Probability and Mathematical Statistics
  RAO • Linear Statistical Inference
  ROBERTSON, WRIGHT, and DYE
  ROGERS and WILLIAMS • Diffus
    II: Îto Calculus
  ROHATGI • Statistical Inference
  ROSS • Stochastic Processes
  RUBINSTEIN • Simulation and the Monte Carlo Method
  RUZSA and SZEKELY • Algebraic Probability Theory
  SCHEFFE • The Analysis of Variance
  SEBER • Linear Regression Analysis
  SEBER • Multivariate Observations
  SEBER and WILD • Nonlinear Regression
  SEN • Sequential Nonparametrics: Invariance Principles and Statistical Inference
  SERFLING • Approximation Theorems of Mathematical Statistics
  SHORACK and WELLNER • Empirical Processes with Applications to Statistics
  STAUDTE and SHEATHER • Robust Estimation and Testing
  STOYANOV • Counterexamples in Probability
  STYAN • The Collected Papers of T. W. Anderson: 1943–1985
  WHITTAKER • Graphical Models in Applied Multivariate Statistics
  YANG • The Construction Theory of Denumerable Markov Processes

Applied Probability and Statistics
  ABRAHAM and LEDOLTER • Statistical Methods for Forecasting
  AGRESTI • Analysis of Ordinal Categorical Data
  AGRESTI • Categorical Data Analysis
  AICKIN • Linear Statistical Analysis of Discrete Data
  ANDERSON and LOYNES • The Teaching of Practical Statistics
  ANDERSON, AUQUIER, HAUCK, OAKES, VANDAELE, and
    WEISBERG • Statistical Methods for Comparative Studies
  ARTHANARI and DODGE • Mathematical Programming in Statistics
  ASMUSSEN • Applied Probability and Queues
 *BAILEY • The Elements of Stochastic Processes with Applications to the Natural
    Sciences
  BARNETT • Interpreting Multivariate Data
  BARNETT and LEWIS • Outliers in Statistical Data, *Second Edition*
  BARTHOLOMEW • Stochastic Models for Social Processes, *Third Edition*
  BARTHOLOMEW and FORBES • Statistical Techniques for Manpower Planning
  BATES and WATTS • Nonlinear Regression Analysis and Its Applications
  BECK and ARNOLD • Parameter Estimation in Engineering and Science
  BELSLEY • Conditioning Diagnostics: Collinearity and Weak Data in Regression
  BELSLEY, KUH, and WELSCH • Regression Diagnostics: Identifying Influential
    Data and Sources of Collinearity
  BHAT • Elements of Applied Stochastic Processes, *Second Edition*
  BHATTACHARYA and WAYMIRE • Stochastic Processes with Applications
  BIEMER, GROVES, LYBERG, MATHIOWETZ, and SUDMAN • Measurement
    Errors in Surveys
  BLOOMFIELD • Fourier Analysis of Time Series: An Introduction
  BOLLEN • Structural Equations with Latent Variables
  BOX • R. A. Fisher, the Life of a Scientist
  BOX and DRAPER • Empirical Model-Building and Response Surfaces
  BOX and DRAPER • Evolutionary Operation: A Statistical Method for Process
    Improvement
  BOX, HUNTER, and HUNTER • Statistics for Experimenters: An
    Introduction to Design, Data Analysis, and Model Building
  BROWN and HOLLANDER • Statistics: A Biomedical Introduction
  BUCKLEW • Large Deviation Techniques in Decision, Simulation, and Estimation
  BUNKE and BUNKE • Nonlinear Regression, Functional Relations and Robust
    Methods: Statistical Methods of Model Building
  BUNKE and BUNKE • Statistical Inference in Linear Models, Volume I
  CHAMBERS • Computational Methods for Data Analysis
  CHATTERJEE and HADI • Sensitivity Analysis in Linear Regression
  CHATTERJEE and PRICE • Regression Analysis by Example, *Second Edition*

*Now available in a lower priced paperback edition in the Wiley Classics Library.

Applied Probability and Statistics (Continued)

CHOW · Econometric Analysis by Control Methods
CLARKE and DISNEY · Probability and Random Processes: A First Course with Applications, *Second Edition*
COCHRAN · Sampling Techniques, *Third Edition*
COCHRAN and COX · Experimental Designs, *Second Edition*
CONOVER · Practical Nonparametric Statistics, *Second Edition*
CONOVER and IMAN · Introduction to Modern Business Statistics
CORNELL · Experiments with Mixtures, Designs, Models, and the Analysis of Mixture Data, *Second Edition*
COX · A Handbook of Introductory Statistical Methods
COX · Planning of Experiments
CRESSIE · Statistics for Spatial Data
DANIEL · Applications of Statistics to Industrial Experimentation
DANIEL · Biostatistics: A Foundation for Analysis in the Health Sciences, *Fourth Edition*
DANIEL and WOOD · Fitting Equations to Data: Computer Analysis of Multifactor Data, *Second Edition*
DAVID · Order Statistics, *Second Edition*
DAVISON · Multidimensional Scaling
DEGROOT, FIENBERG, and KADANE · Statistics and the Law
*DEMING · Sample Design in Business Research
DILLON and GOLDSTEIN · Multivariate Analysis: Methods and Applications
DODGE · Analysis of Experiments with Missing Data
DODGE and ROMIG · Sampling Inspection Tables, *Second Edition*
DOWDY and WEARDEN · Statistics for Research, *Second Edition*
DRAPER and SMITH · Applied Regression Analysis, *Second Edition*
DUNN · Basic Statistics: A Primer for the Biomedical Sciences, *Second Edition*
DUNN and CLARK · Applied Statistics: Analysis of Variance and Regression, *Second Edition*
ELANDT-JOHNSON and JOHNSON · Survival Models and Data Analysis
FLEISS · The Design and Analysis of Clinical Experiments
FLEISS · Statistical Methods for Rates and Proportions, *Second Edition*
FLEMING and HARRINGTON · Counting Processes and Survival Analysis
FLURY · Common Principal Components and Related Multivariate Models
FRANKEN, KÖNIG, ARNDT, and SCHMIDT · Queues and Point Processes
GALLANT · Nonlinear Statistical Models
GIBBONS, OLKIN, and SOBEL · Selecting and Ordering Populations: A New Statistical Methodology
GNANADESIKAN · Methods for Statistical Data Analysis of Multivariate Observations
GREENBERG and WEBSTER · Advanced Econometrics: A Bridge to the Literature
GROSS and HARRIS · Fundamentals of Queueing Theory, *Second Edition*
GROVES · Survey Errors and Survey Costs
GROVES, BIEMER, LYBERG, MASSEY, NICHOLLS, and WAKSBERG · Telephone Survey Methodology
GUPTA and PANCHAPAKESAN · Multiple Decision Procedures: Theory and Methodology of Selecting and Ranking Populations
GUTTMAN, WILKS, and HUNTER · Introductory Engineering Statistics, *Third Edition*
HAHN and MEEKER · Statistical Intervals: A Guide for Practitioners
HAHN and SHAPIRO · Statistical Models in Engineering
HALD · Statistical Tables and Formulas
HALD · Statistical Theory with Engineering Applications
HAND · Discrimination and Classification
HEIBERGER · Computation for the Analysis of Designed Experiments
HELLER · MACSYMA for Statisticians
HOAGLIN, MOSTELLER, and TUKEY · Exploratory Approach to Analysis of Variance
HOAGLIN, MOSTELLER, and TUKEY · Exploring Data Tables, Trends and Shapes
HOAGLIN, MOSTELLER, and TUKEY · Understanding Robust and Exploratory Data Analysis
HOCHBERG and TAMHANE · Multiple Comparison Procedures
HOEL · Elementary Statistics, *Fourth Edition*

Continued on back end papers

*Now available in a lower priced paperback edition in the Wiley Classics Library.

Forecasting with Dynamic Regression Models

Forecasting with Dynamic Regression Models

ALAN PANKRATZ

DePauw University
Greencastle, Indiana

Wiley-Interscience Publication
JOHN WILEY & SONS INC.
New York · Chichester · Brisbane · Toronto · Singapore

In recognition of the importance of preserving what has been
written, it is a policy of John Wiley & Sons, Inc., to have books
of enduring value published in the United States printed on
acid-free paper, and we exert our best efforts to that end.

Copyright © 1991 by John Wiley & Sons, Inc.

All rights reserved. Published simultaneously in Canada.

Reproduction or translation of any part of this work
beyond that permitted by Section 107 or 108 of the
1976 United States Copyright Act without the permission
of the copyright owner is unlawful. Requests for
permission or further information should be addressed to
the Permissions Department, John Wiley & Sons, Inc.

Library of Congress Cataloging in Publication Data:
Pankratz, Alan, 1944–
  Forecasting with dynamic regression models / Alan Pankratz.
    p.   cm. — (Wiley series in probability and mathematical
statistics. Applied probability and statistics, ISSN 0271-6356)
  "A Wiley-Interscience publication."
  Includes bibliographical references and index.
  ISBN 0-471-61528-5
  1. Time-series analysis. 2. Regression analysis. 3. Prediction
theory. I. Title. II. Series.
QA280.P368   1991
519.5'5—dc20   91-12484
                                                  CIP

Printed in the United States of America

10 9 8 7 6 5 4 3 2 1

To Aaron Mark Dietrich and
Sherith Hope Laura

Contents

Preface   xi

Chapter 1   Introduction and Overview   1

1.1   Related Time Series, 1
1.2   Overview: Dynamic Regression Models, 7
1.3   Box and Jenkins' Modeling Strategy, 15
1.4   Correlation, 17
1.5   Layout of the Book, 21
      Questions and Problems, 22

Chapter 2   A Primer on ARIMA Models   24

2.1   Introduction, 24
2.2   Stationary Variance and Mean, 27
2.3   Autocorrelation, 34
2.4   Five Stationary ARIMA Processes, 39
2.5   ARIMA Modeling in Practice, 49
2.6   Backshift Notation, 52
2.7   Seasonal Models, 54
2.8   Combined Nonseasonal and Seasonal Processes, 57
2.9   Forecasting, 59
2.10 Extended Autocorrelation Function, 62
2.11 Interpreting ARIMA Model Forecasts, 64
      Questions and Problems, 69

Case 1 Federal Government Receipts (ARIMA) 72

Chapter 3 A Primer on Regression Models 82

3.1 Two Types of Data, 82
3.2 The Population Regression Function (PRF) with One Input, 82
3.3 The Sample Regression Function (SRF) with One Input, 86
3.4 Properties of the Least-Squares Estimators, 88
3.5 Goodness of Fit ($\bar{R}^2$), 89
3.6 Statistical Inference, 92
3.7 Multiple Regression, 93
3.8 Selected Issues in Regression, 96
3.9 Application to Time Series Data, 103
   Questions and Problems, 113

Case 2 Federal Government Receipts (Dynamic Regression) 115

Case 3 Kilowatt-Hours Used 131

Chapter 4 Rational Distributed Lag Models 147

4.1 Linear Distributed Lag Transfer Functions, 148
4.2 A Special Case: The Koyck Model, 150
4.3 Rational Distributed Lags, 156
4.4 The Complete Rational Form DR Model and Some Special Cases, 163
   Questions and Problems, 165

Chapter 5 Building Dynamic Regression Models: Model Identification 167

5.1 Overview, 167
5.2 Preliminary Modeling Steps, 168
5.3 The Linear Transfer Function (LTF) Identification Method, 173
5.4 Rules for Identifying Rational Distributed Lag Transfer Functions, 184
   Questions and Problems, 193
   Appendix 5A The Corner Table, 194
   Appendix 5B Transfer Function Identification Using Prewhitening and Cross Correlations, 197

Chapter 6 Building Dynamic Regression Models: Model Checking, Reformulation, and Evaluation   202

6.1 Diagnostic Checking and Model Reformulation, 202
6.2 Evaluating Estimation Stage Results, 209
Questions and Problems, 215

Case 4 Housing Starts and Sales   217

Case 5 Industrial Production, Stock Prices, and Vendor Performance   232

Chapter 7 Intervention Analysis   253

7.1 Introduction, 253
7.2 Pulse Interventions, 254
7.3 Step Interventions, 259
7.4 Building Intervention Models, 264
7.5 Multiple and Compound Interventions, 272
Questions and Problems, 276

Case 6 Year-End Loading   279

Chapter 8 Intervention and Outlier Detection and Treatment   290

8.1 The Rationale for Intervention and Outlier Detection, 291
8.2 Models for Intervention and Outlier Detection, 292
8.3 Likelihood Ratio Criteria, 299
8.4 An Iterative Detection Procedure, 313
8.5 Application, 315
8.6 Detected Events Near the End of a Series, 319
Questions and Problems, 320
Appendix 8A BASIC Program to Detect AO, LS, and IO Events, 321
Appendix 8B Specifying IO Events in the SCA System, 322

Chapter 9 Estimation and Forecasting   324

9.1 DR Model Estimation, 324
9.2 Forecasting, 328
Questions and Problems, 340
Appendix 9A A BASIC Routine for Computing the Nonbiasing Factor in (9.2.24), 340

Chapter 10 Dynamic Regression Models in a Vector ARMA Framework 342

10.1 Vector ARMA Processes, 342
10.2 The Vector AR (π Weight) Form, 345
10.3 DR Models in VAR Form, 346
10.4 Feedback Check, 349
10.5 Check for Contemporaneous Relationship and Dead Time, 354
Questions and Problems, 356

Appendix 357

Table A Student's t Distribution, 357
Table B χ^2 Critical Points, 359
Table C F Critical Points, 360

Data Appendix 362

References 376

Index 381

Preface

Single-equation regression models are one of the most widely used statistical forecasting tools. Over the last two decades many ideas relevant to regression forecasting have arisen in the time series literature, starting with the first edition of Box and Jenkins' text, *Time Series Analysis: Forecasting and Control*, Holden-Day (1976). Those who apply regression methods to time series data without knowledge of these ideas may miss out on a better understanding of their data, better models, and better forecasts.

This book is a companion volume to my earlier text, *Forecasting with Univariate Box–Jenkins Models: Concepts and Cases*, Wiley (1983), where I present the Box–Jenkins modeling strategy as applied to ARIMA models. There is more to Box and Jenkins than ARIMA models. They also discuss "combined transfer function–disturbance" models, or what I call dynamic regression models. The purpose of the present text is to pull together recent time series ideas in the Box–Jenkins tradition that are important for the informed practice of single-equation regression forecasting. I pay special attention to possible intertemporal (dynamic) patterns — distributed lag responses of the output series (dependent variable) to the input series (independent variables), and the autocorrelation patterns of the regression disturbance.

This book may be used as a main text for undergraduate and beginning graduate courses in applied time series and forecasting in areas such as economics, business, biology, political science, engineering, statistics, and decision science. It may be used in advanced courses to supplement theoretical readings with applications. And it can serve as a guide to the construction and use of regression forecasting models for practicing forecasters in business and government.

Special features of this book include the following:

- Many step-by-step examples using real data, including cases with multiple inputs, both stochastic and deterministic.
- Emphasis on model interpretation.

- Emphasis on a model identification method that is easily applied with multiple inputs.
- Suggested practical rules for identifying rational form transfer functions.
- Emphasis on feedback checks as a preliminary modeling step.
- Careful development of an outlier and intervention detection method, including a BASIC computing routine.
- A chapter linking dynamic regression models to vector ARMA models.
- Use of the extended autocorrelation function to identify ARIMA models.

While there are other books that cover some of these ideas, they tend to cover them briefly, or at a fairly high level of abstraction, or with few supporting applications. I present the theory at a low level, and I show how these ideas may be applied to real data by means of six case studies, along with other examples within the chapters. Several additional data sets are provided in the Data Appendix. The examples and cases are drawn mainly from economics and business, but the ideas illustrated are also applicable to data from other disciplines.

Those who read the entire text will notice some repetition; I realize that many readers will read only selected chapters and cases, and that most won't already know the contents of this book before they read, so I have tried to repeat some key ideas. I hope my readers will learn enough of the main ideas and practices of dynamic regression forecasting to feel more comfortable and be more insightful when doing their own data analysis, criticizing their own procedures and results, moving ahead to more advanced literature and practices, and when moving back to less advanced methods.

I have successfully used a majority of the material in this book with undergraduates, most of whom have had just one course in statistics. I assume that readers have been exposed to the rudiments of probability, hypothesis testing, and interval estimation. A background in regression methods is helpful but not required; Chapter 3 is an introduction to the fundamentals of regression. I use matrix algebra in a few places, primarily Chapter 10, but a knowledge of matrix algebra is not required to understand the main ideas in the text. For many readers the most difficult part of the text may be the use of backshift notation. If you are one of those readers, to you I say (1) it's only notation, (2) learning to use it is easier than learning to ride a bicycle, and (3) it's immensely useful.

As you read this book, I hope you find parts that are interesting, informative, and helpful. When you do, you should be as grateful as I am to Gregory Hudak, Brooks Elliott, and an anonymous reviewer who provided many excellent comments on a draft of the manuscript. I am also grateful to Fran Cappelletti for his perceptive comments on parts of the manuscript. I wish I could also blame shortcomings and undetected errors on these people, but I can't; I didn't take all of their advice, and any problems you encounter are my responsibility. Various members of the DePauw University Computer

Center staff, my colleagues in the Department of Economics and Management, and those faculty and administrators responsible for DePauw summer grants all facilitated my work in various ways. Paul Koch graciously provided some of the data used in Case 5.

Anyone who knows Bea Shube knows of her competence and wisdom. I was fortunate in starting this project under her editorial guidance. Kate Roach was informative, encouraging, and patient while taking on this project in editorial midstream. Rosalyn Farkas and Ernestine Franco provided outstanding editorial and production assistance.

All of the computational results shown in this text were produced by the SCA System software (developed in collaboration with George E. P. Box, Mervin E. Muller, and George C. Tiao). Contact: Scientific Computing Associates, Lincoln Center, Suite 106, 4513 Lincoln Avenue, Lisle, IL, 60532, (708) 960–1698. Similar results may be obtained with various other packages, including several from Automatic Forecasting Systems, Inc., P.O. Box 563, Hatboro, PA, 19040, (215) 675–0652; SAS Institute Inc., Box 8000, Cary, NC, 27511, (919) 467–8000; or Gwilym Jenkins & Partners Ltd., Parkfield, Greaves Road, Lancaster LA1 4TZ, U.K.

My wife Charity didn't read or type anything this time, but she was her usual good natured and supportive self as I showed all the symptoms of Author's Disease. Though I mentioned it dozens of times, I don't think my children are much aware that I have written another book. But then, they are teenagers, or were while I was writing, with far more important things to think about. Maybe they will notice that this book is dedicated to them, with pride and affection.

ALAN PANKRATZ

Greencastle, Indiana
June 1991

CHAPTER 1

Introduction and Overview

1.1 RELATED TIME SERIES

Comedian George Carlin has noticed that "If there were no such thing as time, everything would happen all at once." Of course, there is such a thing as time, and everything doesn't happen all at once. Events occurring at one time are often *related* to other events occurring at *other times*.

Our daily lives are full of examples. For instance, several years ago my son had a severe headache. The pain was so bad that we took him to the local hospital where a physician gave him some pain medication. After 10 minutes my son felt a little better. After 20 minutes he felt much better. After 30 minutes he felt little pain and was able to sleep. My son's pain relief at any given time was related to the receipt of pain medication at an earlier time.

As another example, suppose a state passes a law requiring all car occupants to wear seat belts as of January 1. Some people who have not worn belts might start wearing them immediately on January 1; others might start wearing them only after many reminders and many days have passed; still others might wear belts only after being arrested or fined; and some might never wear belts. Our ability to understand the effect of the seat belt law, and our ability to forecast the effect of passing a similar law in another state, depend on a correct understanding of the role of time in this situation.

The same holds true in many other contexts: We may misunderstand the effects of public or private policy actions, or the nature of physical, chemical, or biological mechanisms, if we do not understand the special role played by time in the relationship between time-ordered variables. And our ability to forecast time-ordered variables may be impaired if we don't understand the time structure of relationships between variables.

Studying such interrelationships using time-ordered data is called *time series analysis*. The goal in time series analysis is to find a useful way (a "model") to express a time-structured relationship among some variables or events. We may then evaluate this relationship or we may forecast one or more of the variables. That is the focus of this book.

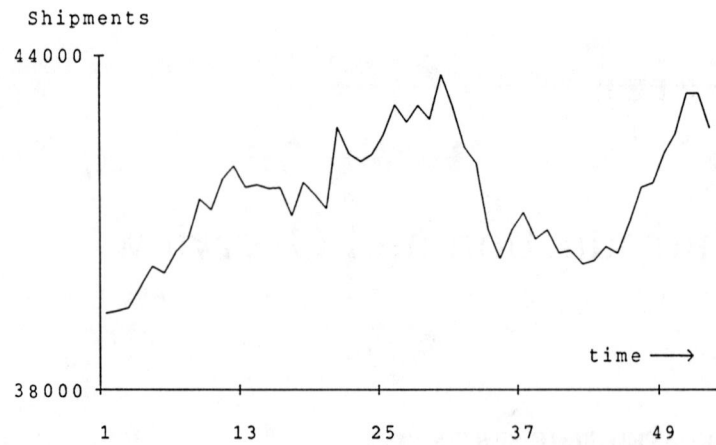

Figure 1.1 Valve shipments, January 1984–May 1988.

Let's consider three more examples of related time series. In each example we have *sequences of observations* on two separate *time-ordered variables* that are (possibly) related to each other.

1.1.1 Example 1: Valve Shipments and Orders

Figure 1.1 shows 53 time-ordered (from left to right) monthly observations of the number of small valves shipped by a manufacturer. The data are recorded from January 1984 through May 1988. Figure 1.2 shows 53 corresponding time-ordered monthly observations of the volume of orders for this valve to be shipped during the same month.

The volume of valve shipments is recorded at the end of the month, while

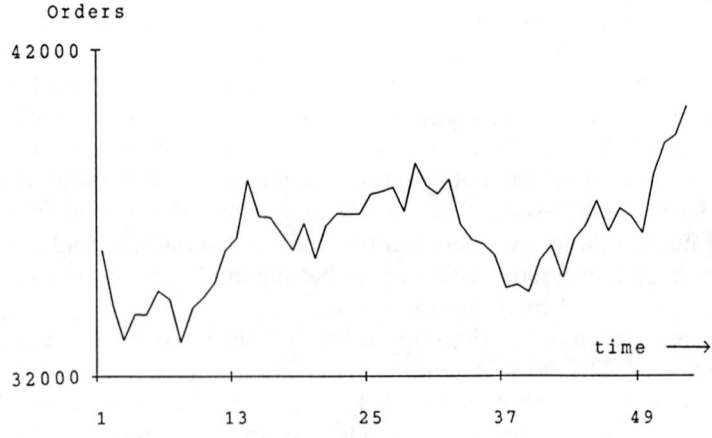

Figure 1.2 Valve orders, January 1984–May 1988.

RELATED TIME SERIES

Table 1.1 Partial Listing of Valve Shipments and Orders Data

Date	Shipments	Orders
January 1984	39,377	35,862
February 1984	39,417	34,165
March 1984	39,475	33,127
.	.	.
.	.	.
.	.	.
April 1988	43,299	39,331
May 1988	42,687	40,203
June 1988	—	38,916

the volume of valve orders for that month is recorded at the start of the month. Thus, the volume of shipments can differ from the volume of orders since the manufacturer accepts late orders, and some orders may be canceled. Since the volume of valve orders is recorded one month earlier than the corresponding shipments, there are 54 observations of orders (rather than 53), and the orders data are recorded through June 1988; Figure 1.2 shows only the first 53 values. The full data set is listed in the Data Appendix as Series 1 (shipments) and Series 2 (orders).

Table 1.1 is a partial listing of the data. We said that the shipments and orders observations are "corresponding." That is, the data occur in time-ordered *pairs*; each value of shipments for a *given month* (recorded at the end of the month) is paired with a corresponding value of orders for the *same month* (recorded at the start of the month). For example, as shown in Table 1.1, the manufacturer began March 1984 with orders for 33,127 valves to be shipped that month, but ended March 1984 with 39,475 valves actually shipped during that month.

It is reasonable to think that the volume of shipments during a given month might be related to the volume of orders on the books at the start of that month. Indeed, inspection of Figures 1.1 and 1.2 suggests that higher shipment levels are associated with higher order levels, and lower shipment levels with lower order levels. Thus knowledge of the volume of orders on the books at the start of a given month might help us to forecast the volume of shipments for that month.

1.1.2 Example 2: Interrupted Saving Rate Series

Figure 1.3 shows 100 time-ordered observations of the saving rate (saving as a percent of income) in the United States from the first quarter (I) of 1955 to the fourth quarter (IV) of 1979. These data are listed in the Data Appendix as Series 3. The observation for 1975 II (the 82nd observation) stands out from the rest of the data: It is markedly higher than both earlier and later

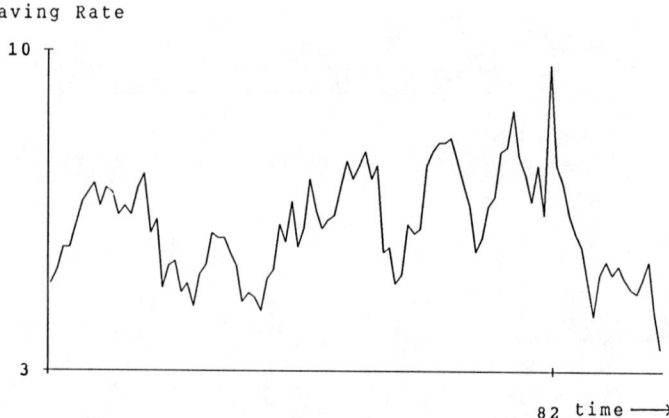

Figure 1.3 U.S. saving rate, 1955 I–1979 IV.

observations. This jump up at the 82nd observation could be just unexplainable random variation in the data. However, during that quarter Congress passed a law granting a one-time tax rebate. According to at least one economic theory (the permanent income hypothesis), such a rebate will lead to a temporary increase in the saving rate. The saving rate series may have been *interrupted by a known event* (at 1975 II) whose effect can be explained.

We can represent the tax rebate event by creating a binary indicator variable (a series consisting of only 0's or 1's) to reflect the "on–off" nature of the rebate. We define this series as having two parts: a set of zeros that stand for "no rebate during this quarter" and a set of ones that stand for "rebate during this quarter." This gives us a set of 100 time-ordered and paired observations for the two series. Table 1.2 is a partial listing of these

Table 1.2 Partial Listing of Saving Rate Data and Tax Rebate Binary Indicator Variable

Date	Saving Rate (%)	Binary Indicator Variable
1955 I	4.9	0
.	.	.
.	.	.
.	.	.
1975 I	6.4	0
1975 II	9.7	1
1975 III	7.5	0
.	.	.
.	.	.
1979 IV	3.5	0

series. The binary indicator variable has the value 0 throughout, except at 1975 II when the tax rebate took place; then it has the value 1.0.

This example is similar to Example 1. However, in this example only the values of the saving rate series are observed; the values of the binary indicator variable are constructed by us rather than observed. In the previous example both the valve shipments and valve orders data were observed, not constructed.

1.1.3 Notation

In the next example we illustrate some helpful notation that we will use throughout this book. The letter Y stands for a variable whose values may depend on one or more other variables or events. In our earlier examples Y is the level of headache pain, the number of car occupants wearing seat belts, the monthly volume of valve shipments, or the saving rate. The letter X stands for a variable that might affect Y. Thus X could be the amount of pain medication received, a binary indicator variable representing the presence or absence of a seat belt law, the volume of valve orders, or a binary indicator variable representing the presence or absence of the tax rebate.

The letter t is used as a subscript to indicate time: Y_t and X_t stand for variables whose possible values are time ordered. Thus, we have $t = 1, 2, 3, \ldots$, where the value of t indicates the place of the corresponding observation (either actual or possible) in the data sequence; that is, Y_1 is the first observation, Y_{69} is the 69th observation, and so forth. Of course, t can correspond to a particular date. For example, for monthly data it could be that $t = 1$ corresponds to December 1979, $t = 2$ corresponds to January 1980, and so on.

Sometimes it is useful to write the time subscript as $t - 1, t - 2, t - 3, \ldots$, or as $t + 1, t + 2, t + 3, \ldots$. The meaning of the positive and negative signs is fairly obvious: A negative refers to a time period before period t, and a positive refers to a time period after period t. For example, if we have daily data and X_t corresponds to Wednesday, then X_{t-1} corresponds to the prior Tuesday, and X_{t-2} corresponds to the prior Monday, while X_{t+1} corresponds to the next Thursday, and X_{t+2} corresponds to the next Friday, and so forth.

1.1.4 Example 3: Sales and Leading Indicator

Figure 1.4 shows 150 time-ordered observations (denoted Y_t) of the sales of a certain product. Figure 1.5 shows 150 corresponding time-ordered observations (denoted X_t) of a leading indicator variable that might be useful in forecasting the sales data. Each horizontal axis is measured in time units ($t = 1, 2, 3, \ldots$); each vertical axis is measured in units of Y_t or X_t. Table 1.3 is a partial listing of these data. For time period $t = 1$, $Y_1 = 200.1$, *and* $X_1 = 10.01$; for $t = 2$, $Y_2 = 199.5$, *and* $X_2 = 10.07$; and so forth. The full sales and

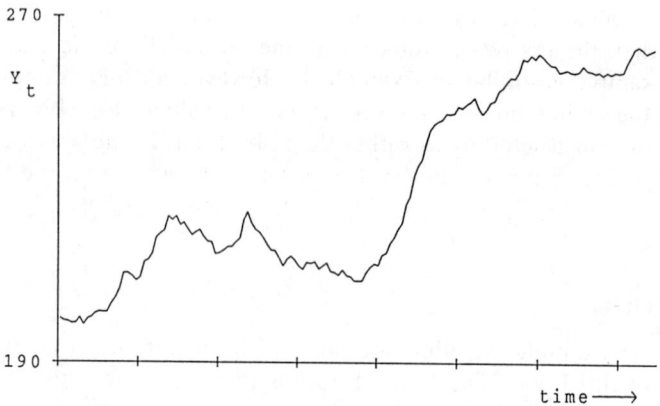

Figure 1.4 Product sales.

leading indicator data sets are listed in the Data Appendix as Series 4 and Series 5, respectively.

The idea here is that changes in X_t may signal current or later (especially later) changes in Y_t since X_t is supposedly a *leading* indicator for Y_t. That is, maybe movements in Y_t reflect movements in X_t, or X_{t-1}, or X_{t-2}, and so forth. Another way to say this is that changes in X_t may lead to changes in Y_t, Y_{t+1}, Y_{t+2}, and so forth. Figures 1.4 and 1.5 suggest that this might be so because higher (lower) values of X_t seem to be associated with higher (lower) values of Y_t. We will consider in detail the possible relationship between X_t and Y_t in this example in Chapters 5, 6, 9, and 10.

1.1.5 Inputs, Output, and Models

In this book we consider situations with *one* time series variable to be forecast or predicted (denoted Y_t), and one or more time series variables (denoted

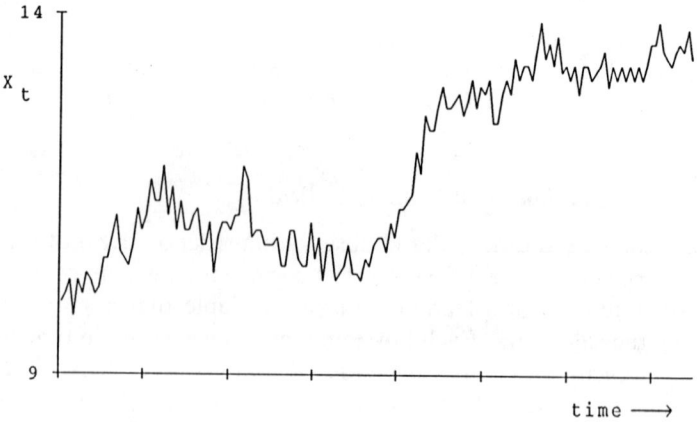

Figure 1.5 Leading indicator.

Table 1.3 Partial Listing of Sales and Leading Indicator Data

t = Time	Y_t = Sales	X_t = Indicator
1	200.1	10.01
2	199.5	10.07
3	199.4	10.32
.	.	.
.	.	.
.	.	.
149	262.2	13.77
150	262.7	13.40

$X_{1,t}, X_{2,t}, \ldots$) that might help to *forecast the future movements* or *explain the past movements of* Y_t. For convenience we will call variable Y_t the *output* and variables $X_{1,t}, X_{2,t}, \ldots$ the *inputs*. In Examples 1, 2, and 3 we had just one input in each case—the valve orders input, the binary tax rebate input, and the leading indicator input, respectively. (Chapter 3 and Cases 2, 3, 5, and 6 in this book show examples with multiple inputs.) As in Example 2, an input may be constructed rather than observed. Thus inputs can be *stochastic* (having some random variation, like the valve orders data in Example 1) or *deterministic* (having no random variation, like the constructed rebate versus no rebate binary variable in Example 2).

Our goal is to *use the available data to build a statistical model* that represents the relationship between the output (Y_t) and the inputs $(X_{1,t}, X_{2,t}, \ldots)$. We may then use this model to *forecast* the output series, *test a hypothesis* about the relationship between the output and the inputs, or *assess the effect of an event*, including a policy action, on the output. In the latter case we may simply want to *understand the historical effect* of a past event on the output. Or, we may want to *predict* the possible effect of a similar future event on the output.

1.2 OVERVIEW: DYNAMIC REGRESSION MODELS

In this book we focus on a *family* of statistical models called *dynamic regression* (DR) models. A DR model states how an output (Y_t) is linearly related to current and past values of one or more inputs $(X_{1,t}, X_{2,t}, \ldots)$. It is usually assumed that observations of the various series occur at *equally spaced time intervals*. (However, a model in which we allow for slightly unequally spaced data is shown in Case 2.) While the output may be affected by the inputs, a crucial assumption is that the *inputs are not affected by the output*. This means that we are limited to *single-equation* models.

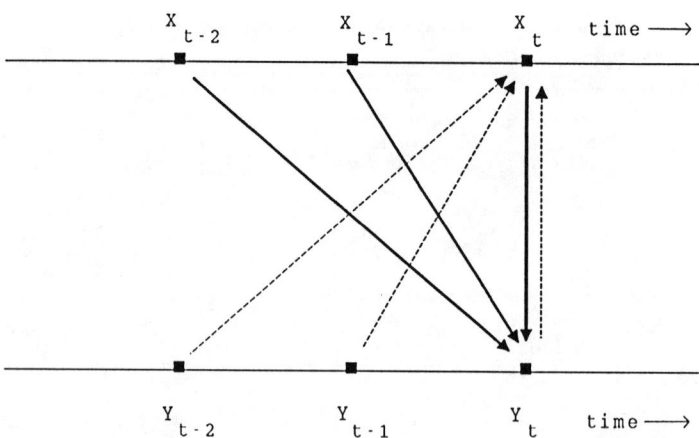

Figure 1.6 Possible time-lagged relationships between Y_t and X_t.

These ideas are illustrated schematically in Figure 1.6. Note that X_t and Y_t are measured at equally spaced time intervals. The solid arrows show that it is acceptable in a *single-equation* DR model to have statistical effects going from current and past values of X_t ($X_t, X_{t-1}, X_{t-2}, \ldots$) to the current value of Y_t. However, it is *not appropriate* to have effects (the dashed line arrows) going from current or past values of Y_t ($Y_{t-1}, Y_{t-2}, \ldots$) to the current X_t.

I assume that most readers have some exposure to the basic ideas of correlation and regression analysis. Section 1.4 reviews the idea of correlation; Chapter 3 reviews regression. To introduce the idea of a DR model, we first consider how we might apply ordinary regression analysis to Examples 1, 2, and 3 in Section 1.1. (An ordinary regression model, when applied to time series data, is a simple example of a DR model.)

1.2.1 Regression

One way to link the valve shipments data to the valve orders data (or the saving rate data to the constructed binary variable, or the sales data to the leading indicator) is to estimate the ordinary regression model

$$Y_t = C + v_0 X_t + N_t \tag{1.2.1}$$

where C is the population intercept, a parameter to be estimated; v_0 is the population slope coefficient, also a parameter to be estimated; and N_t is a stochastic disturbance term. The disturbance is present because we cannot expect variation in the input(s) to explain all of the variation in the output. The disturbance represents variation in the output that is *not associated* with movements in the input(s).

Equation (1.2.1) is a simple example of a DR model with one input. Y_t is the output such as valve shipments, and X_t is the input such as valve orders. According to (1.2.1), valve shipments and valve orders are *contem-*

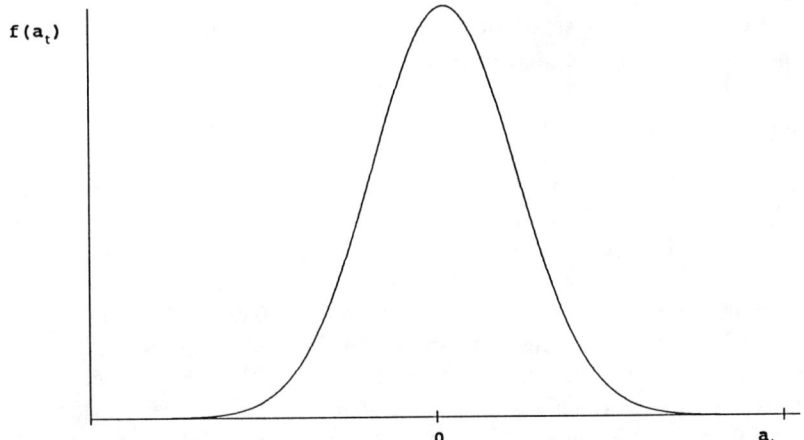

Figure 1.7 Normal distribution for a_t.

poraneously related. That is, if X_t rises by one unit *at time t*, then Y_t changes by v_0 units (other things held constant) *at time t*; there is *no time lag* involved in this response.

In the standard regression model the disturbance term is often assumed to be a set of zero-mean and normally distributed *white-noise* random shocks, denoted in this book as a_t. White noise is defined as a sequence of *mutually independent* and *identically distributed* random variables. Thus if N_t in (1.2.1) is zero mean and normally distributed white noise, then $N_t = a_t$. The normal probability density function for a_t is illustrated in Figure 1.7. Since the a_t values are assumed to be *identically* distributed, they all come from the *same* distribution *for all t*, with the same mean (zero) and the same variance (denoted σ_a^2). The random shocks (a_t) are also assumed to be *independent of* X_t. This is another way of saying that X_t is *not* statistically related to current or past values of Y_t, as we discussed with reference to Figure 1.6.

If a regression model is correctly specified and if the assumptions about the shocks are met, then the ordinary least squares regression estimates of the parameters C and v_0 have certain desirable statistical properties: In repeated sampling they are *unbiased*, *consistent*, and *efficient* estimators. (You may want to review the meanings of these properties now if you don't recall them. Chapter 3 has a brief discussion.)

There are several potential problems in estimating a simple regression model like (1.2.1) with *time series data*. In the next four sections we introduce these problems; they will receive a great deal of attention in this book.

1.2.2 Time-Lagged Relationships

Y_t may be related to X_t *with a time lag*; that is, Y_t may be related to $X_{t-1}, X_{t-2}, X_{t-3}$, and so forth, in addition to (or instead of) being related

to X_t. For example, instead
statement of how Y_t is related

$$Y_t = C + v_0$$

We use C as a generic symbol
value in (1.2.2) than it has in
of the relationship between th ...put, estimating the parameters in (1.2.1) *instead of* (1.2.2) can lead to a variety of problems. For example, in estimating (1.2.1) we lose useful information about the roles of X_{t-1} and X_{t-2} in explaining movements in Y_t. Thus our estimate of the residual variance (the variance of N_t) will be larger than necessary, and our forecasts are likely to be less accurate than they could be. Further, the estimate of the coefficient v_0 attached to the included variable (X_t) in (1.2.1) will be biased and inconsistent *if* the excluded variables (X_{t-1} and X_{t-2}) are correlated with the included variable (X_t). This problem could easily occur with time series data since such data (the input series, in this case) tend to be *self*-correlated; that is, X_t tends to be correlated with its own past values $X_{t-1}, X_{t-2}, \ldots$.

In this book we study methods for identifying the nature of the time-lagged relationship between an output and current and past values of one or more inputs. These methods are designed to help us include all relevant earlier values of the inputs, and exclude all irrelevant ones. These methods are discussed especially in Chapters 5 through 7 and Cases 4 through 6.

1.2.3 Feedback from Output to Inputs

Another potential problem in the use of ordinary regression is that there may be *feedback* from Y_t to X_t. This is illustrated in Figure 1.6 by the dashed arrows. For example, in addition to Equation (1.2.1) or (1.2.2), there might also be *another relationship* between X_t and Y_t, expressed as

$$X_t = C^* + w_1 Y_{t-1} + M_t \tag{1.2.3}$$

where C^* is a constant; w_1 is a slope coefficient parameter; and M_t is a stochastic disturbance. When an input X_t responds in some way to current or past values of Y_t, as in (1.2.3), this is called *feedback* from Y_t to X_t.

If feedback is present, the disturbance N_t in (1.2.1) or (1.2.2) is correlated with one or more of the input variables on the right side of those equations. Estimating (1.2.1) or (1.2.2) using ordinary single-equation regression methods when feedback is present leads to *inconsistent* estimates of the parameters: As the sample size grows, the repeated sampling distributions of the estimates collapse onto values that *differ from the parameters by unknown amounts*. Even if we had all data from the population, applying the usual regression formulas to the data would not give us estimates equal to the

OVERVIEW: DYNAMIC REGRESSION MODELS   11

parameters. Clearly, feedback could lead us to draw incorrect statistical inferences about the relationship between X_t and Y_t. For a more formal discussion of this issue, see Box and MacGregor (1974).

Procedures for detecting feedback are given in Chapters 5 and 10. Sometimes there is sound theoretical reasoning in favor of a one-way relationship from the inputs to the output (no feedback). But *even then* it is important to have a test procedure to detect feedback. Why? Because statistical feedback may exist, even when the underlying system is one way, if the data are *aggregated across time* (Tiao and Wei, 1976; Wei, 1982; 1990). For example, the shipments and orders data in Figures 1.1 and 1.2 are monthly flow data, which means they are sums of daily data. This fact alone means that there might be feedback even though there may not be any direct causal mechanism linking orders to past values of shipments.

When feedback exists, we need a *multiple-equation* framework to properly specify the relationship between inputs and outputs. We could use either a *vector autoregressive moving average (VARMA)* framework or a system of *simultaneous DR* equations. These topics are largely beyond the scope of this book; however, we will introduce a vector ARMA framework for analyzing single-equation DR models in Chapter 10. For more advanced discussion of multiple-equation time series methods, see Tiao and Box (1981), Jenkins and Alavi (1981), Liu (1986b), Liu and Hudak (1986), and Wei (1990).

1.2.4 Autocorrelated Disturbance and ARIMA Processes

A third possible problem in the use of ordinary regression methods is that the disturbance series may be *autocorrelated*, meaning self-correlated; that is, N_t may be *related to its own past values* N_{t-1}, N_{t-2}, and so on. For example, N_t might depend on its own past in this way:

$$N_t = \phi_1 N_{t-1} + a_t \qquad (1.2.4)$$

where ϕ_1 is a coefficient stating how N_t is related to N_{t-1} and a_t is the random shock component of N_t. Or, N_t might depend on its own past in this way:

$$N_t = a_t - \theta_1 a_{t-1} \qquad (1.2.5)$$

where θ_1 is a coefficient stating how N_t is related to a_{t-1}. While a_{t-1} is not a past value of N_t, like N_{t-1}, it is the *random shock component* of N_{t-1}. That is, subtract 1 from each time subscript in (1.2.5) to obtain $N_{t-1} = a_{t-1} - \theta_1 a_{t-2}$. From this expression we see that the past random shock a_{t-1} (on the right side) is part of the past value N_{t-1} (on the left side).

Equations (1.2.4) and (1.2.5) are simple examples of *autoregressive integrated moving average (ARIMA) processes*. An ARIMA process is a useful way of stating the nature of the autocorrelation pattern in a single time series. Chapter 2 introduces the main ideas about ARIMA processes and models. In later chapters we discuss methods for identifying ARIMA models to represent the autocorrelation patterns in a regression disturbance. Including this information in a regression model can help us avoid the problems that arise when a regression disturbance series is autocorrelated.

If the disturbance is autocorrelated, but we ignore that fact in estimating an equation like (1.2.1) or (1.2.2), then the estimate of the residual variance will be larger than necessary. Therefore forecasts from the model will be less accurate than they could be since, in using (1.2.1) or (1.2.2), we are ignoring useful information about past values of N_t. Ignoring autocorrelation in the disturbance also leads to inefficient estimates of the model coefficients and invalid statistical tests. The usual tests of significance in regression analysis rest on the assumption that the disturbance is a sequence of mutually *independent* shocks; an autocorrelated disturbance violates this assumption.

1.2.5 Misleading Correlations

A fourth potential problem in using ordinary regression with time series data arises from the fact that time series data in general tend to be autocorrelated. If there are *common elements in the autocorrelation structure of Y_t and X_t*, then an ordinary regression equation can show a strong relationship between these variables when, in fact, X_t has *no real explanatory power*. For example, if both X_t and Y_t tend to wander away from any fixed overall level (e.g., as happens with many economic time series) but are otherwise unrelated, then an ordinary regression equation is likely to show a mistakenly strong relationship between Y_t and X_t. For a discussion of this phenomenon with some empirical evidence see Granger and Newbold (1986, Section 6.4). Using the methods for building dynamic regression models discussed in this book can help to reduce the chance that we will settle on a misleading model of the relationship between Y_t and X_t.

1.2.6 Applications

Consider the application of the simple bivariate equation (1.2.1), using ordinary least-squares regression estimation, to Examples 1, 2, and 3 in Section 1.1. (In doing this we are creating some straw men; but these examples illustrate in a simple way the possible disadvantages of ignoring the ideas discussed in this book.)

Figure 1.8 is a scatterplot of the paired data for Example 1, Y_t = valve shipments (vertical axis) and X_t = valve orders (horizontal axis). Inspection of this sample plot suggests that there is a useful contemporaneous (same time period) linear relationship between valve shipments and orders. Table 1.4 shows the results of estimating Equation (1.2.1) for these data using ordinary least-squares regression. The resulting least-squares regression line is drawn through the scatter of data in Figure 1.8.

A scatterplot for Example 2, the saving rate data, would not be very

OVERVIEW: DYNAMIC REGRESSION MODELS

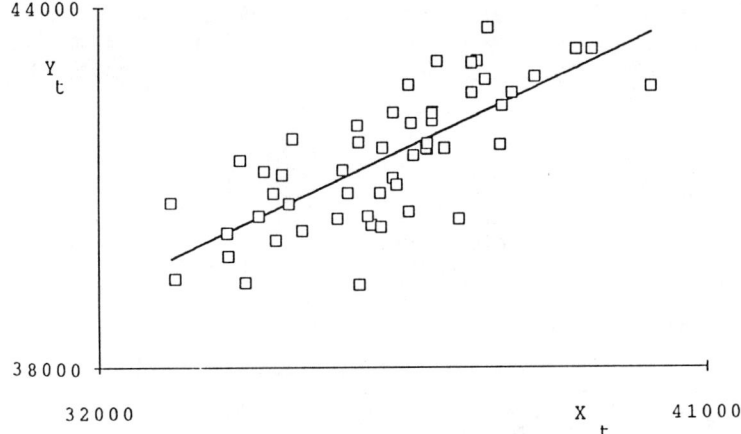

Figure 1.8 Valve shipments (Y_t) vs. valve orders (X_t), with sample regression line.

informative since the constructed binary variable X_t in that example is equal to zero for all time periods except one ($X_{82} = 1$). Table 1.4 shows the ordinary least-squares regression estimation results for these data.

Figure 1.9 is a scatterplot of the paired data for Example 3, Y_t = sales and X_t = leading indicator. Inspection of this sample plot suggests a useful contemporaneous linear relationship between sales and the leading indicator. Table 1.4 shows the ordinary least-squares estimation results for these data. The estimated regression line is drawn through the scatter of data in Figure 1.9.

The important point to notice, in all three examples, is that the input variable (X_t) seems to be important in explaining the movements of the output variable (Y_t). This is suggested by the t value associated with the estimated slope coefficient of each X_t variable; the t values are shown in parentheses in Table 1.4 below each estimated coefficient.

A sample regression coefficient has an associated t value. While there are sometimes pitfalls in using a regression t value, it can often be used appropriately to determine the statistical significance of the associated coefficient.

Table 1.4 Ordinary Regression Estimation Results, Applying Equation (1.2.1) to Examples 1, 2, and 3

	Estimate of C (t value)	Estimate of v_0 (t value)	Residual Standard Error
Example 1 Valve shipments	22308.46 (9.56)	0.529 (8.24)	719.837
Example 2 Saving rate	6.1343 (56.29)	3.5656 (3.27)	1.08427
Example 3 Sales	33.293 (5.81)	16.584 (34.13)	6.6853

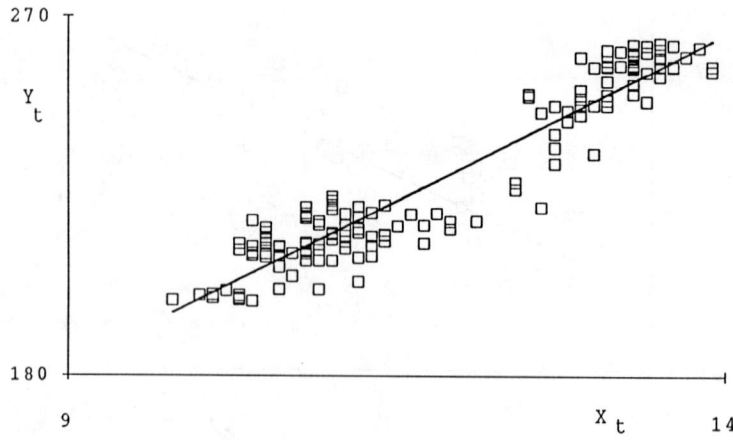

Figure 1.9 Sales (Y_t) vs. leading indicator (X_t), with sample regression line.

A standard regression t value tells us *how many standard errors away from zero* the associated estimated coefficient falls. When a t value is large in absolute value, this suggests that the associated coefficient is significantly different from zero; in turn, this suggests that the associated input variable is important in explaining the movements of the output. As a practical rule, when an absolute t value is about 2.0 or larger, we may conclude that the associated coefficient is significant. This topic is discussed further in Chapter 3. (Table A at the end of the book lists critical t values for selected levels of significance and degrees of freedom.)

Notice in Table 1.4 that the t value of each estimate of coefficient v_0 is large in absolute value: Each estimated coefficient falls more than three standard errors away from zero. This implies that each estimated coefficient is significantly different from zero, and that each input is important in explaining the behavior of the corresponding output.

It turns out that the results for each equation in Table 1.4 are *misleading* to varying degrees for various reasons. We will consider the valve shipments and orders data (Example 1) further in Chapter 5. *Despite the evidence in Figure 1.8 and Table 1.4 we will see that there is no statistically reliable relationship between sales and orders, either contemporaneously or with a time lag.* In addition, there is a strong *autocorrelation pattern in the disturbance* when Equation (1.2.1) is estimated for these data. The *t test for the estimated coefficient v_0 in Table 1.4 is therefore invalid.*

In the case of the saving rate data (Example 2), things are not quite so bad. We consider these data again in Chapters 7 and 8. We will see that there is an *autocorrelation pattern in the disturbance* for Equation (1.2.1) when it is estimated for these data. This invalidates the t test in Table 1.4. However, when the autocorrelation pattern of the disturbance is taken into account, we find the following: (1) the binary input variable (X_t) continues to be statistically important; but (2) its associated coefficient is different from that shown in Table 1.4; also (3) the statistical significance of its impact (the coefficient t value) is different; and (4) the overall equation goodness of fit (residual standard error) is improved substantially when the disturbance autocorrelation pattern is taken into account.

We will examine the sales and leading indicator data (Example 3) more carefully in Chapters 5 and 6. There we will see that sales are *not much related to the leading indicator on a contemporaneous basis*, despite the evidence in Figure 1.9 and Table 1.4. There is, however, a fairly strong relationship between current sales and *past values* of the leading indicator. The results in Table 1.4 do not properly show the *time-lagged* (dynamic) nature of the relationship between sales and the leading indicator. We will also see in Chapter 5 that there is a strong *autocorrelation pattern in the disturbance* when Equation (1.2.1) is estimated for these data; therefore *the t test in Table 1.4 is invalid*.

1.2.7 Dynamic Regression

The three applications we have just considered illustrate how applying a simple regression model like (1.2.1) to time series data can produce badly misleading results. In this book we consider how to improve on simple regression by using DR models for time series data. A DR model shows how the output is related to the inputs, *taking into account* (1) the possible time-lagged relationship between the output and the inputs, and (2) the possible time structure (autocorrelation pattern) of the disturbance series.

"Dynamic" means that we pay special attention to (1) the *time structure* of the input–output relationship and (2) the *time structure* of the disturbance series. By using an appropriate modeling strategy and statistical estimation procedure, we hope to reduce or avoid the problems that could arise if we use simple regression methods inappropriately. In Chapters 5 and 6 we consider modeling procedures that can reveal (1) possible feedback effects from past values of Y_t to current values of X_t; (2) time lags in the relationship between Y_t and X_t; and (3) the autocorrelation structure of the disturbance series.

1.3 BOX AND JENKINS' MODELING STRATEGY

In building DR models we use a three-stage model building strategy associated with the names of George E. P. Box and the late Gwilym M. Jenkins (the so-called Box–Jenkins strategy). While the general strategy was used by practicing analysts before the work of Box and Jenkins appeared, these authors are perhaps most responsible for formalizing and popularizing the strategy. They have also made important contributions to the related statistical theory and practice. Their book (Box and Jenkins, 1976) is a standard reference. Another important developer of some of the material presented in this book is George C. Tiao. In fact, some analysts refer to "Box–Jenkins–Tiao" models and analysis. Many further contributions to the theory and practice of time series analysis have come from students of these three people.

In building a DR model, we first use *whatever theory* is relevant (e.g., economic or biological) to *choose the input variable(s)*. This theory may also suggest something about the *form* of the relationship between the output and the input(s). Then we apply the modeling strategy. The modeling strategy has three parts: (1) model identification, (2) model estimation, and (3) model checking. If a satisfactory model is found, we may then forecast and monitor the performance of the model.

Identification

At this stage we first attempt to get a feel for the nature of both the input–output relationship and the possible time-structured pattern in the disturbance by "letting the data talk to us." This involves estimating a rather large number of parameters (regression coefficients and autocorrelation coefficients). We then choose from the family of DR models a *tentative model* that summarizes these patterns in a *parsimonious* (compact) way. That is, we attempt to identify a model that seems to be *consistent with the data* and that has *the fewest coefficients* needed to adequately explain the behavior of the data. DR model identification is discussed especially in Chapter 5 and Cases 2–6.

Estimation

At this stage we obtain estimates of the parameters of our tentative DR model. Here we make more efficient use of the data than we can at the identification stage since here we estimate just a few parameters. These few parameters are (1) regression-type coefficients that represent the input–output relationship and (2) ARIMA model coefficients that represent the disturbance series autocorrelation pattern.

In Chapter 3 we discuss the idea of least-squares regression estimation. Throughout this book we use a maximum-likelihood estimation method developed by Hillmer and Tiao (1979) to estimate DR models. This method has been implemented as part of the SCA System software, which is used for all computational procedures in this book. DR model estimation is discussed in more detail in Chapter 9.

Checking

At this stage we examine the model to see if it is adequate. We want to see if we have left out some terms that seem called for, or included some that play no important role in the model. We also want to see if the model violates commonsense rules about what constitutes a good model. DR model checking and reformulation are discussed especially in Chapter 6.

CORRELATION   **17**

Table 1.5 Sample Data on Sons' and Fathers' Heights (inches); $n = 25$

i	X_i	Y_i	i	X_i	Y_i
1	68	69	14	73	73
2	72	71	15	78	75
3	73	71	16	70	72
4	70	71	17	72	69
5	72	69	18	72	70
6	69	68	19	71	70
7	75	73	20	69	70
8	72	67	21	71	73
9	69	72	22	68	64
10	69	68	23	69	72
11	75	73	24	75	74
12	72	68	25	75	75
13	69	72			

We may also check the model residual series for outliers (unusual values). A study of outliers may lead us to a better understanding of the data, better model identification, better estimated model coefficients, and better forecasts. A method for detecting and treating outliers is discussed in Chapter 8.

Parsimony

Box and Jenkins emphasize the important principle of *parsimony*. According to this principle, we want to find the simplest adequate model—one that contains the fewest coefficients needed to adequately explain the behavior of the observed data. A parsimonious model makes good use of a limited sample and tends to give more accurate forecasts. See Ledolter and Abraham (1981) for a study of the importance of parsimony in ARIMA model forecasting.

1.4 CORRELATION

We will use the idea of correlation in many places in this book. In this section we summarize and illustrate the most basic ideas about correlation.

Consider the 25 pairs of sample observations in Table 1.5. The random variable X_i is the height, in inches, of a sample of fathers. The random variable Y_i is the height, also in inches, of the college-age sons of these fathers. Each father's height is paired with his son's height. For example, the first father in the list is $X_1 = 68$ inches tall; this father's son is $Y_1 = 69$ inches tall. These are not time-sequenced data of the sort that are of special

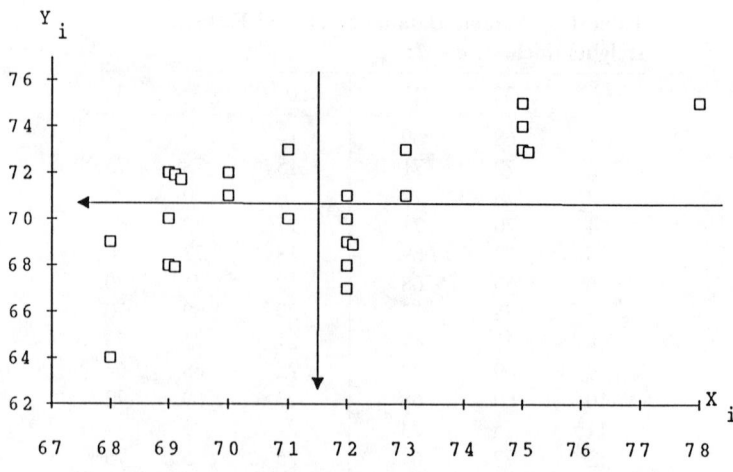

Figure 1.10 Son's height (Y_i) vs. father's height (X_i).

interest to us in this book. However, the ideas we develop using this example are also relevant to time series data.

Because of genetic factors, we expect to see a positive relationship between father's height and son's height: It seems likely that taller fathers will tend to have taller sons and that shorter fathers will tend to have shorter sons. Figure 1.10 is a scatterplot of the heights data in Table 1.5. Each point plotted in Figure 1.10 represents *both* an X_i value (on the horizontal axis) *and* its corresponding Y_i value (on the vertical axis). This scatterplot tends to confirm our expectation that Y_i and X_i are positively related: Higher (lower) X_i values tend to correspond to higher (lower) Y_i values. Of course, there are likely to be other factors that also affect a son's height. For example, the mother's height and the quality of the nutrition received by the son during his growing years would probably affect the son's height.

Correlation measures the strength and direction of the linear relationship between two random variables such as father's height and son's height. Of course, the data set in Table 1.5 (and Figure 1.10) is only one *sample* of 25 pairs drawn from the unobserved *population* of all possible pairs of father's height and son's height. In classical statistical inference, our task is to say something in probability terms about an unknown population parameter based on an observed sample statistic.

1.4.1 Population Correlation Coefficient

In the present case the parameter is the *population correlation coefficient* for two random variables (father's height and son's height, in our example), defined as

$$\rho = \text{cov}(X, Y)/\sigma_x \sigma_y \tag{1.4.1}$$

CORRELATION

where σ_x and σ_y are the population standard deviations of random variables X and Y, respectively, and $\text{cov}(X, Y)$ is the population covariance of X and Y. As usual, σ_x is the square root of the population variance σ_x^2, which is defined as the expected value

$$\sigma_x^2 = E(X_i - \mu_x)^2 \quad \text{distance from pop mean} \tag{1.4.2}$$

The symbol μ_x is the expected value (population mean) of the random variable X,

$$\mu_x = E(X_i) \tag{1.4.3}$$

σ_y, σ_y^2, and μ_y are similarly defined for random variable Y. Dividing $\text{cov}(X,Y)$ by the product $\sigma_x \sigma_y$ in (1.4.1) conveniently standardizes the covariance so that ρ has values only in the inclusive range between -1 and $+1$.

The covariance of X and Y is defined as the expected value,

$$\text{cov}(X, Y) = E[(X_i - \mu_x)(Y_i - \mu_y)] \tag{1.4.4}$$

If higher Y_i values tend to go with higher X_i values, and if lower Y_i values tend to go with lower X_i values, then $\text{cov}(X, Y)$ is positive and X and Y are positively correlated ($\rho > 0$). But if higher Y_i values tend to go with lower X_i values, and if lower Y_i values tend to go with higher X_i values, then $\text{cov}(X, Y)$ is negative and X and Y are negatively correlated ($\rho < 0$). Thus, inspection of Figure 1.10 suggests that the correlation between son's height and father's height may be positive. (But remember that Figure 1.10 is only sample information, not population information.)

We can understand covariance more clearly by considering the definition of $\text{cov}(X, Y)$ in (1.4.4). Suppose that Y_i tends to lie above μ_y when X_i is above μ_x, and that Y_i tends to lie below μ_y when X_i is below μ_x; then the product $(X_i - \mu_x)(Y_i - \mu_y)$, and therefore $\text{cov}(X, Y)$, will tend to be positive. Alternatively, suppose that Y_i tends to lie below μ_y when X_i is above μ_x, and that Y_i tends to lie above μ_y when X_i is below μ_x; then the product $(X_i - \mu_x)(Y_i - \mu_y)$, and therefore $\text{cov}(X, Y)$, will tend to be negative.

Figure 1.10 illustrates these ideas; but remember that Figure 1.10 is a sample rather than a population. The vertical arrow points to the sample mean of the X_i values,

$$\overline{X} = \frac{1}{n} \sum_{i=1}^{n} X_i = 71.52 \text{ inches}$$

where $n = 25$ in this example. Similarly, the horizontal arrow points to the sample mean ($\overline{Y} = 70.76$ inches) of the Y_i values. Notice that X_i values above

$\overline{X}$ tend to be paired with $\overline{Y}_i$ values above $\overline{Y}$, while X_i values below $\overline{X}$ tend to be paired with Y_i values below $\overline{Y}$. This suggests that $\text{cov}(X, Y)$ is positive, and therefore that ρ is positive for these two variables. However, the population quantities $\text{cov}(X, Y)$ and ρ are unobserved. We want sample statistics that estimate these values.

1.4.2 Sample Correlation Coefficient

The *sample correlation coefficient*, which provides an estimate of ρ, is computed as

$$r = c(X, Y)/s_x s_y \quad \text{stan dev} \tag{1.4.5}$$

where s_x and s_y are the sample standard deviations of X_i and Y_i, respectively, and $c(X, Y)$ is the sample covariance of X and Y. As usual, s_x is the square root of the sample variance, which is computed as

$$s_x^2 = (n-1)^{-1} \sum_{i=1}^{n} (X_i - \overline{X})^2 \quad \text{distance from sample mean}$$

s_y is obtained similarly. The sample covariance is computed as

$$c(X, Y) = (n-1)^{-1} \sum_{i=1}^{n} (X_i - \overline{X})(Y_i - \overline{Y})$$

For the data in Table 1.5, we find $s_x = 2.61598$, $s_y = 2.63439$, $c(X, Y) = 4.21071$, and $r = 0.611$. Inspection of the sample data in Figure 1.10 suggests that the unobserved population value ρ is positive. The formal measure of sample correlation for these data, $r = 0.611$ from Equation (1.4.5), supports this conclusion.

1.4.3 Testing Correlation Coefficient for Significance

Of course, r is only a sample measure and is thus subject to sampling variation. But $r = 0.611$ may be considered significantly different from zero since it is more than two standard errors away from zero. We can determine this by first finding the standard error of r, denoted $s(r)$, as

$$\begin{aligned} s(r) &= [(1 - r^2)/(n-2)]^{1/2} \\ &= [(1 - 0.611^2)/(25 - 2)]^{1/2} \\ &= 0.16506 \end{aligned} \tag{1.4.6}$$

LAYOUT OF THE BOOK 21

Dividing r by $s(r)$ gives $0.611/0.16506 = 3.70$; this says that $r = 0.611$ lies 3.70 standard errors above zero.

We can make our decision about the significance of r more formally using hypothesis testing. Let's test the null hypothesis that there is no linear association in the population between father's height and son's height. That is, we hypothesize

$$H_0: \rho = 0 \qquad (1.4.7)$$

with alternate hypothesis $H_a: \rho \neq 0$. If the sample value r is far enough from zero at our chosen α level (risk of type I error), we reject H_0. If H_0 is true, then in repeated sampling the test statistic $t = r/s(r) = 0.611/0.16506$ is t distributed with $n - 2$ degrees of freedom. For our example, with $\alpha = 0.05$ and $n - 2 = 25 - 2 = 23$ degrees of freedom, the critical t value is ± 2.069. The computed test statistic is $t = 3.70$; because this falls beyond the critical t value, we reject H_0 at the 5% level.

1.5 LAYOUT OF THE BOOK

You may think of this book as coming in two "parts," bound together for your convenience, with the two parts interspersed. Part I (Chapters 1–10) contains the main ideas that are needed to understand DR models. While these chapters are somewhat theoretical, the theory is at a low level. These chapters have a practical orientation and flavor, with examples to clarify the ideas. In Part II (Cases 1–6) the theory is applied to real data to illustrate how well DR analysis works out in practice (sometimes).

Chapter 2 is a discussion of ARIMA models. Chapter 3 is a review of the basic ideas about regression. In Chapter 4 we consider some special ways of writing models when the output is related to an input with a time lag. This leads to Chapter 5 where we consider the practical matter of model building. (How can we check for feedback? What steps do we follow when trying to find a useful DR model for a real data set? Why follow those steps?) Chapter 6 discusses DR model diagnostic checking, reformulation, and evaluation. Chapter 7 extends Chapters 4 through 6 to include *intervention analysis*, where the input variables are binary variables constructed to represent an event (like the tax rebate event in Example 2 in Section 1.1). Chapter 8 is an application of earlier ideas to the problem of *detecting* and treating *outliers* (unusual values) in time series data. Chapter 9 is a discussion of DR model estimation and forecasting procedures. Finally, Chapter 10 explains how DR models are related to vector ARMA models.

QUESTIONS AND PROBLEMS

1. Can you think of any two logically related events that do not occur exactly simultaneously, but instead with a time lag?

2. Can you think of any two logically related events that occur exactly simultaneously, with no time lag at all? Which question is more difficult to answer, this one or Question 1? Why?

3. Develop a headache. Or, imagine that you have one. Let the output variable be Y_t = headache pain level, measured subjectively. Suppose your initial pain level is indicated by the level of the horizontal line on the graph in part (b). The arrow indicates the time when you take a headache pain remedy.
   a. Fill in appropriate numbers on the time axis, starting with $t = 0$ at the arrow. (Equal distances on the time axis might represent 2-minute intervals or 10-minute intervals.)
   b. Extend the path of the pain level line across the rest of the graph.

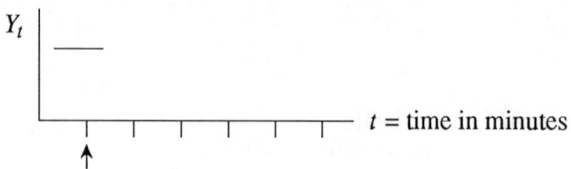

   Y_t

   t = time in minutes

   c. Is there an input variable X_t in this situation? If so, how is it measured? Construct a graph, similar to the preceding, showing the behavior of X_t over time.
   d. Are Y_t and X_t stochastic or deterministic? How does your answer show up in the appearance of the graphs you have drawn in parts (b) and (c)?
   e. How would the graph of Y_t be different if t were measured in seconds? ... if t were measured in hours? ... if you were immune to the pain remedy? ... if you were allergic to the pain remedy?

4. For the following data compute the sample correlation (r) between Y_t and X_t.

t	X_t	Y_t
1	5	11
2	2	9
3	4	8
4	8	8
5	6	14
6	3	12

t	X_t	Y_t
7	6	8
8	8	11
9	3	12
10	2	9
11	7	7
12	5	12
13	4	9
14	4	9
15	6	8
16	6	10
17	1	11
18	2	6
19	5	9
20	4	10

a. Is r significantly different from zero? What is your standard for determining the significance of r?

b. Plot Y_t separately against X_t, X_{t-1}, and X_{t-2}. What linear equation, similar to Equation (1.2.1) or (1.2.2), seems appropriate for expressing the possible relationship between the output and the input?

5. Under what circumstances might a researcher be willing to work with a biased estimator or with an inconsistent estimator? For an introductory discussion, see Kennedy (1985, Chapter 2).

CHAPTER 2

A Primer on ARIMA Models

2.1 INTRODUCTION

In Chapter 1 we introduced the idea of an autoregressive integrated moving average (ARIMA) model as a way of describing how a time series variable is related to its own past values. In this chapter we discuss this idea in more detail. For a more complete discussion of ARIMA models at an introductory level see Pankratz (1983). For a more advanced presentation see Box and Jenkins (1976).

2.1.1 Weighted Average Forecasts and ARIMA Models

The solid line in Figure 2.1 is 60 quarterly observations of the change in business inventories for the United States, from 1958 I to 1969 IV. The data are listed in the Data Appendix as Series 6. Suppose the present time is $t = 52$ so that only the first $n = 52$ observations (denoted as $z_1, z_2, \ldots, z_{52}$) are available; we will treat the remaining 8 observations ($z_{53}, z_{54}, \ldots, z_{60}$) as unknown "future" values. Suppose we want to forecast these 8 future values using only the 52 available observations of z_t, and no other information.

How might we use these past observations to forecast? One possibility is to use the sample arithmetic mean of the past values, $\bar{z} = (\frac{1}{52}) \Sigma_{t=1}^{52} z_t = 5.7$, as the forecast for all eight future time periods. This constant value is plotted on the right side of Figure 2.1 as the horizontal dashed line. Using $\bar{z}$ as the forecast of a time series is appealing only if the data return regularly to a fixed overall level.

The arithmetic mean is a special case of a weighted mean,

$$\bar{z}_c = \frac{\sum\limits_{t=1}^{n} c_t z_t}{\sum\limits_{t=1}^{n} c_t} \qquad (2.1.1)$$

INTRODUCTION

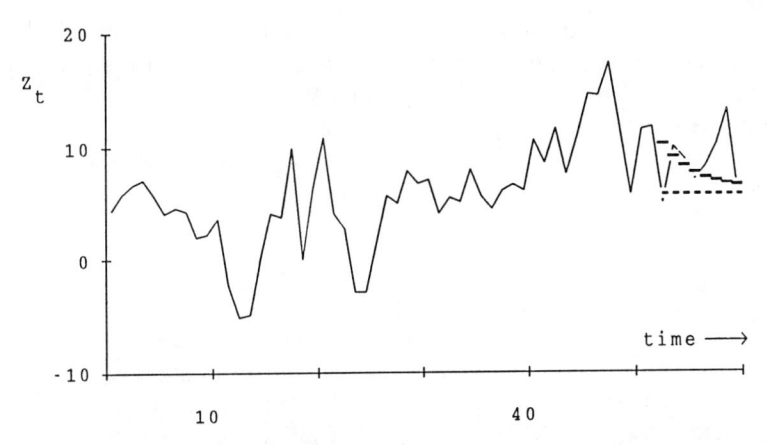

Figure 2.1 Change in business inventories, 1955 I–1969 IV.

where c_t is the numerical weight given to each observation z_t. For the ordinary arithmetic mean, implicitly $c_t = 1.0$ for all t, and the sum of the weights in the denominator of (2.1.1) is n. Perhaps a weighted average of the available observations, with weights different from 1.0, would produce a better set of forecasts than the ordinary arithmetic mean. In particular, it seems intuitively appealing to give more weight to the most recent observations and less weight to observations in the distant past.

In this chapter we discuss how to use the *Box–Jenkins modeling strategy* to produce the (theoretically) *best weighted average* forecasts for a single time series. These forecasts are produced by the appropriate *single equation* (*univariate*) *autoregressive integrated moving average* (*ARIMA*) *model* for that series. A single equation ARIMA model states how any value in a *single time series* is linearly related to *its own past* values. Any forecast from an ARIMA model is a *weighted average of past values* of the series. The Box–Jenkins modeling strategy is designed to help us choose an appropriate model. In theory, if we have the "correct" model, the ARIMA weighted average forecasts are "best" (the mean of the squared forecast errors is minimized). The steplike horizontal lines at the right of Figure 2.1 are forecasts from an ARIMA model for this data series. In this example these ARIMA weighted average forecasts track the eight future values better than the arithmetic mean forecast $\bar{z} = 5.7$ (the horizontal dashed line).

2.1.2 Example: An ARIMA Process and Model

An ARIMA *process* corresponds to the *population mechanism* that generates the time series. A *model* is *based on sample data*. We hope that any ARIMA model we build is a *useful approximation* of the true but unobservable underlying process. If a model is a good approximation of a process, the model tends to mimic the behavior of the process. Thus forecasts from the model may provide useful information about future values of the series. As a rough rule, we need about $n = 50$ or more observations (at equally spaced time intervals) to build an ARIMA model.

Many different ARIMA processes and models are possible, but just a few tend to occur often in practice. A simple example of an ARIMA process for a series z_t is

$$z_t = C + \phi_1 z_{t-1} + a_t \quad (2.1.2)$$

C and ϕ_1 are parameters to be estimated, and a_t is a random shock term. Process (2.1.2) says that any value z_t is related to the immediately previous value z_{t-1} through coefficient ϕ_1. The values of the random shock a_t can't be predicted from the ARIMA process. However, a_t is assumed to take on different values according to a certain probability density function: a_t is assumed to be zero-mean and normally distributed white noise (defined in Chapter 1). In addition, a_t is assumed to be independent z_{t-1}.

Consider again the business inventories data in Figure 2.1. Using only the first 52 observations, I found that the following ARIMA model provides a good fit to the data:

$$z_t = \underset{(2.30)}{1.65} + \underset{(7.39)}{0.73 z_{t-1}} + \hat{a}_t \quad (2.1.3)$$

This estimated model corresponds to the process in (2.1.2). The estimated constant is $\hat{C} = 1.65$; the estimate of ϕ_1 is $\hat{\phi}_1 = 0.73$; and a_t is written with a circumflex (^) to represent *model residuals* in contrast to process random shocks. The numbers in parentheses below each estimated coefficient are the corresponding t values (ratio of coefficient to standard error). Since both t values are more than 2.0 in absolute value, we conclude that both coefficients are significantly different from zero.

There is no guarantee that these data really were generated by process (2.1.2); but model (2.1.3) seems to provide a useful approximation to the historical behavior of the first 52 observations. Later in this chapter we discuss how to choose an ARIMA model for a given data set and how to produce forecasts from an ARIMA model. The steplike horizontal lines at the right of Figure 2.1 are eight forecasts (for $t = 53, 54, \ldots, 60$) produced by model (2.1.3), using the change in business inventories data, observations z_1 through z_{52}.

2.1.3 ARIMA Analysis and DR Models

Why do we study ARIMA processes and models in connection with dynamic regression (DR) models? There are at least five reasons:

1. Suppose we want to forecast Y_t using a DR model. If we first build a preliminary ARIMA model for Y_t, we establish a *baseline* model and set of forecasts. We can then compare our DR model and forecasts for Y_t with the ARIMA model and forecasts for Y_t. If the DR model cannot fit or forecast the data any better than an ARIMA model, we may not want to go to the extra trouble of working with a DR model.
2. As discussed in Chapter 1, the disturbance series (N_t) in a dynamic regression will often be autocorrelated. Often we can represent that pattern very well with an ARIMA model for the disturbance; this will improve the DR model and its forecasts.
3. To forecast Y_t using a DR model, we may also need forecasts of the inputs ($X_{1,t}, X_{2,t}, \ldots$). Often we will be able to produce forecasts for the inputs conveniently with ARIMA models for these series.
4. Often an ARIMA model for a DR model output or input series will reveal something interesting or useful about the data. This may influence the way we choose to model the data when we build a DR model.
5. Finally, ARIMA models for stochastic inputs are needed to perform diagnostic checks of the DR model's adequacy (discussed in Chapter 6) and to compute standard errors for the DR model forecasts (discussed in Chapter 9).

Before discussing ARIMA processes and models in more detail, we first develop two key ideas in time series analysis: (1) *stationary series* and (2) *autocorrelation*.

2.2 STATIONARY VARIANCE AND MEAN

Standard ARIMA analysis rests on the simplifying assumption that the process that generated a single time series is *stationary*. For our purposes *a stationary process is one whose mean, variance. and autocorrelation function are constant through time*. In this section we focus on the variance and mean. We discuss the autocorrelation function in Section 2.3. Our notion of stationarity is a *weak form*. There is also a *strong form*, which requires that the entire probability distribution function for the process is independent of time. For practical reasons it is common to work with the weak form. If the random shocks a_t are normally distributed, then the two forms are identical.

The stationarity assumption is a practical one. It permits us to develop both a relatively simple theoretical framework and some useful sample statistics. For example, if the process mean is constant, then we can use all n sample observations to estimate it. If the process mean were different each time period, we could not get useful estimates of its values since, in practice, we have only one observation per time period.

If an observed data series is not stationary, we can often modify the data

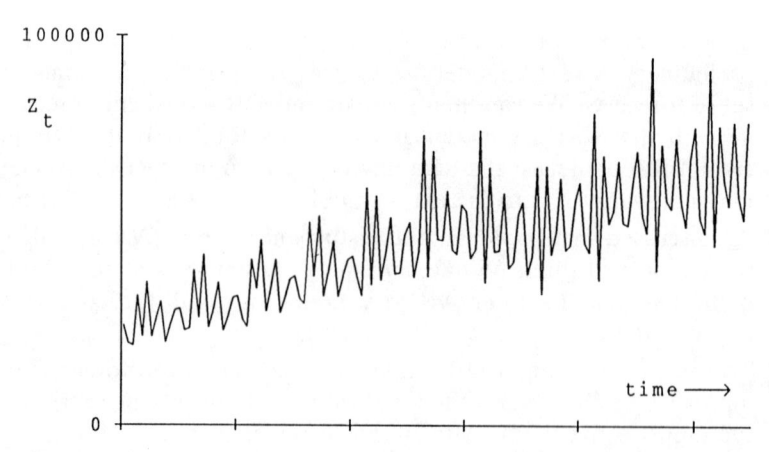

Figure 2.2 U.S. government receipts, January 1976–December 1986.

to produce a stationary series. Since the modifications are known to us, we can later reverse them so that the resulting forecasts are comparable to the original data. We first consider methods for achieving a stationary variance. Then we consider methods for achieving a stationary mean.

2.2.1 Stabilizing the Variance

Modifications to induce a stationary variance should be applied (if needed) *before* any further modifications or analysis of the data. The series in Figure 2.2 is monthly U.S. federal government budget receipts in billions of dollars, from January 1976 to December 1986. The data are listed in the Data Appendix as Series 7, along with 12 additional observations (for 1987) that will be used to evaluate forecasts. The variance of the series in Figure 2.2 seems to increase as its overall level increases. We may be able to induce a constant variance by transforming the data. For example, (1) if the *standard deviation* of a series is *proportional to its level*, taking the *natural logarithms* yields a new series with a constant variance; or (2) if the *variance* of the original series is *proportional to its level*, taking the *square root* induces a constant variance. Many other transformations are possible, but these two (especially the log transformation) are often useful in practice. The log transformation is both common and *interpretable*: the *changes* in a log value are relative (percent) changes in the original metric.

Usually we assume the data are normally distributed. The usual inferential procedures (such as t tests on model coefficients) depend on the normality assumption. Often a transformation to induce a constant variance will also bring the data closer to normality, but this will not always happen.

STATIONARY VARIANCE AND MEAN 29

Box–Cox Transformation

The log and square root transformations are members of a *family* of power transformations called the *Box–Cox transformation* (Box and Cox, 1964). With this transformation we define a new (transformed) series z'_t as

$$z'_t = \frac{z_t^\lambda - 1}{\lambda} \qquad (2.2.1)$$

where λ is a real number. Note that z_t must not be negative. If some values of z_t are negative, we add a positive constant to z_t so all values are positive. After modeling the data, we may then return forecasts of this series to the correct overall level by subtracting the same constant from the forecasts. (Getting back to the original metric can sometimes be a bit complicated; see the discussion in Section 9.2.3.)

For the case $\lambda = \frac{1}{2}$, for example, (2.2.1) gives the square root transformation since $z_t^{1/2}$ is the square root of z_t. Subtracting 1 from z_t^λ and dividing the result by λ will not alter the time structure of the variance of the series; these steps do, however, give (2.2.1) several attractive properties. (See Johnston, 1984, Chapter 5, for a more detailed discussion.) These steps (1) preserve the ordering of the data when λ is negative; (2) make (2.2.1) continuous in the limiting case $\lambda \to 0$; and (3) lead to the natural log transformation in the limiting case $\lambda \to 0$.

Inspection of a plot of the data may suggest an appropriate transformation. If the variance tends to *rise* as the level of the series *rises*, setting $\lambda < 1$ is called for. If the variance tends to *fall* as the level of the series *rises*, setting $\lambda > 1$ is called for. Further inspection of plots of the data after various transformations may confirm which transformation seems best. Visual inspection of data plots is a relatively simple approach to choosing a power transformation. However, in most cases the careful use of visual inspection seems to give results that are as good as more complicated approaches. Relatively simple or interpretable transformations are preferred; thus, we would tend to choose $\lambda = 0.5$ (the square root transformation) instead of its close neighbor $\lambda = 0.54$, and choose $\lambda = 0$ (the natural log transformation) instead of its close neighbor $\lambda = -0.08$.

The natural logs of the budget receipts series are plotted in Figure 2.3. The variance of the series is now roughly constant at all levels of the data. Therefore we will analyze the series $z'_t = \ln(z_t)$, where ln denotes natural logarithm. The resulting model will produce forecasts in the natural log metric, which may not be useful to the user of the forecasts. We can retransform the log metric forecasts to the original metric by taking antilogs of the forecasts. Further discussion of retransforming forecasts to the original metric appears in Chapter 9.

11.6

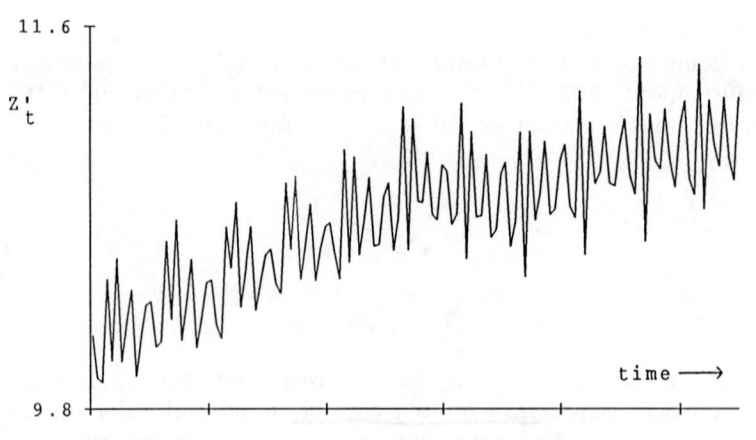

z'_t

time →

9.8

Figure 2.3 Log U.S. government receipts.

2.2.2 Differencing

Inspection of Figure 2.3 shows that z'_t does not fluctuate around a constant mean. Of course, we can *compute* a single mean for this series; but a series with a *stationary* mean *returns fairly quickly* to a constant mean. When a series does not return quickly to a single overall mean, usually we can create a new series with a constant mean by *differencing* the data. That is, we compute the *successive changes* in the series for all t, as follows:

$$w_t = z_t - z_{t-1} \qquad (2.2.2)$$

(If a variance stabilizing transformation has been used, we difference series z'_t instead of z_t.) Performing this calculation once, for all t, is called *first differencing*. If the resulting series does not yet have a constant overall mean, we then compute the first differences *of the first differences* for all t. That is, denote the first differences of z_t as w_t^*. Then the first differences of the w_t^* series are

$$\begin{aligned} w_t &= w_t^* - w_{t-1}^* \\ &= (z_t - z_{t-1}) - (z_{t-1} - z_{t-2}) \end{aligned} \qquad (2.2.3)$$

The resulting series is called the *second differences* of z_t. Let d denote the degree of differencing. For first differencing $d = 1$. For second differencing $d = 2$. If the original data lack a constant mean, usually setting $d = 1$ will create a new (differenced) series with a constant mean; setting $d > 2$ is almost never needed.

STATIONARY VARIANCE AND MEAN 31

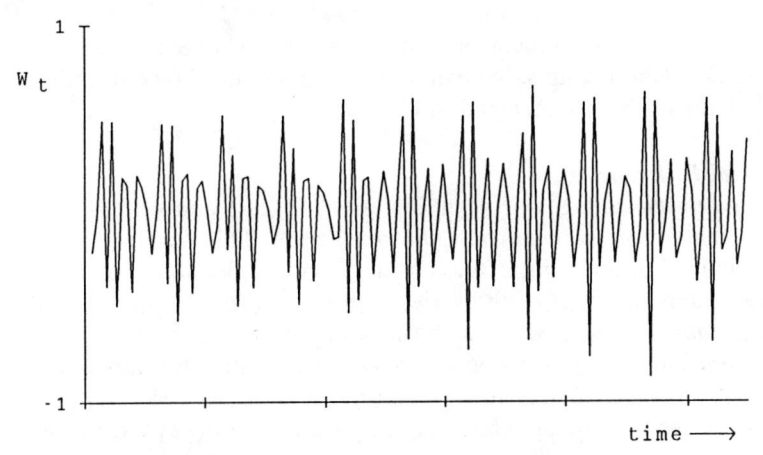

Figure 2.4 Log U.S. government receipts, $d = 1$.

Example
For the logs of the budget receipts data in Figure 2.2, before any differencing we have $d = 0$. To find the first differences of the log series we compute $w_t = z'_t - z'_{t-1}$. The following table shows the first four values of the log series:

t	z'_t
1	10.152
2	9.945
3	9.925
4	10.415

We cannot compute the differenced value $w_1 = z'_1 - z'_0$ because z'_0 is not observed. But we can find the remaining w_t values as follows:

$$w_2 = z'_2 - z'_1 = 9.945 - 10.152 = -0.207$$
$$w_3 = z'_3 - z'_2 = 9.925 - 9.945 = -0.020$$
$$w_4 = z'_4 - z'_3 = 10.415 - 9.925 = 0.490$$

Figure 2.4 shows the first differences ($d = 1$) of the logs of the budget receipts series. The data now seem to return quickly to a constant overall mean; it seems for the moment that no further differencing is needed. We discuss differencing further in the next section.

One way to make a decision about the value of d is to inspect a plot of the data. When the data show a strong uptrend (as in Figure 2.3) or downtrend, it is a fairly safe bet that $d > 0$ is needed. But a series can be nonstationary in the mean without showing a persistent uptrend or downtrend. As we will see in Section 2.3, we also get information about the need for differencing

Seasonal Differencing

The computed changes of z_t in Equations (2.2.2) and (2.2.3) are between values that are *separated by just one time period*. This is called "nonseasonal" or "regular" differencing. But consider Figure 2.3 again: The wavy pattern suggests that the level of the series might be *shifting regularly* according to which month is observed. And consider Figure 2.4 (with $d = 1$) again: While the *overall* mean seems to be constant, observations for *certain months* are *regularly above or below* this overall mean. For example, June values are always above that mean, while February values are always below it. It seems that the level of the series is shifting in a *periodic (or seasonal)* fashion: the level of any observation (w_t) is similar to the level of other observations that are separated from it by $s = 12$ time periods (w_{t+12}, w_{t-12}), $2s = 24$ time periods (w_{t-24}, w_{t+24}), $3s = 36$ time periods (w_{t-36}, w_{t+36}), and so forth.

Usually, *seasonal differencing* induces a constant mean in a series that shifts in a seasonal fashion. To perform seasonal differencing, we compute the *successive changes between observations separated by s time periods*, where s is the number of seasons. For quarterly data, $s = 4$; for monthly data, $s = 12$; and so forth. A series may be differenced nonseasonally only, seasonally only, or both ways. Let D denote the degree of seasonal differencing (the number of times seasonal differencing is performed). If $d = 0$, a seasonally differenced series ($D = 1$) is computed for all t as

$$w_t = z_t - z_{t-s} \tag{2.2.4}$$

(If a variance stabilizing transformation has been used, seasonal differencing is applied to z'_t.) Almost always, setting $D = 1$ removes any large seasonal shifts in the level of a series. If both nonseasonal and seasonal differencing are used, either one may be done first; the result is always the same.

Example

For the logs of the budget receipts series we found the regular first differences (denoted now as w^*_t), $w^*_t = z'_t - z'_{t-1}$, plotted in Figure 2.4. Then the regularly *and* seasonally differenced series ($d = 1$, $D = 1$, with $s = 12$) for all t is

$$w_t = w^*_t - w^*_{t-12}$$
$$= (z'_t - z'_{t-1}) - (z'_{t-12} - z'_{t-13}) \tag{2.2.5}$$

Following are the first 15 values of the nonseasonal first differences w^*_t, for $t = 2, 3, \ldots, 16$, for the budget receipts data:

STATIONARY VARIANCE AND MEAN   33

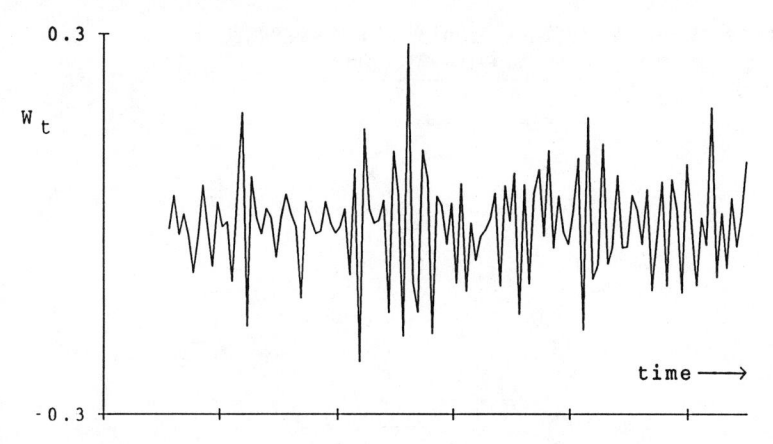

[Figure: plot of w_t vs time, ranging from -0.3 to 0.3]

Figure 2.5 Log U.S. government receipts, $d = D = 1$.

t	w_t^*
2	−0.207
3	−0.020
4	0.490
5	−0.386
6	0.484
7	−0.488
8	0.191
9	0.149
10	−0.412
11	0.201
12	0.137
13	0.016
14	−0.214
15	0.026
16	0.473

We cannot compute $w_t = w_t^* - w_{t-12}^*$ for $t < 14$ for these data since w_t^* is not available before $t = 2$. But starting with $t = 14$, we get the following results for w_t:

$$w_{14} = w_{14}^* - w_2^* = -0.214 - (-0.207) = -0.007$$
$$w_{15} = w_{15}^* - w_3^* = 0.026 - (-0.020) = 0.046$$
$$w_{16} = w_{16}^* - w_4^* = 0.473 - 0.490 = -0.017$$

The resulting series is plotted in Figure 2.5. This series returns quickly to the overall mean (approximately zero), *and* it does not shift around the overall mean in an obvious seasonal fashion. For this example we have

Table 2.1 Partial List of Change in Business Inventories Data (z_t)

t	z_t	z_{t+1}
1	4.4	5.8
2	5.8	6.7
3	6.7	7.1
4	7.1	5.7
⋮	⋮	⋮
50	5.8	11.5
51	11.5	11.7
52	11.7	—

created a new series w_t that has both a stationary variance and a stationary mean.

As in choosing the value of d, we might choose the value of D by inspecting a plot of the data. When the data show strong seasonal variation (as in Figures 2.3 or 2.4), seasonal differencing is a fairly safe bet. But we also get information about the need for seasonal differencing from (1) the sample autocorrelation function of the data and (2) the ARIMA model estimation results. These matters are discussed further later.

2.3 AUTOCORRELATION

In Chapter 1 we reviewed the idea of correlation between two random variables. With time series data we are often interested in *auto*correlation patterns: We want to measure how any current value of a series (z_t) is related to *its own* future values ($z_{t+1}, z_{t+2}, \ldots$), or, equivalently, to its own past values ($z_{t-1}, z_{t-2}, \ldots$). Thus autocorrelation measures the direction (positive or negative) and strength of the relationship among observations *within a single time series* z_t when the observations are *separated by k time periods*, for $k = 1, 2, 3, \ldots, K$. In this case we treat series z_t as one random variable and series z_{t+k} as another random variable. We then consider the correlation coefficient for the two random variables z_t and z_{t+k} in the same way that we considered the correlation coefficient for the two random variables X_i and Y_i in Section 1.4. A study of the autocorrelation patterns in a data series often can lead us to identify an *ARIMA* model for that series.

2.3.1 Autocorrelation Coefficients

The second column in Table 2.1 is a partial list of observations for the inventory changes data (plotted in Figure 2.1) under the heading z_t. We will continue to use only the first $n = 52$ observations for our example. Suppose

AUTOCORRELATION

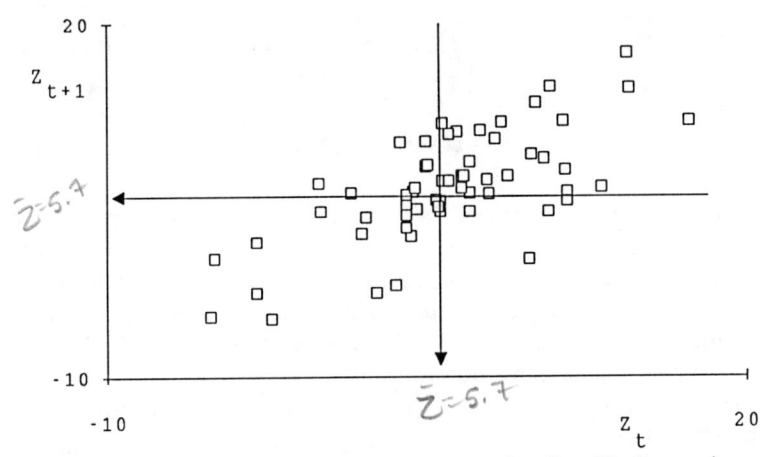

Figure 2.6 Inventory changes, z_{t+1} vs. z_t, using first 52 observations.

we want to consider the correlation between z_t and z_{t+1} ($k = 1$). That is, we want to associate each observation at any time t with the corresponding observation at time $t + 1$. This is illustrated in Table 2.1: We create the column labeled z_{t+1} by offsetting the column labeled z_t by one time period (moving column z_t up by one time period). Notice that $z_t = z_{52} = 11.7$ is not paired with a corresponding value $z_{t+1} = z_{53}$ because z_{53} is not observed (we are treating $z_{53}, z_{54}, \ldots, z_{60}$ as if they were unobserved). However, each remaining observation z_t for $t = 1, 2, \ldots, 51$, is now paired with the corresponding value z_{t+1}, for $t + 1 = 2, 3, \ldots, 52$.

These data are plotted in Figure 2.6, with z_t on the horizontal axis and z_{t+1} on the vertical axis. The vertical and horizontal arrows each represent the sample mean ($\bar{z} = 5.7$) of the first $n = 52$ observations. In Figure 2.6, when z lies above $\bar{z}$, we see that z_{t+1} also tends to lie above $\bar{z}$. And when z_t lies below $\bar{z}$, we see that z_{t+1} also tends to lie below $\bar{z}$. This pattern suggests that z_t and z_{t+1} are *positively correlated*.

We could create tables and data plots similar to Table 2.1 and Figure 2.6 for other values of $k > 0$. For each k we offset the column of z_t observations by k time periods to create the column z_{t+k}; this creates two columns of $n - k$ ($= 52 - k$ for this example) ordered pairs of observations. Thus we can have *many autocorrelation coefficients* for a single data series z_t, one for each k. Since we lose another observation on z_{t+k} each time k increases by one, the maximum useful value of k (denoted by K) is somewhat less than n; a *rough* rule is to choose $K \leq n/4$, where n is the sample size.

The data in Figures 2.1 and 2.6 and Table 2.1 are just a sample (a realization) drawn from the population consisting of all possible observations on the quarterly change in business inventories in the United States. From sample data we can compute sample autocorrelation coefficients. We use sample data to obtain information about the *population autocorrelation coefficient* at various lags $k = 1, 2, \ldots, K$. This theoretical coefficient is defined as

$$\rho_k = \text{cov}(z_t, z_{t+k})/\sigma_z^2 \quad (2.3.1)$$

where the population variance σ_z^2 is defined as the expected value $\sigma_z^2 = E(z_t - \mu_z)^2$; μ_z is the expected value (population mean) of the random variable z_t, $\mu_z = E(z_t)$; and $\text{cov}(z_t, z_{t+k})$ is defined as the expected value $\text{cov}(z_t, z_{t+k}) = E[(z_t - \mu_z)(z_{t+k} - \mu_z)]$. For a *stationary* series $\text{cov}(z_t, z_{t+k})$, and therefore ρ_k, are independent of t; they depend only on k, the number of time periods separating z_t and z_{t+k}.

The *sample autocorrelation coefficient*, which provides an estimate of ρ_k, is usually computed as

$$r_k = \frac{\sum_{t=1}^{n-k}(z_t - \bar{z})(z_{t+k} - \bar{z})}{\sum_{t=1}^{n}(z_t - \bar{z})^2} \quad (2.3.2)$$

This and other formulas for r_k are discussed by Jenkins and Watts (1968). For the first 52 observations of the business inventories data in Figure 2.1, Equation (2.3.2) gives

$$r_1 = \frac{(z_1 - \bar{z})(z_2 - \bar{z}) + (z_2 - \bar{z})(z_3 - \bar{z}) + \cdots + (z_{51} - \bar{z})(z_{52} - \bar{z})}{(z_1 - \bar{z})^2 + (z_2 - \bar{z})^2 + \cdots + (z_{52} - \bar{z})^2}$$

$$= \frac{(4.4 - 5.7)(5.8 - 5.7) + (5.8 - 5.7)(6.7 - 5.7) + \cdots + (11.5 - 5.7)(11.7 - 5.7)}{(4.4 - 5.7)^2 + (5.8 - 5.7)^2 + \cdots + (11.7 - 5.7)^2}$$

$$= 0.71$$

Values of r_k for $k = 2, 3, \ldots, K$ are computed in a similar fashion. The resulting set of values is the sample autocorrelation function, abbreviated SACF.

Significance Tests for Autocorrelation Coefficients

Any r_k is only a sample value that might differ from zero just because of sampling variation. We can get some idea of the importance of this sample statistic by comparing it with its standard error. An approximate standard error for r_k, due to Bartlett (1946), is

$$s(r_k) = \left(1 + 2\sum_{j=1}^{k-1} r_j^2\right)^{1/2} n^{-1/2} \quad (2.3.3)$$

AUTOCORRELATION

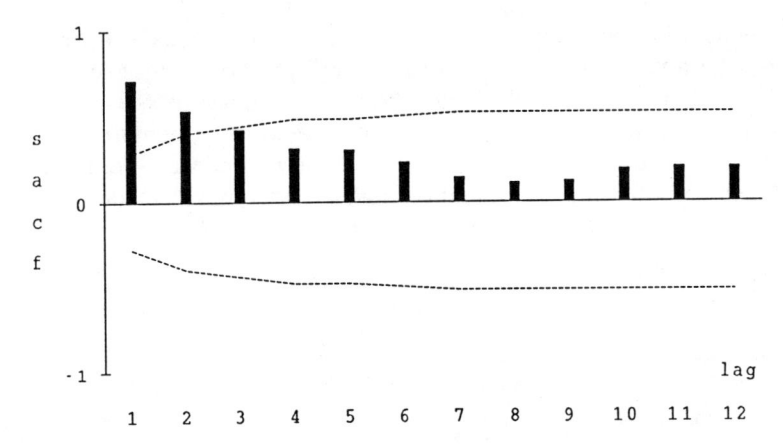

Figure 2.7 Sample ACF for the change in business inventories data in Figure 2.1 using the first 52 observations.

For example, for $k = 1$ for the business inventories data, we have $s(r_k) = (52)^{-1/2} = 0.14$. To test for a linear association in the population between z_t and z_{t+k}, we test the null hypothesis

$$H_0: \rho_k = 0 \tag{2.3.4}$$

against the alternate $H_a: \rho_k \neq 0$. We then compute the approximate t statistic, $t = (r_k - \rho_k)/s(r_k)$. Note that t is the ratio of the statistic r_k to its standard error $s(r_k)$ since ρ_k is hypothesized to be zero. In our example with $k = 1$, $t = (r_1 - \rho_1)/s(r_1) = (0.71 - 0)/0.14 = 5.07$; that is, $r_1 = 0.71$ falls more than 5 standard errors above zero. This t-value is considered significant at better than the 5% level since it exceeds the approximate 5% critical value of 2.0.

Graphical Form of SACF
It is convenient to present the SACF (the set of r_k values for $k = 1, 2, \ldots, K$) in graphical form. The SACF for the first 52 observations in Figure 2.1 is shown in Figure 2.7. For $k = 1$, for example, the value of $r_1 = 0.71$ is represented on the graph by the length of the bar at lag 1. The dashed lines on either side of zero are two standard errors from zero. Notice that r_1 falls well beyond the two standard error limit on the positive side of the graph. This is consistent with our computations showing that r_1 for this example is more than five standard errors above zero.

Stationary Mean and the SACF
Is the overall mean of the first 52 observations in Figure 2.1 stationary? *The SACF can help to answer this question.* If the mean is stationary, then the SACF will tend to decay *quickly* toward zero. In practice a quick decay means that the autocorrelation coefficients are well below their two standard

error limits by about lag 5 or 6: The ratio of the coefficients to their standard errors (the approximate t values) should fall to about 1.6 or less by about lag 5 or 6. In Figure 2.7 the autocorrelation coefficients fall to about one standard error above zero by lag 5 or 6. We conclude tentatively that the mean of the inventory changes series is stationary, and we do not difference. However, as we will see, estimation stage results may provide further evidence about whether differencing is needed.

2.3.2 Partial Autocorrelation Coefficients

Another useful measure of autocorrelation for stationary series is the *partial autocorrelation coefficient*. One way to think of this coefficient is to consider the set of K regression equations:

$$z_t = C_1 + \phi_{11} z_{t-1} + e_{1,t}$$
$$z_t = C_2 + \phi_{21} z_{t-1} + \phi_{22} z_{t-2} + e_{2,t}$$
$$\vdots$$
$$z_t = C_K + \phi_{K1} z_{t-1} + \phi_{K2} z_{t-2} + \cdots + \phi_{KK} z_{t-K} + e_{K,t} \quad (2.3.5)$$

(The main ideas about regression are reviewed in Chapter 3.) The *population* partial autocorrelation coefficient at lag $k = 1, 2, \ldots, K$ is the *last* coefficient (ϕ_{kk}) in each equation. Each population coefficient is estimated for a given data set by its *sample* counterpart $\hat{\phi}_{kk}$. The resulting set of values is the sample partial autocorrelation function, abbreviated SPACF. Recall that in computing r_k we considered only the two random variables z_t and z_{t+k}, and we ignored the intervening random variables $z_{t+k-1}, z_{t+k-2}, \ldots, z_{t+1}$. But in computing $\hat{\phi}_{kk}$ we simultaneously take into account the role of these intervening random variables, as seen in each equation in (2.3.5). Computationally efficient formulas for computing $\hat{\phi}_{kk}$ values are available (see Pankratz, 1983, p. 40).

We can gauge the significance of each $\hat{\phi}_{kk}$ by comparing it with its standard error,

$$s(\hat{\phi}_{kk}) = n^{-1/2} \quad (2.3.6)$$

It is convenient to present the SPACF (the set of estimates of the $\hat{\phi}_{kk}$ values for $k = 1, 2, \ldots, K$) in graphical form. Figure 2.8 shows the SPACF based on the first 52 observations of the inventory changes data. For $k = 6$, for example, the value $\hat{\phi}_{66} = -0.07$ is represented by the length of the bar at lag 6. The dashed lines on either side of zero at each lag are two standard error limits.

From (2.3.6) the standard error of the partial autocorrelations at each

FIVE STATIONARY ARIMA PROCESSES

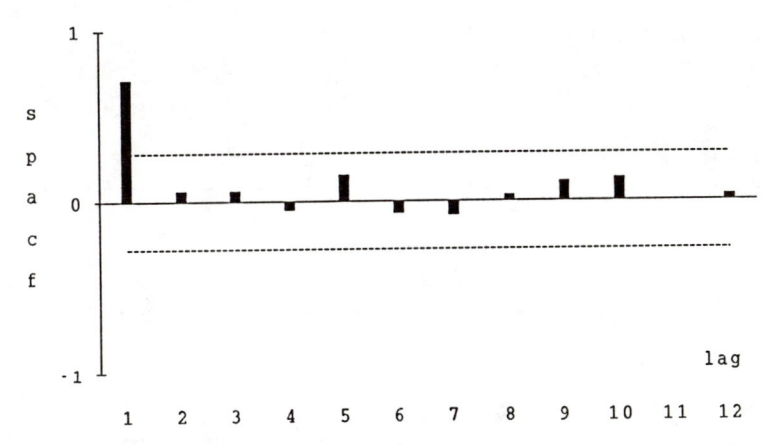

[PACF plot with bars at lags 1 through 12, y-axis labeled "s p a c f" ranging from -1 to 1, dashed lines indicating two-standard-error bounds]

Figure 2.8 Sample PACF for the change in business inventories data in Figure 2.1 using the first 52 observations.

lag for the inventory changes data is $n^{-1/2} = (52)^{-1/2} = 0.14$. The SPACF coefficient at lag 6 (-0.07) for these data is much smaller in absolute value than this standard error. This is seen in Figure 2.8, where the bar length of the SPACF coefficient at lag 6 falls far short of the two-standard-error dashed line on the negative side of the graph.

2.4 FIVE STATIONARY ARIMA PROCESSES

An ARIMA *model* is based on the available data; its theoretical (population) counterpart is an ARIMA *process*. Each ARIMA process has a *theoretical autocorrelation function (ACF)* and *partial autocorrelation function (PACF)* associated with it. To *identify* an ARIMA model in practice we first construct the SACF and SPACF for a given data series. Then we compare the SACF and SPACF with some common theoretical ACFs and PACFs. Upon finding a reasonable match, we choose the ARIMA process associated with the matching theoretical ACF and PACF as a tentative ARIMA model for the data. In this section we present some common *stationary* ARIMA processes and their associated theoretical ACFs and PACFs.

2.4.1 Autoregressive Processes

Two common and fairly simple ARIMA processes are

$$z_t = C + \phi_1 z_{t-1} + a_t \tag{2.4.1}$$

$$z_t = C + \phi_1 z_{t-1} + \phi_2 z_{t-2} + a_t \tag{2.4.2}$$

[We saw process (2.4.1) earlier in Equation (2.1.2), and a corresponding

(a) AR(1), $0 < \phi_1 = 0.8$

(b) AR(1), $0 > \phi_1 = -0.8$

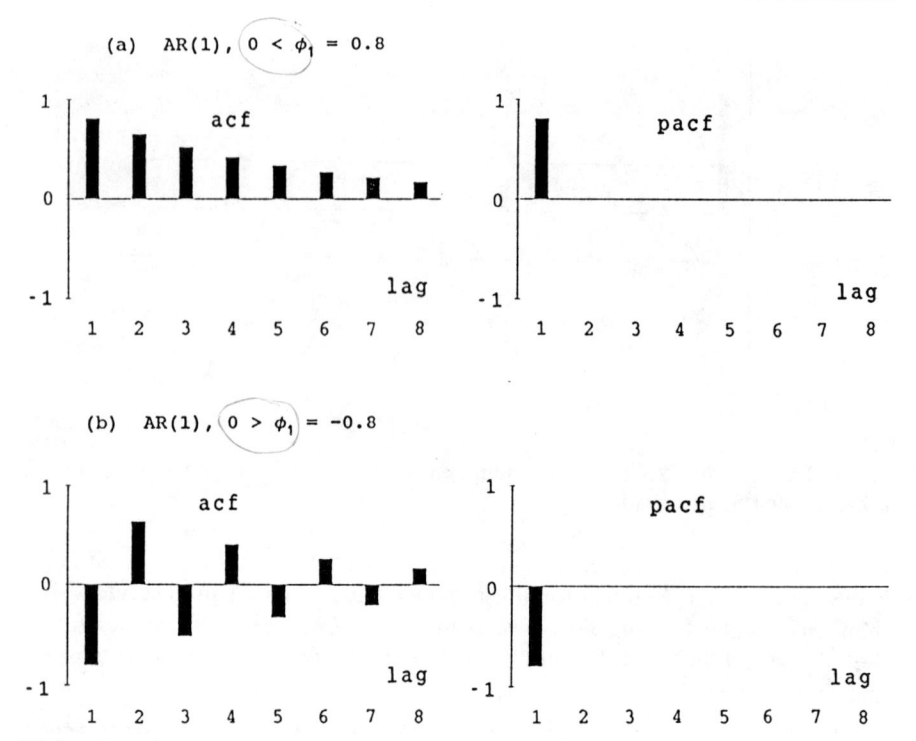

Figure 2.9 Two examples of theoretical ACFs and PACF's for AR(1) processes. (*a*) AR(1), $0 < \phi_1 = 0.8$. (*b*) AR(1), $0 > \phi_1 = -0.8$.

model in (2.1.3). An example of a model corresponding to process (2.4.2) appears in Case 1 following this chapter.] In these processes past z values $(z_{t-1}, z_{t-2}, \ldots)$ with their associated coefficients are called *autoregressive* (*AR*) terms. The *order* (*p*) of an AR process (or model) is the highest lag length of the AR terms. For example, (2.4.1) is a first-order ($p = 1$) AR process, denoted AR(1), while (2.4.2) is an AR(2) process. C is a constant that is related to both the mean (μ_z) of the series z_t and the AR coefficients (the ϕ's) as follows:

$$C = \mu_z \left(1 - \sum_{i=1}^{p} \phi_i \right) \qquad (2.4.3)$$

As discussed in Chapter 1, the random shock a_t is assumed to be zero mean and normally distributed white noise.

Theoretical ACFs and PACFs
Figure 2.9 shows examples of theoretical ACFs and PACFs associated with the AR(1) process in Equation (2.4.1). (We will not derive these theoretical ACFs and PACFs. See Pankratz, 1983, or Box and Jenkins, 1976, for deri-

FIVE STATIONARY ARIMA PROCESSES

vations.) In both cases the ACF decays exponentially to zero, while the PACF has one spike at lag 1. When $\phi_1 > 0$ as in Figure 2.9(a), the ACF decays all on the positive side, and the PACF spike is positive. When $\phi_1 < 0$ in Figure 2.9(b), the ACF decays with alternating signs starting on the negative side at lag $k = 1$, and the PACF spike is negative.

Figure 2.10 shows examples of ACFs and PACFs associated with the AR(2) process in Equation (2.4.2). The AR(2) has a greater variety of ACF and PACF patterns than the AR(1): the AR(2) ACF may show a compound exponential decay or a damped sine wave decay. However, in all cases the ACF *decays* to zero, and the PACF has *spikes* at lags 1 and 2.

Stationary AR processes of order $p > 2$ have ACF and PACF patterns similar to those in Figures 2.9 and 2.10. To summarize, *stationary AR processes* of order p have these characteristics:

1. The theoretical ACF *decays*, either exponentially or with a damped sine wave pattern or with both of these patterns.
2. The theoretical PACF has *spikes* through lag p, then all zeros. (Some values before lag p could be zero; the main point is that the last nonzero value occurs at lag p.)

Remember how we *use* theoretical ACFs and PACFs in practice: If we see a *sample* ACF and PACF pair that look much like one of the theoretical pairs shown in Figures 2.9 or 2.10, we identify an AR(1) or AR(2) model, respectively, as a reasonable tentative model for the available data. The model we choose corresponds to the process whose theoretical ACF and PACF match the SACF and SPACF. Then we estimate and check the tentatively identified model to see if it is adequate. We will illustate this modeling procedure in Section 2.5.

Stationarity Conditions

The ϕ coefficients satisfy certain conditions if the mean of the process is stationary. For an AR(1), stationarity requires that

$$|\phi_1| < 1 \tag{2.4.4}$$

For an AR(2), the following three conditions must all be met:

$$|\phi_2| < 1$$
$$\phi_2 + \phi_1 < 1 \tag{2.4.5}$$
$$\phi_2 - \phi_1 < 1$$

Stationarity conditions for the general AR(p) case are discussed in Section

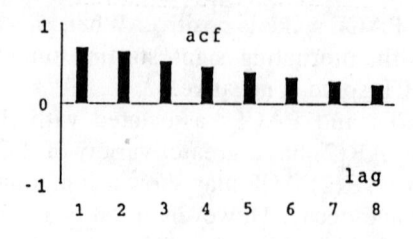

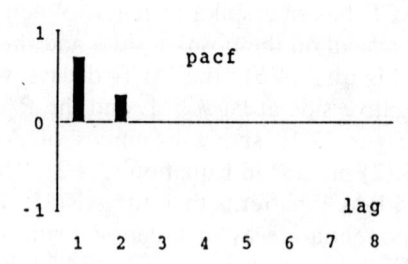

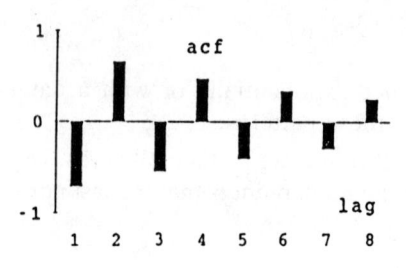

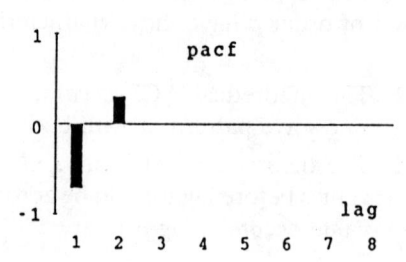

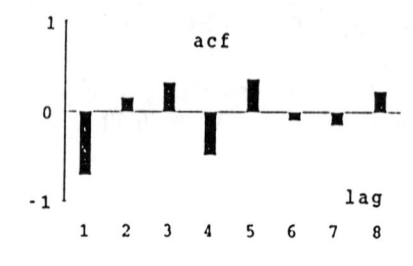

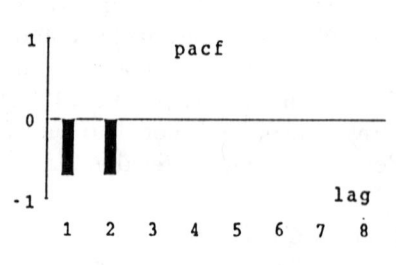

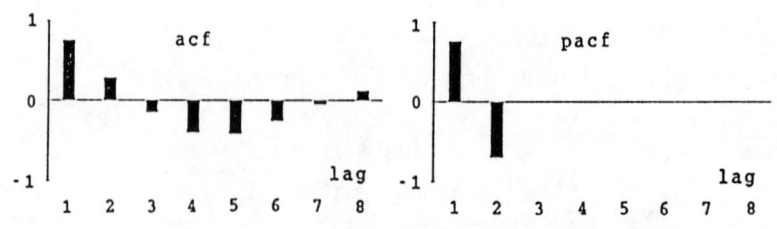

Figure 2.10 Four examples of theoretical ACFs and PACF's for AR(2) processes. (a) AR(2), $\phi_1 = 0.5$, $\phi_2 = 0.3$. (b) AR(2), $\phi_1 = -0.5$, $\phi_2 = 0.3$. (c) AR(2), $\phi_1 = -1.2$, $\phi_2 = -0.7$. (d) AR(2), $\phi_1 = 1.3$, $\phi_2 = 0.7$.

FIVE STATIONARY ARIMA PROCESSES

2.8. In practice we don't observe the AR process coefficients; instead we see if the estimated model coefficients satisfy the stationarity conditions.

2.4.2 Moving Average and Mixed Processes

Three other common processes are

$$z_t = C - \theta_1 a_{t-1} + a_t \qquad (2.4.6)$$

$$z_t = C - \theta_1 a_{t-1} - \theta_2 a_{t-2} + a_t \qquad (2.4.7)$$

$$z_t = C + \phi_1 z_{t-1} - \theta_1 a_{t-1} + a_t \qquad (2.4.8)$$

Past random shocks (a_t) with their associated coefficients are called *moving average* (*MA*) terms. A past random shock a_{t-k} is not a past value of z_t, but it is the *random shock component of* z_{t-k}; thus an MA term represents *part of* a past value of z_t. For example, the MA term in (2.4.6) is $-\theta_1 a_{t-1}$; here a_{t-1} is not the past value z_{t-1}, but it is a part of z_{t-1}. To see this, subtract 1 from each time subscript in (2.4.6) to obtain $z_{t-1} = C - \theta_1 a_{t-2} + a_{t-1}$. Thus, for this process a_{t-1} (on the right side of this expression) is a component of z_{t-1} (on the left side of this expression). As a convention, MA terms are written with negative signs.

The *order* (*q*) of an MA process (or model) is the highest lag length of the MA terms; for example, (2.4.6) is a first-order ($q = 1$) MA process, also referred to as MA(1), while (2.4.7) is an MA(2) process. Process (2.4.8) is a *mixed* [*both AR* (*autoregressive*) *and MA*] process with $p = 1$ and $q = 1$, referred to as ARMA(1, 1) (autoregressive moving average). C is a constant that is related to both μ_z and the AR coefficients (the ϕ's) if any are present, as shown in Equation (2.4.3).

Theoretical ACFs and PACFs

Figure 2.11 shows examples of theoretical ACFs and PACFs associated with the MA(1) process shown in (2.4.6). (For derivations of these theoretical functions, see Pankratz, 1983, or Box and Jenkins, 1976.) In both cases the ACF cuts off to zero after lag 1, while the PACF decays. When $\theta_1 > 0$ as in Figure 2.11(*a*), the ACF spike is negative, and the PACF decays all on the negative side. When $\theta_1 < 0$ as in Figure 2.11(*b*), the ACF spike is positive, and the PACF decay alternates in sign starting from the positive side at lag $k = 1$.

Figure 2.12 shows examples of theoretical ACFs and PACFs associated with the MA(2) process in (2.4.7). The MA(2) has a greater variety of ACF and PACF patterns than the MA(1). However, in all cases the ACF cuts off to zero after lag 2, and the PACF decays.

Figure 2.13 shows an example of a theoretical ACF and a PACF for an ARMA(1, 1) process. The distinguishing characteristic is that *both* the ACF and PACF *decay*. Mixed process of higher orders ($p > 1$ and/or $q > 1$) also

(a) MA(1), $0 < \theta_1 = 0.8$

(b) MA(1), $0 > \theta_1 = -0.8$

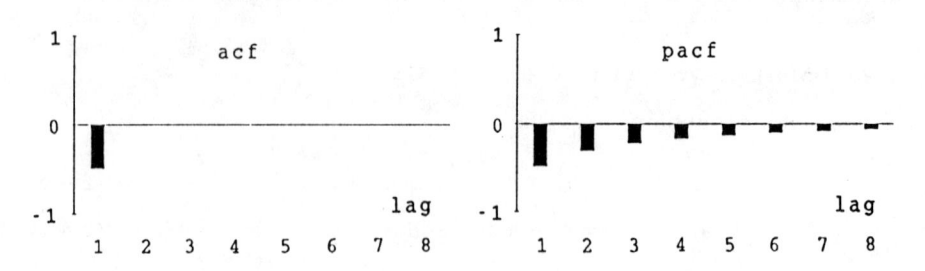

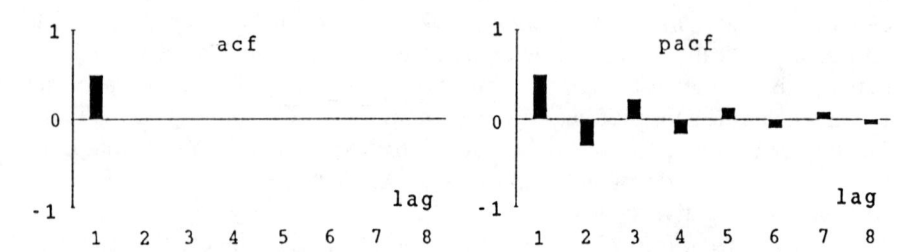

Figure 2.11 Two examples of theoretical ACFs and PACFs for MA(1) processes. (*a*) MA(1), $0 < \theta_1 = 0.8$. (*b*) MA(1), $0 > \theta_1 = -0.8$.

have a decay in *both* the ACF and PACF. Identifying the exact orders of a mixed model is somewhat more difficult than identifying the order of a pure AR or pure MA. An identification tool known as the *extended autocorrelation function* (EACF) is especially useful in identifying the orders of mixed models. The EACF is discussed in Section 2.10.

Other stationary MA processes of order q and mixed processes of orders p and q have ACF and PACF patterns similar to those in Figures 2.11, 2.12, and 2.13. To summarize,

1. An MA theoretical ACF has spikes through lag q, then all zeros. (Some values before lag q could be zero; the main point is that the last nonzero value occurs at lag q.)
2. An MA theoretical PACF *decays*.
3. A mixed theoretical ACF and PACF both *decay*.

Remember how we *use* theoretical ACFs and PACFs in practice: if we see a *sample* ACF and PACF pair that look much like one of the theoretical pairs in Figures 2.11, 2.12, or 2.13, we identify an MA(1), MA(2), or ARMA(1, 1) model, respectively, as a reasonable tentative model for the

FIVE STATIONARY ARIMA PROCESSES

(a) MA(2), $\theta_1 = 0.5$, $\theta_2 = 0.3$

(b) MA(2), $\theta_1 = -0.8$, $\theta_2 = 0.5$

(c) MA(2), $\theta_1 = -0.8$, $\theta_2 = -0.5$

(d) MA(2), $\theta_1 = 0.8$, $\theta_2 = -0.5$

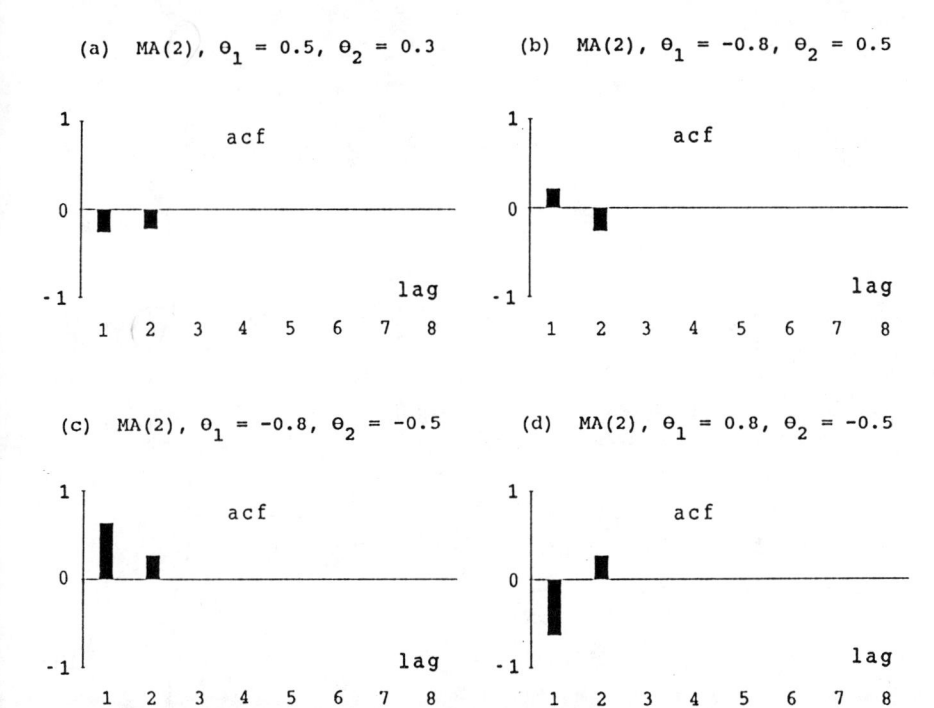

Figure 2.12 Four theoretical ACFs for MA(2) processes. (*a*) MA(2), $\theta_1 = 0.5$, $\theta_2 = 0.3$. (*b*) MA(2), $\theta_1 = 0.8$, $\theta_2 = 0.5$. (*c*) MA(2), $\theta_1 = -0.8$, $\theta_2 = -0.5$. (*d*) MA(2), $\theta_1 = 0.8$, $\theta_2 = -0.5$.

available data. The model we choose corresponds to the process whose theoretical ACF and PACF match the sample ACF and PACF. Then we estimate and check the tentatively identified model to see if it is adequate.

Invertibility

Earlier we discussed stationarity conditions for AR processes. MA processes have similar conditions that are required for *invertibility*. The common sense of the invertibility requirement is explained in two steps. (1) An MA process has an equivalent AR form (though this form is not parsimonious). This is discussed in more detail in Section 2.11.2. (2) Invertibility ensures that the absolute values of the *implied* weights on past z's in this equivalent AR form become smaller as the lag length on the past z's increases. Having declining weights on past z's satisfies a commonsense criterion: It seems sensible that more recent data should be more important for understanding the present than data in the distant past. (This may not always be correct; in psychoanalysis, for example, data from the distant past are often deemed to be more important than more recent data.)

It can be shown that the θ coefficients will satisfy certain conditions if the

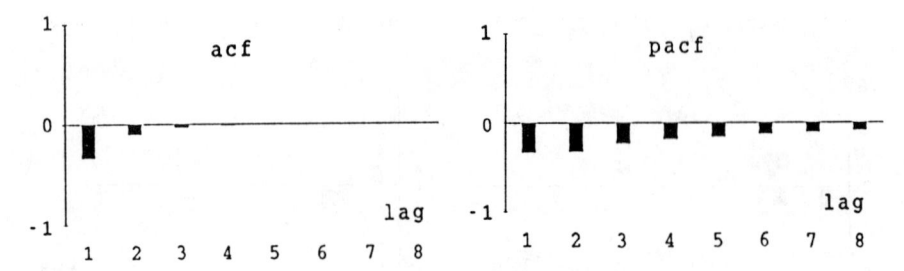

Figure 2.13 Example of theoretical ACF and PACF for ARMA(1,1) process with $\phi_1 = 0.3$, $\theta_1 = 0.9$.

process is invertible. For an MA(1) invertibility requires that

$$|\theta_1| < 1 \qquad (2.4.9)$$

For an MA(2) the following three conditions must all be met:

$$|\theta_2| < 1$$
$$\theta_2 + \theta_1 < 1 \qquad (2.4.10)$$
$$\theta_2 - \theta_1 < 1$$

Invertibility conditions for the general MA(q) case are discussed further in Section 2.8. In practice we don't observe the MA process coefficients; instead we see if the estimated model coefficients satisfy the invertibility conditions.

2.4.3 Integrated Processes

Equations (2.4.1), (2.4.2), and (2.4.6) through (2.4.8) are processes for series with a *stationary mean*. How are processes for nonstationary series written? The only change is that we start by writing the model for the *differenced* series w_t.

For example, suppose $d = 1$ and $D = 0$, and suppose the differenced series w_t is described by an ARMA(0, 1) process. This gives an ARIMA(p, d, q) = ARIMA(0, 1, 1) process for the original series z_t. This simple integrated process deserves special attention because it occurs so often in practice. Each forecast from an ARIMA(0, 1, 1) is an *exponentially weighted moving average* (EWMA) of the available data. For a one-step ahead forecast, for example, based on data through $t = n$, a certain weight is given to z_n, a smaller weight to z_{n-1}, an even smaller weight to z_{n-2}, and so forth. These weights decline in value exponentially. This aspect of the ARIMA(0, 1, 1) is discussed further in Section 2.11.2.

Let's write out the ARIMA(0, 1, 1). With $d = 1$ we are considering the

FIVE STATIONARY ARIMA PROCESSES

differenced series $w_t = z_t - z_{t-1}$. With $p = 0$ and $q = 1$, we write this process as

$$w_t = C - \theta_1 a_{t-1} + a_t \qquad (2.4.11)$$

where C is given by Equation (2.4.3) except that μ_z is replaced by μ_w, the mean of the differenced series w_t. As seen in (2.4.3), C is also related to any ϕ coefficients that may be present; since (2.4.11) has no AR terms, for this example we have $C = \mu_w$.

While process (2.4.11) is written for w_t, it *implies* a process for the original series z_t. We can see this by using the definition $w_t = z_t - z_{t-1}$. Substitute this definition into (2.4.11) and add z_{t-1} to both sides of the resulting equation to get this corresponding model for z_t:

$$z_t = z_{t-1} + C - \theta_1 a_{t-1} + a_t \qquad (2.4.12)$$

By adding z_{t-1} to both sides, after substituting for w_t, we are doing the opposite of differencing: We are integrating, and this accounts for the I in the acronym ARIMA. A process that includes differencing is referred to as an *integrated process* since we must integrate to get back to the process for the original series. In the present example, differencing introduces the term z_{t-1} on the right side of the process for z_t in (2.4.12). This looks like an AR(1) term, but its coefficient is implicitly 1.0. In practice this coefficient is not estimated from the data but is imposed by the differencing and integration procedures; thus we do not refer to this or similar terms as AR terms even though they are past values of z_t.

If a process for a differenced series always implies a corresponding integrated process for the original series, why do we difference? Why not just study the original series? After all, it is the original series that we want to forecast. The answer is a practical one: The SACF for an integrated series decays *slowly* and thus *obscures the rest of the patterns* that may be present. Differencing permits us to uncover those other patterns.

To illustrate this idea, consider the data plotted in Figure 2.14. This series is the natural logs of the repair parts demand series listed in the Data Appendix as Series 8. The SACF for this series (Figure 2.15) decays to zero quite slowly, suggesting that the data do not return quickly to a fixed mean. This is consistent with the behavior of the data in Figure 2.14. What values of p and q do you identify for this series based on the SACF in Figure 2.15? No answer is possible since this SACF is dominated by the slow decay caused by the nonstationary mean; this slow decay pattern overwhelms information about other possible autocorrelation patterns.

The SACF for the first differences of the log series (Figure 2.16) is more easily interpreted. While this SACF pattern is somewhat irregular, the significant spike at lag 1 stands out. It appears that the differenced series has an MA(1) pattern, so that the original log series has an ARIMA(0, 1, 1)

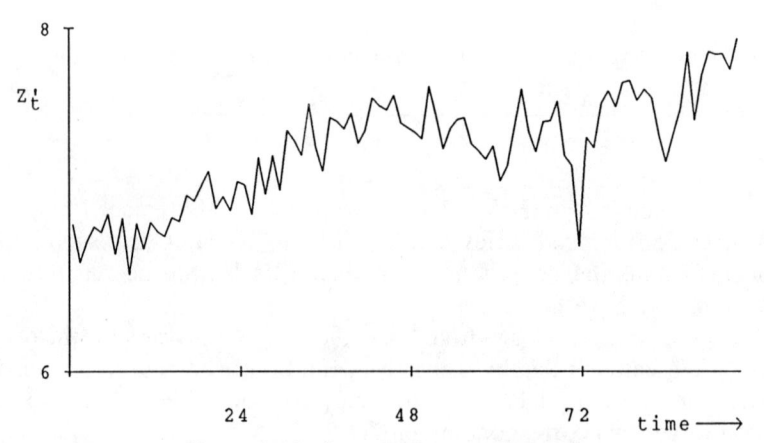

Figure 2.14 Log repair parts demand.

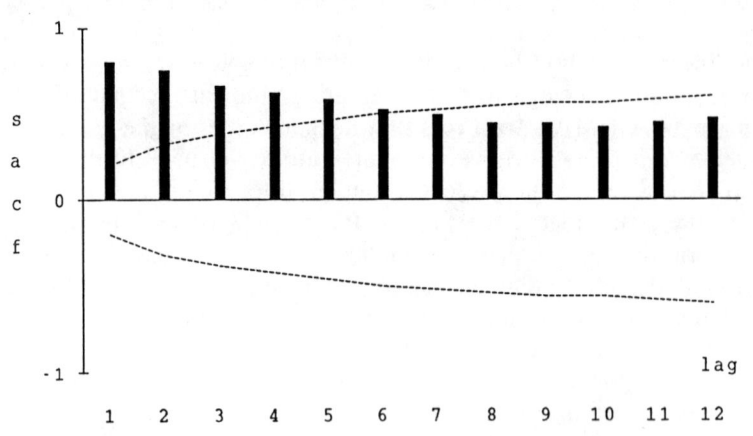

Figure 2.15 SACF for log repair parts demand, $d = 0$.

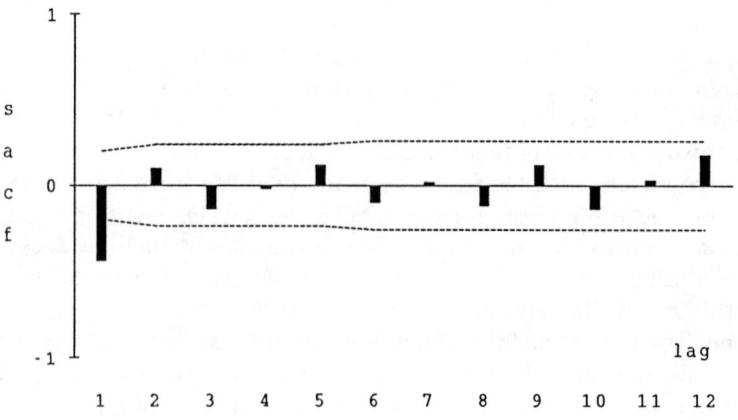

Figure 2.16 SACF for log repair parts demand, $d = 1$.

pattern. The main point is that differencing has allowed us to uncover the possible MA(1) component.

2.5 ARIMA MODELING IN PRACTICE

If we see a *sample* ACF and PACF that look much like one of the pairs of theoretical ACFs and PACFs shown in Figures 2.9 through 2.13, we tentatively *identify* an AR(1), AR(2), MA(1), MA(2), or ARMA(1, 1) model as a reasonable tentative model for the available data. We choose the process whose theoretical ACF and PACF best match the SACF and SPACF for the data.

In using this identification procedure we apply the *principle of parsimony*. According to this principle we try to choose a model with as *few* coefficients as are needed to adequately explain the behavior of the data. Parsimonious models tend to give more accurate forecasts. Having identified a tentative model, we then *estimate* and *check* the model to see if it is adequate.

Practical Modeling Rules
The SACF and SPACF can reveal much about the patterns in a data series. But they are based only on sample information, so we must make allowance for sampling variation. We will get the best results by focusing on lags 1, 2, and 3, and lags s, $2s$, and $3s$ in the SACF and SPACF. In practice we should pay special attention to nonseasonal r_k values that are about 1.6 or more times their standard errors in absolute value, and seasonal r_k values that are about 1.25 or more times their standard errors in absolute value. In the SPACF pay special attention to $\hat{\phi}_{kk}$ values that are two or more times their standard errors. These are not rigid rules; they are guidelines, or warning values.

Along with these practical rules, we should consider the *overall pattern* of the r_k and $\hat{\phi}_{kk}$ coefficients. For example, only the first three r_k values in Figure 2.7 exceed 1.6 times their standard errors in absolute value. However, the overall pattern suggests a decay. Even though the r_k coefficients after lag 3 are relatively insignificant, they are part of a possible overall decay pattern that we should at least consider.

Not every moderately large spike or wave in an SACF or SPACF is worth our attention since SACF coefficients are often correlated with each other. Thus an SACF can show moderately large waves that reflect only sampling variation.

Example
Consider the SACF and SPACF for·the change in business inventories in Figures 2.7 and 2.8. These resemble the AR(1) theoretical ACF and PACF in Figure 2.9(a), with $\phi_1 > 0$. Of course, the match is not perfect since the SACF and SPACF are based on sample information, but it is fairly close.

We can estimate the coefficients (C and ϕ_1) in an AR(1) model for these data using a statistical software package. The SCA System gives the following approximate maximum-likelihood estimates [also shown in (2.1.3)]:

$$z_t = 1.65 + 0.73 z_{t-1} + \hat{a}_t$$
$$\quad\quad (2.30) \quad (7.39) \quad\quad\quad\quad (2.5.1)$$

$$\hat{\sigma}_a = 3.28$$

According to (2.5.1), for these data we have $\hat{C} = 1.65$ and $\hat{\phi}_1 = 0.73$. Each figure in parentheses in equation (2.5.1) is an approximate t value (ratio of each estimated coefficient to its standard error). We may use each t value to test the null hypothesis, H_0: parameter (C or ϕ_1) = 0. Each estimated coefficient ($\hat{C}$ or $\hat{\phi}_1$) is more than twice its standard error in absolute value (each absolute t value is greater than 2.0); therefore we conclude that each estimated coefficient is significantly different from zero, and we reject each H_0 at about the 5% level.

The standard deviation (σ_a) of the shock term series (a_t) is unknown; we estimate it with the sample standard deviation of the residual series, $\hat{\sigma}_a = 3.28$. Notice that $|\hat{\phi}_1| = 0.73 < 1$; therefore the stationarity condition for an AR(1) stated in (2.4.4) is satisfied; we do not have evidence that differencing is needed.

Model Checking

If (2.5.1) is an adequate ARIMA representation of the autocorrelation patterns in the available business inventories data, then there should be *no further significant autocorrelation pattern left in the residual series* $\hat{a}_t$. The residual series is computed from the data by solving model (2.5.1) for $\hat{a}_t$ to get $\hat{a}_t = z_t - 1.65 - 0.73 z_{t-1}$. Then we construct the SACF *of the residual series*; the residual autocorrelation coefficient at lag k is denoted by $r_k(\hat{a})$. The residual SACF for model (2.5.1) in Figure 2.17 shows that each residual autocorrelation is small relative to its standard error: Each falls well short of its two standard error limits (the dashed lines). This suggests that model (2.5.1) adequately represents the autocorrelation patterns in the data.

We may also perform a *joint test* with the null hypothesis H_0: $\rho_1(a) = \rho_2(a) = \cdots = \rho_K(a) = 0$. If H_0 is true, then the following statistic is approximately χ^2 distributed with $K - m$ degrees of freedom (m is the number of coefficients estimated in the model):

$$Q^* = n(n+2) \sum_{k=1}^{K} (n-k)^{-1} r_k^2(\hat{a}) \quad\quad (2.5.2)$$

For a detailed discussion of the properties of this statistic, see Ljung and Box (1978). (Table B at the end of the book lists χ^2 critical points for selected levels of significance and degrees of freedom.)

ARIMA MODELING IN PRACTICE 51

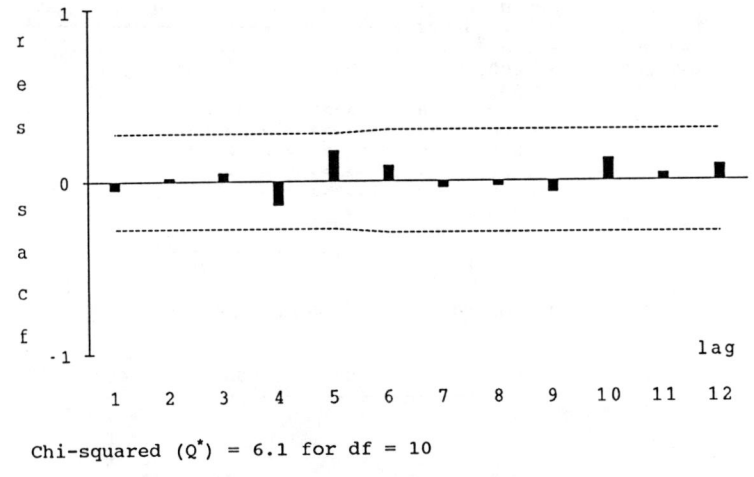

```
r    1
e
s
s         ─────────────────────────────────────
a    0  ─┬──┬──┬──┬──┬──┬──┬──┬──┬──┬──┬──┬──
c         ─────────────────────────────────────
f   -1
                                              lag
     1  2  3  4  5  6  7  8  9 10 11 12
```

Chi-squared (Q*) = 6.1 for df = 10

Figure 2.17 Residual SACF for model (2.5.1).

Using the set of sample residual autocorrelations in Figure 2.17, with $K = 12$ and $m = 2$, applying (2.5.2) we obtain $Q^* = 6.1$ for 10 degrees of freedom (shown at the bottom of the residual SACF in Figure 2.17). We do not reject H_0 since Q^* is less than the 5% critical χ^2 value (18.3) for 10 degrees of freedom. The result of this joint test suggests that model (2.5.1) has adequately captured the autocorrelation patterns in the data.

A standard assumption is that the random shocks (a_t) are normally distributed. This permits us to perform approximate t tests on coefficient significance at the estimation stage. One way to check for normality is to examine a histogram of the residuals. Another is to plot the model residuals in a *normal probability plot*. Both of these procedures give a helpful graphical representation of the data; they do not, however, provide a formal test for normality. To create the normal probability plot, the residuals are standardized (divided by their estimated standard deviation) and ordered from lowest to highest values. These values are then plotted against a set of theoretical ordered normal values. If the residuals are normally distributed, the resulting plot is a straight line. For further discussion of the normal probability plot, see Weisberg (1980) or Liu and Hudak (1986).

The histogram of the residuals from model (2.5.1) is shown in Figure 2.18. This distribution is a bit irregular, but it does not strongly suggest that the shocks are not normal. The normal probability plot for the residuals from model (2.5.1) in Figure 2.19 is roughly a straight line, suggesting that the shocks are approximately normal.

The standardized residuals ($\hat{a}_t/\hat{\sigma}_a$, the number of standard deviations that each residual falls away from zero) are shown in Figure 2.20. All of these residuals fall well within three standard deviations of zero. Thus we have no outliers that call for special attention. (The formal outlier detection method

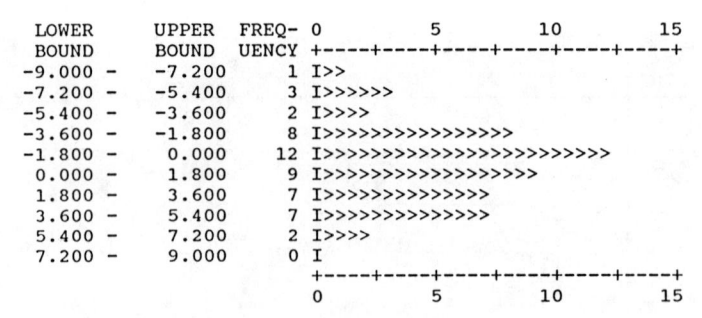

```
LOWER    UPPER    FREQ-  0              5             10             15
BOUND    BOUND    UENCY  +----+----+----+----+----+----+
-9.000 -  -7.200    1    I>>
-7.200 -  -5.400    3    I>>>>>>
-5.400 -  -3.600    2    I>>>>
-3.600 -  -1.800    8    I>>>>>>>>>>>>>>>>
-1.800 -   0.000   12    I>>>>>>>>>>>>>>>>>>>>>>>>
 0.000 -   1.800    9    I>>>>>>>>>>>>>>>>>>
 1.800 -   3.600    7    I>>>>>>>>>>>>>>
 3.600 -   5.400    7    I>>>>>>>>>>>>>>
 5.400 -   7.200    2    I>>>>
 7.200 -   9.000    0    I
                         +----+----+----+----+----+----+
                         0              5             10             15
```

Figure 2.18 Histogram of residuals from model (2.5.1).

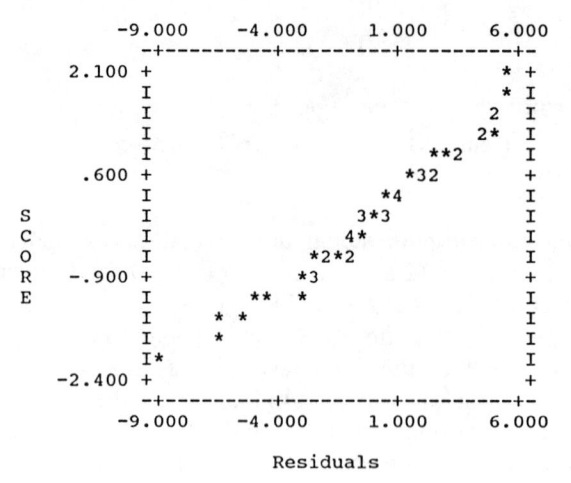

Figure 2.19 Normal probability plot of residuals from model (2.5.1).

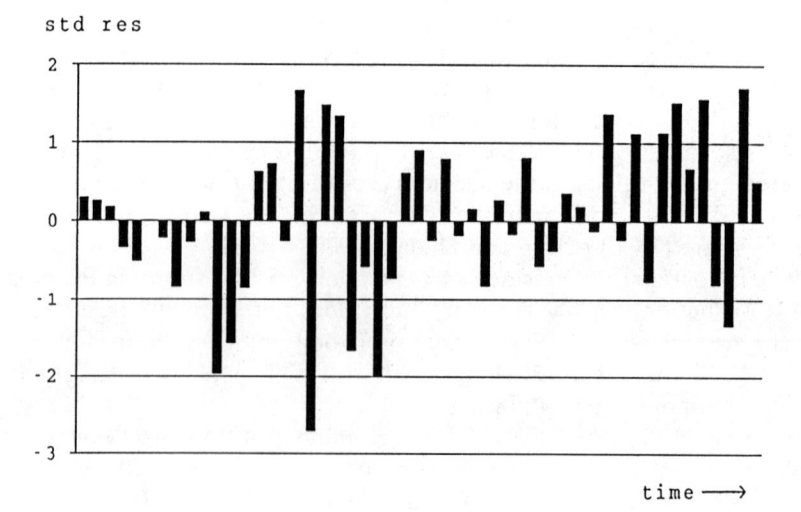

Figure 2.20 Standardized residuals from model (2.5.1).

2.6 BACKSHIFT NOTATION

Consider the ARIMA(1, 1, 1) process, written as

$$w_t = C + \phi_1 w_{t-1} - \theta_1 a_{t-1} + a_t \quad (2.6.1)$$

Substituting $z_t - z_{t-1}$ for w_t, and adding z_{t-1} to both sides of the resulting equation gives

$$z_t = z_{t-1} + C + \phi_1(z_{t-1} - z_{t-2}) - \theta_1 a_{t-1} + a_t \quad (2.6.2)$$

While the ARIMA(1, 1, 1) is a fairly simple process, with just one AR term and one MA term after differencing, writing it out in the form of (2.6.2) can be somewhat awkward. To simplify the business of writing out ARIMA processes and models (and for other reasons), we will often write them using *backshift notation*. This notation is used extensively in time series literature.

We use the backshift operator B^i defined as follows: When B^i multiplies any *time-subscripted* variable, the time subscript is shifted back by i time periods. Thus we have $B^i z_t = z_{t-i}$, $B^i a_t = a_{t-i}$, and $B^i C = C$. We can use backshift notation to write expressions for differenced series. For example, a series differenced $d = 1$ time nonseasonally is written as $w_t = (1 - B)z_t$. To see this, multiply out the right side of the equation to get $w_t = z_t - Bz_t$, then apply the definition of B to get $w_t = z_t - z_{t-1}$. Similarly, a series differenced $D = 1$ time seasonally is written as $w_t = (1 - B^s)z_t$. In general, a series differenced d times nonseasonally and D times seasonally is written as $w_t = (1 - B^s)^D (1 - B)^d z_t$.

The AR(1) process in backshift form is

$$(1 - \phi_1 B)z_t = C + a_t \quad (2.6.3)$$

Multiplying out (2.6.3) gives $z_t - \phi_1 B z_t = C + a_t$. Using the definition of B^i, this is rewritten as $z_t - \phi_1 z_{t-1} = C + a_t$. Rearranging this expression leads to (2.4.1). Thus (2.6.3) is just another way of writing an AR(1) process.

You should verify that the following processes in backshift form are equivalent to the AR(2) in (2.4.2), the MA(1) in (2.4.6), the MA(2) in (2.4.7), the ARMA(1, 1) in (2.4.8), the ARIMA(0, 1, 1) in (2.4.11) and (2.4.12), and the ARIMA(1, 1, 1) in (2.6.1) and (2.6.2), respectively:

$$(1 - \phi_1 B - \phi_2 B^2)z_t = C + a_t \quad (2.6.4)$$

$$z_t = C + (1 - \theta_1 B)a_t \qquad (2.6.5)$$
$$z_t = C + (1 - \theta_1 B - \theta_2 B^2)a_t \qquad (2.6.6)$$
$$(1 - \phi_1 B)z_t = C + (1 - \theta_1 B)a_t \qquad (2.6.7)$$
$$(1 - B)z_t = C + (1 - \theta_1 B)a_t \qquad (2.6.8)$$
$$(1 - \phi_1 B)(1 - B)z_t = C + (1 - \theta_1 B)a_t \qquad (2.6.9)$$

General ARIMA(p, d, q) Process

In practice most ARIMA processes and models are of low order, similar to the examples that we have used so far. Sometimes, however, it is useful to work with a general ARIMA(p, d, q) process. This may be written in a compact way using the following definitions:

$$\nabla^d = (1 - B)^d \quad \text{(the } d\text{-order differencing operator)}$$
$$\phi(B) = (1 - \phi_1 B - \phi_2 B^2 - \cdots - \phi_p B^p) \quad \text{(the } p\text{-order AR operator)}$$
$$\theta(B) = (1 - \theta_1 B - \theta_2 B^2 - \cdots - \theta_q B^q) \quad \text{(the } q\text{-order MA operator)}$$

Using these definitions, the general ARIMA(p, d, q) process is

$$\phi(B)\nabla^d z_t = C + \theta(B)a_t \qquad (2.6.10)$$

You should see that (2.6.10) is just a generalization of the examples shown in Equations (2.6.4) through (2.6.9).

2.7 SEASONAL MODELS

ARIMA models can also represent seasonal (or other periodic) patterns. For example, consider the logs of the budget receipts data (Figure 2.3); the wavy pattern in this plot suggests that the data have a seasonal pattern. Figure 2.4 shows that, after regular differencing ($d = 1$), the data seem to move around a constant overall level; but certain observations are *regularly above* (e.g., June) or *regularly below* (e.g., February) this overall mean. That is, the level of the series shifts in a seasonal fashion. After seasonal length differencing ($D = 1$, with $s = 12$), the data no longer show the same obvious seasonal shifting of the level, as shown in Figure 2.5. However, there might still be some remaining seasonal pattern, even after seasonal differencing.

For monthly data z_t may be related to z_{t-12}, z_{t-24}, z_{t-36}, and so on; that is, a given month may be similar to the same month 1 year earlier, 2 years earlier, 3 years earlier, and so on. In general, for a seasonal pattern of length s, we expect to see a relationship between z_t and z_{t-s}, z_{t-2s}, z_{t-3s}, and so on. If this pattern shows up as a pronounced shifting of the level of the series

SEASONAL MODELS

by season, then seasonal differencing is called for, perhaps in addition to nonseasonal differencing. A seasonal pattern might also appear in the differenced, stationary series w_t. The orders of a purely seasonal ARIMA process are expressed as ARIMA$(P, D, Q)_s$, where Ps is the maximum lag length on seasonal AR terms, and Qs is the maximum lag length on seasonal MA terms. For example, an ARIMA$(1, 0, 2)_{12}$ has an AR term at lag $s = 12$, and MA terms at lags $s = 12$ and $2s = 24$.

The most common seasonal model by far is the ARIMA$(0, 1, 1)_s$. This simple model is written in two equivalent forms:

$$z_t = z_{t-s} + C - \theta_s a_{t-s} + a_t \qquad (2.7.1)$$

$$\nabla_s z_t = C + (1 - \theta_s B^s)a_t \qquad (2.7.2)$$

where $\nabla_s = (1 - B^s)$. Notice that the ARIMA$(0, 1, 1)_s$ in (2.7.1) has the same form as the nonseasonal ARIMA$(0, 1, 1)$ in (2.4.12); however in (2.7.1) the differencing is by length s instead of length 1, and the MA term appears at lag s instead of at lag 1. For exercise, show that (2.7.2) is equivalent to (2.7.1) by using the definition of B^i.

Each forecast from the ARIMA$(0, 1, 1)_s$ is an *exponentially weighted moving average* of a portion of the available data. For example, consider such a model for monthly data ($s = 12$). A one-step ahead forecast from this model (say, for period t, an October) gives a certain weight to the observation in the same season (October) *1 year ago* ($z_{t-s} = z_{t-12}$), a smaller weight to the October observation *2 years ago* ($z_{t-2s} = z_{t-24}$), an even smaller weight to the October observation *3 years ago* ($z_{t-3s} = z_{t-36}$), and so on. These weights decline exponentially. We discuss this feature of the ARIMA$(0, 1, 1)_s$ further in Section 2.11.2. An example of such a model for real data appears in Case 1 following this chapter.

Theoretical ACFs and PACFs

In practice the seasonal and nonseasonal patterns occur together within a time series and in the SACF and SPACF. But is it usually helpful to try separating these two parts in your mind's eye when examining SACFs and SPACFs. Purely seasonal processes of orders P and Q have theoretical ACFs and PACFs that are *identical* to those of nonseasonal processes of orders p and q, with one exception: For the purely seasonal process the patterns occur at lags $s, 2s, 3s, \ldots$, instead of at lags $1, 2, 3, \ldots$.

Thus purely seasonal theoretical ACFs and PACFs have the following patterns:

1. The ACF *decays slowly* at multiples of lag s ($s, 2s, 3s, \ldots$) if seasonal differencing is needed.

2. For a stationary purely seasonal AR process of order P, the ACF

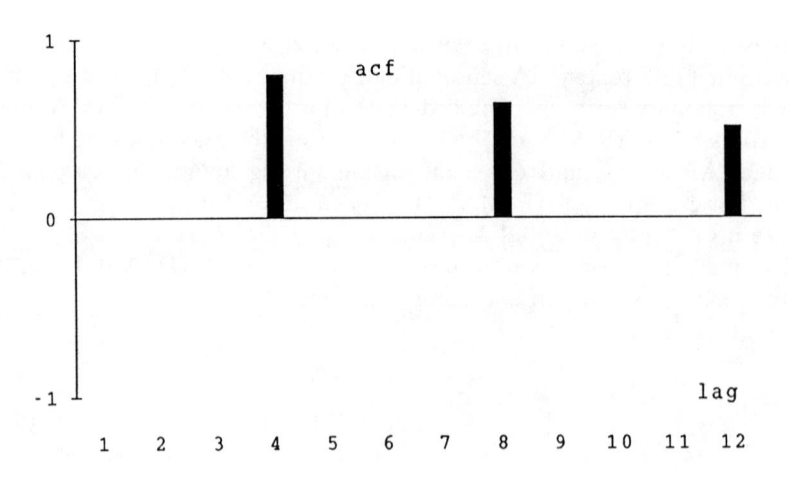

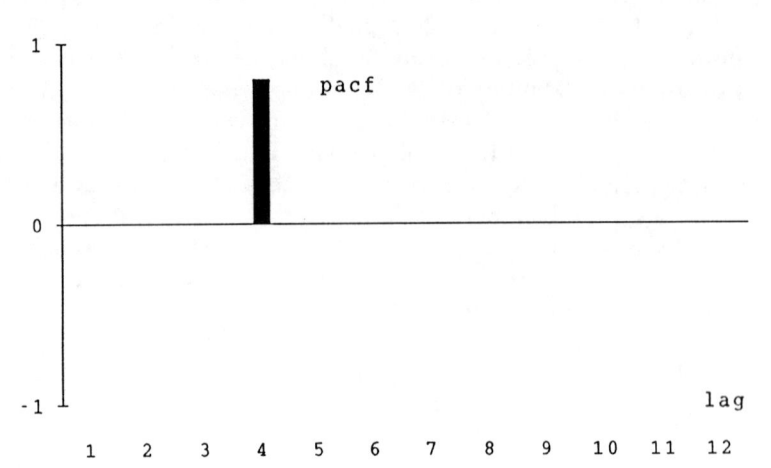

Figure 2.21 Example of theoretical ACF and PACF for seasonal AR(1)$_4$ process with $\phi_4 = 0.8$.

decays at multiples of lag s; the PACF has *spikes* at multiples of lag s and then a cutoff to zeros, with the last nonzero spike at lag Ps.

3. For a stationary purely seasonal MA process of order Q, the *ACF* has *spikes* at multiples of lag s and then a cutoff to zeros, with the last nonzero spike at lag Qs; the PACF *decays* at multiples of lag s.
4. For a stationary purely seasonal mixed process of orders P and Q, both the ACF and PACF *decay* at multiples of lag s.

For example, Figure 2.21 shows a theoretical ACF and PACF for a purely seasonal AR process of order 1, denoted AR(1)$_s$, with $s = 4$. Figure 2.22

COMBINED NONSEASONAL AND SEASONAL PROCESSES 57

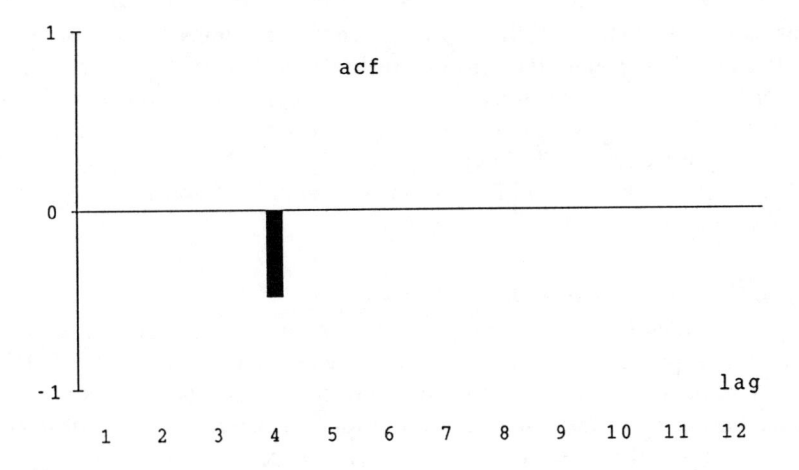

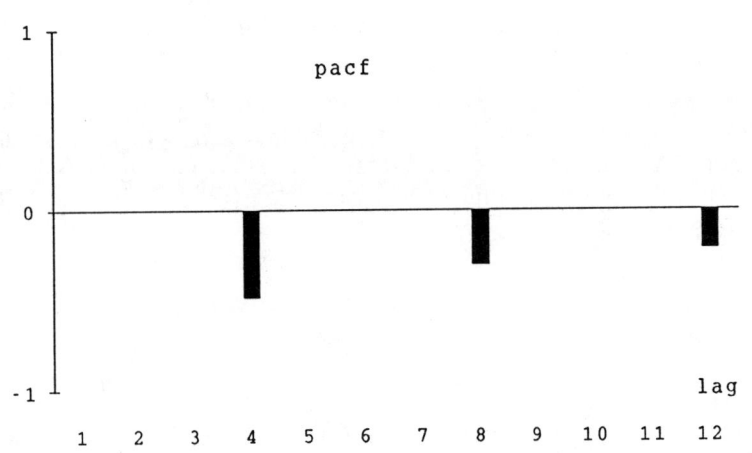

Figure 2.22 Example of theoretical ACF and PACF for seasonal MA(1)$_4$ process with $\theta_4 = 0.8$.

shows a theoretical ACF and PACF for a purely seasonal MA(1)$_s$, with $s = 4$.

2.8 COMBINED NONSEASONAL AND SEASONAL PROCESSES

An ARIMA process or model may have both seasonal and nonseasonal elements. For example, in Case 1 following this chapter we will find the following model for the log of government receipts data (z'_t) in Figure 2.3:

$$(1 - \phi_1 B - \phi_2 B^2)(1 - B^{12})(1 - B)z'_t = (1 - \theta_{12} B^{12})a_t \quad (2.8.1)$$

This is an ARIMA(2, 1, 0)(0, 1, 1)$_{12}$, where the nonseasonal orders are $(p, d, q) = (2, 1, 0)$, and the seasonal orders are $(P, D, Q)_s = (0, 1, 1)_{12}$. This model says that, after both nonseasonal and seasonal differencing of degree one ($d = D = 1$), the logs of the data have a nonseasonal AR(2) pattern ($p = 2$), and a seasonal MA(1)$_{12}$ pattern ($Q = 1$). For exercise, try writing (2.8.1) without the backshift notation; this may lead you to appreciate the simplicity of the backshift form.

General ARIMA(p, d, q)(P, D, Q)$_s$ Process

Sometimes it is useful to write a general form for combined seasonal and nonseasonal processes. Suppose we have the nonseasonal part of an ARIMA pattern expressed as in (2.6.10), but suppose the seasonal part is not represented yet in the process. Therefore the random shock series is not uncorrelated but instead is a series (denoted b_t) that contains a seasonal pattern. Thus we can write the nonseasonal part as

$$\phi(B)\nabla^d z_t = C^* + \theta(B)b_t \qquad (2.8.2)$$

where $C^* = \mu(1 - \sum_{i=1}^{p} \phi_i)$; if $d = 0$, then $\mu = \mu_z$; if $d > 0$, then $\mu = \mu_w$.

Now suppose the seasonal pattern in the series b_t can be represented by AR and MA terms at the seasonal lags up to some maximum AR seasonal lag (Ps) and some maximum MA seasonal lag (Qs). Define the following operators:

$\nabla_s^D = (1 - B^s)^D$ (the D-order seasonal differencing operator)

$\phi(B^s) = (1 - \phi_s B^s - \phi_{2s} B^{2s} - \cdots - \phi_{Ps} B^{Ps})$
(the P-order seasonal AR operator)

$\theta(B^s) = (1 - \theta_s B^s - \theta_{2s} B^{2s} - \cdots - \theta_{Qs} B^{Qs})$
(the Q-order seasonal MA operator)

Now suppose that the seasonal behavior of b_t can be described as

$$\phi(B^s)\nabla_s^D b_t = \theta(B^s)a_t \qquad (2.8.3)$$

Solving (2.8.3) for b_t gives $b_t = [\theta(B^s)/\phi(B^s)\nabla_s^D]a_t$; substitute this expression for b_t into (2.8.3) and rearrange to get the *combined multiplicative seasonal and nonseasonal* ARIMA(p, d, q)(P, D, Q)$_s$ process

$$\phi(B^s)\phi(B)\nabla_s^D \nabla^d z_t = C + \theta(B^s)\theta(B)a_t \qquad (2.8.4)$$

where

$$C = \phi(B^s)C^* = \mu\left(1 - \sum_{i=1}^{p} \phi_i\right)\left(1 - \sum_{i=1}^{P} \phi_{is}\right)$$

If $d = D = 0$ then $\mu = \mu_z$; otherwise $\mu = \mu_w$, the mean of the differenced series $w_t = \nabla_s^D \nabla^d z_t$. In practice all the orders (p, d, q, P, D, Q) tend to be small, often no more than 1 or 2. Notice in (2.8.4) that the nonseasonal and seasonal AR operators multiply each other, and the nonseasonal and seasonal MA operators multiply each other. These elements may also be treated as additive; see Pankratz (1983) for further discussion.

Stationarity and Invertibility

We can now discuss stationarity and invertibility in more general terms. Here we *treat B as an ordinary algebraic variable*, and consider $\phi(B)$ and $\phi(B^s)$ as polynomials in B. Formally, stationarity requires that all the roots of the *characteristic equations* $\phi(B) = 0$ and $\phi(B^s) = 0$ lie outside the unit circle in the complex plane. Likewise, invertibility requires that all roots of the characteristic equations $\theta(B) = 0$ and $\theta(B^s) = 0$ lie outside the unit circle. For the AR(1), AR(2), MA(1), and MA(2) cases, it is easy to show that these requirements are equivalent to the conditions shown earlier in equations (2.4.4), (2.4.5), (2.4.9), and (2.4.10). For $p > 2$, $P > 2$, $q > 2$, or $Q > 2$, it is easiest to use a numerical routine to solve for the roots of the relevant characteristic equation and then examine their distances from the origin; all of these distances must exceed 1.0 in absolute value for the model to be both stationary and invertible.

For the business inventories data the only polynomial in B in the AR(1) model (2.5.1) is $(1 - 0.73B)$, and the only characteristic equation is $(1 - 0.73B) = 0$. A first-degree polynomial like $(1 - 0.73B) = 0$ has only one root. This root is real, and its value is equal to the distance from the origin. Using the SCA System, we find that this root is 1.37, and it lies 1.37 units from the origin. Since the distance from the origin is greater than 1.0 in absolute value, we conclude that the root lies outside the unit circle and the model is stationary.

2.9 FORECASTING

A correct ARIMA model gives minimum mean squared error forecasts among all linear univariate (single-series) models with fixed coefficients. We prove this result in Chapter 9 where the optimal nature of DR forecasts is proven; since an ARIMA model is a special case of a DR model, the proof in Chapter 9 amounts to a proof of the optimality of ARIMA forecasts.

For each time period we can produce a single-value forecast, called a *point* forecast. We can also construct a confidence interval around each point forecast to give us an *interval forecast*. Interval forecasts are especially useful because they convey the possible degree of error associated with the point forecast.

Producing forecasts from an ARIMA model is straightforward, but the computations can be tedious. Fortunately, software packages with an ARIMA option will do the computations. We will illustrate the idea here with a simple example, model (2.5.1). For this model we have $n = 52$ observations available. First, we want to forecast $z_{n+1} = z_{53}$ (one step ahead) based on information available through period 52 (the forecast "origin" is $t = n = 52$). Entering the appropriate time subscripts in model (2.5.1), the one-step ahead forecast (denoted $\hat{z}$) is

$$\hat{z}_{53} = 1.65 + 0.73 z_{52} + \hat{a}_{53}$$
$$= 1.65 + 0.73(11.7) + 0$$
$$= 10.2 \qquad (2.9.1)$$

At time $t = 52$ the value of a_{53} is unknown, and we have no estimate of it, (no value for $\hat{a}_{53}$), so we replace it with the expected value $E(a_{53}) = 0$. However, z_{52} ($= 11.7$) is known at time $t = 52$. Coefficients C and ϕ_1 are unknown, but we have estimates (1.65 and 0.73, respectively) from the model. The resulting point forecast is $\hat{z}_{53} = 10.2$.

Next we want to forecast $z_{n+2} = z_{54}$ (two steps ahead), based on information available through period 52 (the forecast origin is still $t = n = 52$). Entering the appropriate time subscripts in model (2.5.1), the two-step ahead forecast is

$$\hat{z}_{54} = 1.65 + 0.73 z_{53} + \hat{a}_{54}$$
$$= 1.65 + 0.73 \hat{z}_{53} + \hat{a}_{54}$$
$$= 1.65 + 0.73(10.2) + 0$$
$$= 9.1 \qquad (2.9.2)$$

At time $t = 52$ the value of a_{54} is unknown, and we have no estimate of it (no value for $\hat{a}_{54}$), so we replace it with the expected value $E(a_{54}) = 0$. z_{53} is unknown at time $t = 52$, but we have an estimate of it [the one-step ahead forecast computed in (2.9.1), $\hat{z}_{53} = 10.2$]. The resulting point forecast is $\hat{z}_{54} = 9.1$. Table 2.2 shows eight point forecasts from model (2.5.1), all computed by the SCA System from origin $t = 52$. These forecasts are shown as the steplike horizontal lines in Figure 2.1 and as the square symbols in Figure 2.23.

It is also possible to compute the standard deviation of the forecast error. This allows us to place confidence intervals around our point forecasts. See Pankratz (1983, Chapter 10), or Box and Jenkins (1976) for more discussion. Further, Chapter 9 of this text discusses forecast error variance and confi-

FORECASTING

Table 2.2 Eight Forecasts and Standard Errors from Origin $t = 52$ Using Model (2.5.1)

Time	Forecast	Standard Error	Actual
53	10.2	3.3	5.0
54	9.1	4.1	10.0
55	8.3	4.4	8.9
56	7.7	4.6	7.1
57	7.3	4.7	8.3
58	7.0	4.8	10.2
59	6.7	4.8	13.3
60	6.6	4.8	6.2

dence limits for DR models. Since ARIMA models are special cases of DR models, the results in Chapter 9 apply to ARIMA models. Software packages with an ARIMA option generally provide forecast error variances or standard errors.

Table 2.2 shows the forecast standard error computed by the SCA System for each of the eight forecasts from model (2.5.1). Let f denote a forecast and s its standard error; then an approximate 95% confidence interval for each forecast is $f \pm 2s$. Figure 2.23 shows the approximate 95% intervals (dashed lines) around the eight forecasts (square symbols), and the actual future observations that we have treated as unobserved (solid line after $t = 52$).

For a *stationary* model the forecasts *converge to the mean* of the series. How quickly or slowly this happens depends on the nature of the model and on how close the most recent observations are to the mean. For the stationary model (2.5.1) the forecasts in Table 2.2 converge toward the estimated mean

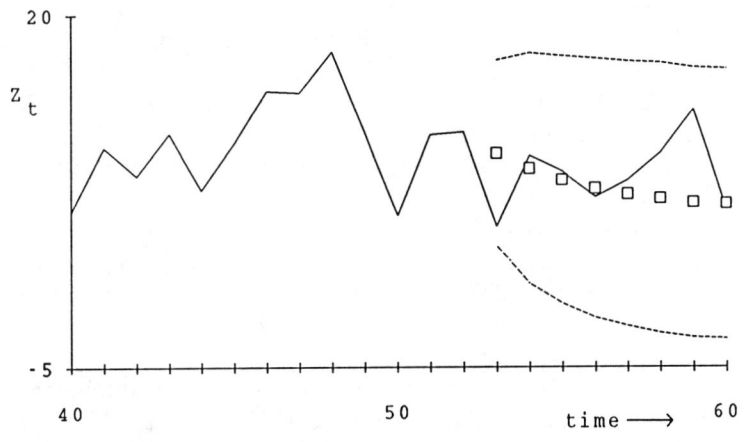

Figure 2.23 Forecasts with two standard error limits from model (2.5.1), and observed values.

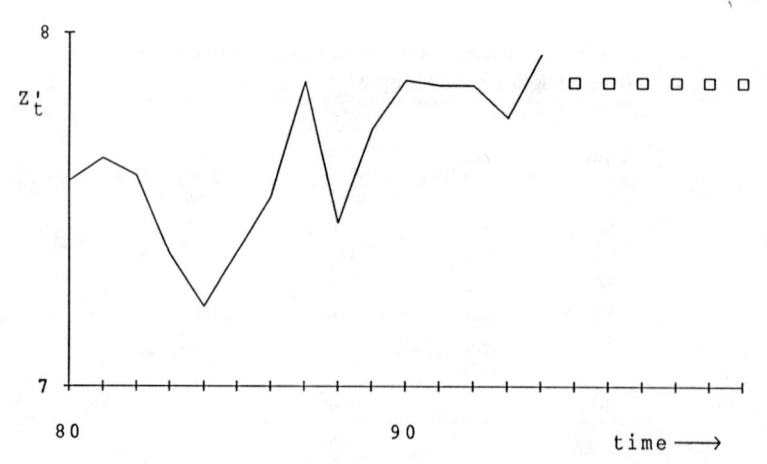

Figure 2.24 Forecasts from ARIMA(0,1,1) model for log repair parts demand.

($\bar{z}$ = 5.7). For a *nonstationary* model the forecasts *do not converge to the mean*; this makes sense since a nonstationary series does not have a constant mean. For example, Figure 2.24 shows six forecasts (square symbols) from origin t = 94 from the ARIMA(0, 1, 1) model identified for the log of the repair parts demand data in Figure 2.14. These forecasts do not converge to the computed mean of the available data.

2.10 EXTENDED AUTOCORRELATION FUNCTION

Identifying the order of a pure AR or a pure MA model is relatively easy. The order of a pure AR(p) is equal to the lag length of the last significant spike in the SPACF. The order of a pure MA(q) is equal to the last significant spike in the SACF. However, it can be difficult to identify p and q from the SACF and SPACF for a mixed (both AR and MA) model. The *extended autocorrelation function* (EACF) and its sample counterpart (ESACF) are especially useful for identifying the orders of mixed models. (The ESACF is difficult to use to identify seasonal orders; we will focus on the nonseasonal orders p and q.) The ESACF could also be used by itself to identify pure AR or MA models; but experience suggests that it is best to use all three identification tools (SACF, SPACF, and ESACF).

We will not work through the details underlying the EACF. The interested reader may consult Wei (1990) for an introductory account or Tsay and Tiao (1984) for a detailed discussion. We will show only the results and discuss how the ESACF is used.

The EACF (and ESACF) may be presented in simplified tabular form, as shown in Table 2.3. Cells marked X are theoretically nonzero or unpatterned (significant in the ESACF), and cells marked 0 are theoretically zeros (insignificant in the ESACF). We search the ESACF for a *triangle of zeros* in the

EXTENDED AUTOCORRELATION FUNCTION

Table 2.3 Theoretical EACF for an ARIMA(2, 0, 1), or ARIMA(1, 1, 1), or ARIMA(0, 2, 1)[a]

($q \to$)	0	1	2	3	4	5	6	7	8
($p' = 0$)	X	X	X	X	X	X	X	X	X
($p' = 1$)	X	X	X	X	X	X	X	X	X
($p' = 2$)	X	0	0	0	0	0	0	0	0
($p' = 3$)	X	X	0	0	0	0	0	0	0
($p' = 4$)	X	X	X	0	0	0	0	0	0
($p' = 5$)	X	X	X	X	0	0	0	0	0

[a]Simplified EACF: X = nonzero or unpatterned; 0 = zero.

lower right part of the table. (In practice the zeros may form a rectangle rather than a triangle.) The upper left vertex of the triangle (or corner of the rectangle) corresponds to the AR order p' at the left of the table and to the MA order q at the top of the table. Table 2.3 shows the theoretical EACF for an ARMA(2, 1) process: The upper left vertex of the triangle of zeros occurs in the $p' = 2$, $q = 1$ cell.

Table 2.4 shows the ESACF for the data in Figure 2.1, based on the first 52 observations. It appears that there is a triangle of zeros in the lower right with an upper left vertex in the $p' = 1$, $q = 0$ cell; therefore we tentatively identify an AR(1) model for these data. This is the same model we found by examining the SACF and SPACF (Figures 2.7 and 2.8). In this case we have used the ESACF to identify a pure AR model rather than a mixed model. However, the great strength of the ESACF is its ability to help in identifying mixed models.

Another special feature of the EACF must be mentioned: It is not necessary to difference data to use the ESACF to identify the orders of a model. The EACF incorporates the differencing order d into the autoregressive order p'. Thus if p' is the AR order given by the EACF, then $p' = p + d$. For the inventory changes example, we have $d = 0$; therefore the ESACF AR order $p' = 1$ is the same as the AR order $p = 1$. Suppose, however, that we have data generated by an ARIMA(1, 1, 0) process:

Table 2.4 Sample EACF for the Inventory Changes Data in Figure 2.1[a]

($q \to$)	0	1	2	3	4	5	6
($p' = 0$)	X	X	0	0	0	0	0
($p' = 1$)	0	0	0	0	0	0	0
($p' = 2$)	X	0	0	0	0	0	0
($p' = 3$)	X	0	0	0	0	0	0
($p' = 4$)	0	0	0	0	0	0	0
($p' = 5$)	0	X	0	0	0	0	0
($p' = 6$)	X	X	0	0	0	0	0

[a]Simplified sample EACF: X = significant at 5% level.

$$(1 - \phi_1 B)\nabla z_t = C + a_t \tag{2.10.1}$$

If we do not difference the data, we expect the ESACF to identify an AR order $p' = 2$; this is composed of $p = 1$ and $d = 1$, so that $p' = p + d = 1 + 1 = 2$. This occurs because the EACF incorporates the nonstationary nature of the series mean into the AR order. The nonstationary feature presumably will be seen at the estimation stage when we check the AR coefficient(s) for stationarity.

To see how the order d can be considered part of an AR order, multiply the AR operator $(1 - \phi_1 B)$ and the differencing operator $\nabla = (1 - B)$ in our ARIMA(1, 1, 0) example in (2.10.1) to get $(1 - \phi_1^* B - \phi_2^* B^2)$, where $\phi_1^* = 1 + \phi_1$ and $\phi_2^* = -\phi_1$. These ϕ^* coefficients do not satisfy the stationarity condition $\phi_2^* + \phi_1^* < 1$. Thus the ARIMA(1, 1, 0) can be thought of as a *nonstationary* ARIMA(2, 0, 0); this is how such a process is understood in the EACF framework, assuming that z_t is not differenced. Thus the theoretical EACF example in Table 2.3 applies to an ARIMA(2, 0, 1), or an ARIMA(1, 1, 1), or an ARIMA(0, 2, 1). In practice we would start with an ARIMA(2, 0, 1) and then check the stationarity conditions using the estimated AR coefficients.

2.11 INTERPRETING ARIMA MODEL FORECASTS

The most general interpretation of the forecasts from an ARIMA model is quite simple: As we stated at the start of this chapter, an ARIMA forecast is a *best weighted average of the past observations*. (This interpretation is somewhat loose since some of the weights may be negative; however, they will sum to 1.0.) There are several other interesting and useful interpretations of ARIMA model forecasts that arise in special cases. We will consider three such cases:

1. The exponentially weighted moving average (EWMA) forecast.
2. The stochastic cycle forecast.
3. The deterministic trend forecast.

These latter three interpretations may apply to an entire ARIMA model or to just certain parts of the model.

2.11.1 Weighted Average Forecasts

Before considering the three special cases noted, we first use a simple example to illustrate how an ARIMA forecast is a weighted average of the available data. Consider the one-step ahead forecast from origin $t = n$ from a stationary AR(1) model:

INTERPRETING ARIMA MODEL FORECASTS

$$\hat{z}_{n+1} = \hat{C} + \hat{\phi}_1 z_n \qquad (2.11.1)$$

where, adapting (2.4.3) to sample information, $\hat{C} = \bar{z}(1 - \hat{\phi}_1)$. Thus (2.11.1) may be written as

$$\hat{z}_{n+1} = \bar{z}(1 - \hat{\phi}_1) + \hat{\phi}_1 z_n \qquad (2.11.2)$$

Within $\bar{z}$ each available z_t observation is given a weight of $1/n$. Thus the first term in (2.11.2) gives a weight of $(1 - \hat{\phi}_1)/n$ to each observation. Since there are n of these observed values, the sum of these n weights in this component of the forecast is $n[(1 - \hat{\phi}_1)/n] = 1 - \hat{\phi}_1$. Then in the second term of (2.11.2) an additional weight of $\hat{\phi}_1$ is given to z_n. Thus the sum of the weights given to all observed z_t values is $1 - \hat{\phi}_1 + \hat{\phi}_1 = 1$. Therefore the sum of the weights given to the observed z_t values is 1.0, and the ARIMA forecast of z_{n+1} is a weighted average of the n available observations. A similar result can be shown for multistep ahead forecasts from the AR(1) and for the forecasts from other ARIMA models.

2.11.2 ARIMA(0, 1, 1) and Exponentially Weighted Moving Average Forecasts

ARIMA(0, 1, 1) models and ARIMA(0, 1, 1)$_s$ models, or models with these components, arise often in practice. As pointed out earlier in this chapter, their forecasts have a special interpretation: They are *exponentially weighted moving averages* of the available data. To show this we begin by writing the ARIMA(0, 1, 1) in backshift form where, for simplicity, $\hat{C} = 0$:

$$(1 - B)z_t = (1 - \hat{\theta}_1 B)\hat{a}_t \qquad (2.11.3)$$

Now we will treat B as an ordinary algebraic variable in addition to using it as the backshift operator. We will first show that the MA(1) part of (2.11.3) has an *equivalent AR form* of infinitely high order. Multiply both sides of (2.11.3) by $(1 - \hat{\theta}_1 B)^{-1}$ to get

$$(1 - \hat{\theta}_1 B)^{-1}(1 - B)z_t = \hat{a}_t \qquad (2.11.4)$$

If $|\hat{\theta}_1| < 1$ (the model is invertible), then from a theorem about infinite series (or using a Taylor series expansion), we may write (2.11.4) as

$$(1 + \hat{\theta}_1 B + \hat{\theta}_1^2 B^2 + \hat{\theta}_1^3 B^3 + \cdots)(1 - B)z_t = \hat{a}_t \qquad (2.11.5)$$

where the first factor in (2.11.5) is an infinite series. This factor is equivalent to an AR operator of infinitely high order, where the AR coefficients have the *special pattern* $\hat{\phi}_i = -\hat{\theta}_1^i$. Thus the MA(1) part of (2.11.3) has an equivalent AR form of infinitely high order. This is an example of a more general result:

All invertible MA(q) models have an AR form of infinitely high order, and all stationary AR(p) models have an MA form of infinitely high order.

Now we show that the one-step ahead forecast from (2.11.5) is an exponentially weighted moving average of the available data. To see this, multiply the first two factors on the left side of (2.11.5), use the definition of B^i, do the algebra required to isolate z_t on the left side, and write the result as a forecast for period $t = n + 1$:

$$\hat{z}_{n+1} = (1 - \hat{\theta}_1)z_n + \hat{\theta}_1(1 - \hat{\theta}_1)z_{n-1} + \hat{\theta}_1^2(1 - \hat{\theta}_1)z_{n-2} + \cdots \quad (2.11.6)$$

As seen in (2.11.6), the implied weights given to the past z's in the one-step ahead forecast from (2.11.3) are $\hat{\theta}_1^{k-1}(1 - \hat{\theta}_1)$, where $k = 1, 2, 3, \ldots$ is the number of time periods into the past (prior to $n + 1$). When $|\hat{\theta}_1| < 1$, these weights decline exponentially as k increases. For example, for $\hat{\theta}_1 = 0.5$ the first three weights in (2.11.6) are

$$(1 - \hat{\theta}_1) = (1 - 0.5) = 0.5$$
$$\hat{\theta}_1(1 - \hat{\theta}_1) = 0.5(1 - 0.5) = 0.25$$
$$\hat{\theta}_1^2(1 - \hat{\theta}_1) = 0.5^2(1 - 0.5) = 0.125$$

In practice we would not use the form (2.11.6) to forecast; instead we use the parsimonious form (2.11.3). But form (2.11.6) is convenient for showing the EWMA nature of the forecasts from an ARIMA(0, 1, 1).

For the nonseasonal ARIMA(0, 1, 1) the implied weights on past z's apply to observations that are 1 period, 2 periods, 3 periods, and so forth, before the period being forecast. For a seasonal ARIMA(0, 1, 1)$_s$, the implied weights on past z's are $\hat{\theta}_s^{k-1}(1 - \hat{\theta}_s)$, where $k = 1, 2, 3, \ldots$ is the number of *years* into the past; these implied weights apply to observations that are 1 year, 2 years, 3 years, and so forth, before the period being forecast. An example of a model with a seasonal EWMA component is shown in Case 1 following this chapter.

2.11.3 Stochastic Cycles

Some AR(2) and AR(2)$_s$ models generate *stochastic cycles* in the forecast pattern. Examples of models with stochastic cycle components appear in Cases 1 and 5. A stochastic cycle pattern can be thought of as a *distorted sine wave* pattern in the forecast pattern: It is a sine wave with a stochastic (probabilistic) period, amplitude, and phase angle. Such models are especially useful for describing data that have a business cycle component, for example.

INTERPRETING ARIMA MODEL FORECASTS

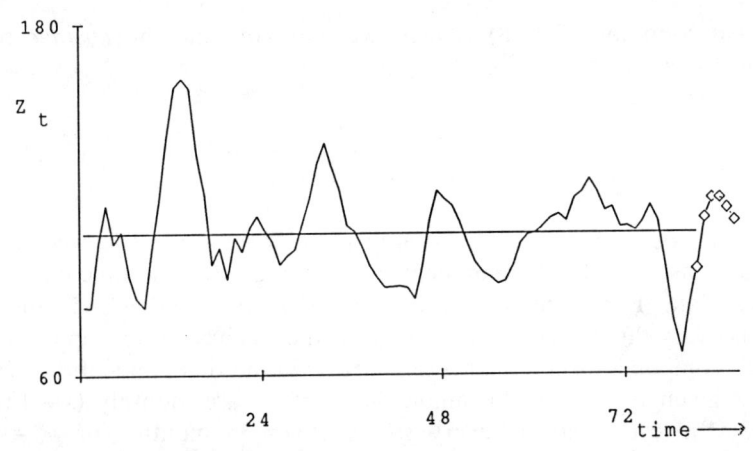

Figure 2.25 Housing permits data with forecasts from model (2.11.7).

The irregular solid line (prior to the diamond symbols) in Figure 2.25 is a sequence of 82 observed values of quarterly housing permits issued in the United States from 1947 I to 1967 II. The data are listed in the Data Appendix as Series 9. The horizontal line is the arithmetic mean ($\bar{z} = 108.1$) of the 82 observations. Notice that the data move around the overall mean in a pattern that could be interpreted as a distorted sine wave.

The model identification and checking methods described in this chapter produce the following ARIMA(2, 0, 2) model (with $\hat{\theta}_1 = 0$) for these data:

$$(1 - 1.20B + 0.54B^2)z_t = 36.85 + (1 + 0.47B^2)\hat{a}_t \qquad (2.11.7)$$

The diamond symbols at the right of Figure 2.25 are five forecasts from model (2.11.7) from time origin $t = n = 82$. These forecasts seem to reproduce the distorted sine wave pattern of the historical data: The forecasts move up through $\bar{z}$, reach a peak, and then move back down toward $\bar{z}$. (To what value would the forecasts at longer lead times eventually converge?)

It turns out that model (2.11.7) does, indeed, contain a stochastic cycle component; therefore model (2.11.7) has the capacity to forecast possible turning points in a time series, as indicated by the forecasts in Figure 2.25. But *not all AR(2) models produce stochastic cycles* in the forecast pattern; a nonseasonal AR(2) model generates a stochastic cycle when the following condition is met:

$$\phi_1^2 + 4\phi_2 < 0 \qquad (2.11.8)$$

In the case of a seasonal AR(2)$_s$, use (2.11.8) to test for the presence of a stochastic cycle but replace ϕ_1 with ϕ_s, and ϕ_2 with ϕ_{2s}. In practice the process coefficients are unknown, and we use the estimated model coefficients.

When condition (2.11.8) is met, we may compute the *average* period (p^*) of the cycle as

$$p^* = \frac{360°}{\cos^{-1}[\phi_1/2(-\phi_2)^{1/2}]} \quad (2.11.9)$$

where the cosine inverse is stated in degrees. If the stochastic cycle appears in a nonseasonal AR(2) component, p^* is measured in the time units of the original data. For example, if the data are *monthly* and $p^* = 3$, the cycle repeats every three *months* on average. If the stochastic cycle appears in an AR(2)$_s$ component, then the average period in the time units of the original data is given by sp^*. For example, if the data are monthly ($s = 12$) and $p^* = 3$, the cycle repeats every $sp^* = 12(3) = 36$ months (or $p^* =$ three years) on average.

While stochastic cycles are associated with AR(2) or AR(2)$_s$ models, we should be alert to their possible presence within higher-ordered AR models. Sometimes the AR operator of a higher-ordered AR model can be factored approximately into an AR(2) or AR(2)$_s$ component and one or more other AR components; the AR(2) or AR(2)$_s$ component may represent a stochastic cycle forecast pattern. For example, suppose we have an AR(3) model with estimated coefficients $\hat{\phi}_1 = -0.58$; $\hat{\phi}_2 = -0.63$; and $\hat{\phi}_3 = -0.21$. The estimated AR(3) operator in this case can be factored approximately into an AR(2) component and an AR(1) component:

$$(1 + 0.58B + 0.63B^2 + 0.21B^3) \approx (1 + 0.2B + 0.5B^2)(1 + 0.4B)$$

The AR(2) part of this approximation has $\hat{\phi}_1 = -0.2$ and $\hat{\phi}_2 = -0.5$. From (2.11.8) we find $\hat{\phi}_1^2 + 4\hat{\phi}_2 = 0.04 - 2.0 = -1.96 < 0$. Thus the estimated AR(3) model has a stochastic cycle component in the forecast pattern.

2.11.4 Deterministic Trends

When $d > 0$ and/or $D > 0$ *and* an ARIMA model contains a constant term, the constant is a *deterministic time trend* component. For example, we showed that model (2.4.11) could be written as $z_t = z_{t-1} + C - \theta_1 a_{t-1} + a_t$, as shown in (2.4.12). To produce a one-step ahead forecast of z_{n+1} from this model, based on information available through period n, we start from the overall level of z_n. Then we *add C units to that level*, and then we make a further adjustment based on the MA term in the model. To produce a two-step ahead forecast for z_{n+2} from origin n, we start from the forecast $\hat{z}_{n+1}$. Then we *add C units to that level*.

The important point is this: As we increase the forecast lead time, we keep *adding the fixed amount C* to the previous observed (z_t) or forecast ($\hat{z}_t$) level of the series. This effect *accumulates* to produce a continuous upward (if $C > 0$) or downward (if $C < 0$) *trend component* in the forecast pattern. Since C is a constant, this trend is *fixed*, or *deterministic*.

The following points summarize some special interpretations of deterministic trend elements. Suppose that we are working with monthly data; then,

1. If $d = 1$ and $D = 0$, then C is a *monthly linear* deterministic trend element. (The forecasts will rise or fall by C units *per month*, in addition to the effects of other terms in the model.)
2. If $d = 0$ and $D = 1$, then C is an *annual linear* deterministic trend element. (The forecasts will rise or fall by C units *per year*, in addition to the effects of other terms in the model.)
3. If $d = 2$ and $D = 0$, then C introduces a *monthly quadratic* deterministic trend into the forecasts.
4. If the data are in the log metric, then C represents a *proportional* deterministic trend in the original metric. For example, if $d = 0$ and $D = 1$, then $C = 0.02$ represents a 2% annual deterministic trend growth in the original metric.

Care should be taken when including a deterministic trend component (a constant term after differencing) in an ARIMA model. Such components imply that the series will perpetually rise or fall (depending on the sign of C) in a deterministic fashion (in addition to the effects of the other terms in the model). This could produce large forecast errors if the historical deterministic trend fails to continue. This problem can be especially serious if we are using the model to produce forecasts over a long time horizon. For this reason we may choose to exclude the constant term from an integrated model even when it is statistically significant. See Case 6 for an example of a DR model with a deterministic trend component.

QUESTIONS AND PROBLEMS

1. Write the following ARIMA processes in backshift form [e.g., like process (2.6.9)] and in ordinary algebraic form [e.g., like process (2.6.2)]:
   a. ARIMA(2, 0, 1)
   b. ARIMA(0, 1, 3), with $\theta_2 = 0$
   c. ARIMA(1, 1, 1)(0, 1, 1)$_{12}$

   (Which of the three forms seems most informative? Which seems to be easiest to work with?)

2. (For the student with calculus.) Show using L'Hospital's rule that the Box–Cox transformation leads to the natural log transformation for the limiting case $\lambda \to 0$.

3. Based on the SACFs in Figures 2.15 and 2.16, I suggested that we might identify an ARIMA(0, 1, 1) model for the natural logs of Series 8 in the Data Appendix.
   a. Compare a plot of the original data, the Box–Cox transformed data with $\lambda = 0.5$, the natural logs of the original data, and the Box–Cox transformed data with $\lambda = -1$. Do you agree that studying the natural logs is reasonable?
   b. Study the SPACF and ESACF of the log series with $d = 0$. Do you learn anything useful from the SPACF? From the ESACF? Answer the same questions for the log series with $d = 1$. Do you find any evidence of seasonality? Explain.
   c. Based on your answer to part (b), what model(s) do you tentatively identify for the log series?
   d. Estimate and check an ARIMA(0, 1, 1) model for the logs of the data. Do you find that this model is adequate? Explain.
   e. Compute four forecasts in the log metric (1 step, 2 steps, 3 steps, and 4 steps ahead) from origin $t = n = 94$. Interpret the forecasts from this model.

4. Consider model (2.11.7) based on Series 9 in the Data Appendix.
   a. Is this model stationary? Invertible? Explain.
   b. Use (2.11.8) to show that (2.11.7) contains a stochastic cycle component.
   c. What is the *average* period of the stochastic cycle component of model (2.11.7)? In what units is your answer measured?
   d. Study the SACF, SPACF, and ESACF of Series 9. Identify, estimate, and check an alternative model. Do you find that it is easy to identify an adequate model for this series? For a discussion of model identification for this series, see Pankratz (1983, Case 4).

5. (Read Case 1 following this chapter before doing this problem.) Consider Series 10 in the Data Appendix. Try to construct an ARIMA model for the natural logarithms of this series.
   a. Inspect a plot of the data. Do you think that the log transformation is justified? Do you think that regular differencing is justified? Seasonal differencing? Explain.
   b. Examine the SACF and SPACF of the natural logs with $d = D = 0$. What value of d is suggested? Why? Is your answer consistent with the one you gave in part (a)?
   c. Study the SACF and SPACF with $d = 1$ and $D = 0$. What value of D is suggested? Why? Is your answer consistent with the one you gave in part (a)?

d. Study the SACF and SPACF with $d = 1$ and $D = 1$. Do you see why an ARIMA$(2, 1, 0)(0, 1, 0)_{12}$ might be justified?

e. Estimate an ARIMA$(2, 1, 0)(0, 1, 0)_{12}$ model. How can you decide if this model is satisfactory? What is your conclusion? Do you find any remaining seasonal pattern? How might you modify the model?

f. Is there an EWMA component to the model you estimated in part (e)? Is there a stochastic cycle component to this model? Do the data show evidence of trading day variation? Explain your answers.

CASE 1

Federal Government Receipts (ARIMA)

In this example we study the monthly U.S. federal government receipts data shown earlier in Figure 2.2 and listed in the Data Appendix as Series 7. This case study illustrates the ARIMA modeling procedures discussed in Chapter 2. It also provides examples of two interpretations of certain ARIMA models discussed in Chapter 2: (1) the exponentially weighted moving average (EWMA) interpretation and (2) the stochastic cycle interpretation. We will study these data again in Case 2 where we build a DR model. We use the first $n = 132$ observations to build a model and hold out the last 12 observations to check the forecasts.

As discussed in Chapter 2 the original series (z_t) in Figure 2.2 has a variance that increases with the level of the data. The natural logarithms of this series in Figure 2.3 have a variance that is roughly constant. Thus we will build our ARIMA model for the transformed series $z'_t = \ln(z_t)$.

C1.1 IDENTIFICATION

Figure C1.1 shows the SACF for z'_t with $d = D = 0$ (no differencing). The key feature of this SACF is that it decays *slowly* to zero. The autocorrelations are much more than twice their standard errors well past lag 6. This suggests that the overall mean of series z'_t is not stationary. This is consistent with the behavior of the data in Figure 2.3 where the series trends upward over time instead of moving around a fixed overall mean.

The nonseasonal first differences $w^*_t = \nabla z'_t$ ($d = 1$) are plotted in Figure 2.4. As noted in Chapter 2 this series has a constant *overall* mean. But the data in Figure 2.4 show a strong *seasonal variation*. For example, the June values are all above the computed overall mean of the differenced series, while the February values are all below that mean. The SACF with $d = 1$ is shown in Figure C1.2. This SACF drops toward zero (insignificant) values fairly quickly; this reinforces our conclusion that the overall mean is now stationary. However, the r_k values at lags 12 and 24 are large, and the decay from lag 12 to 24 is not very quick. This reinforces our observation that the

IDENTIFICATION

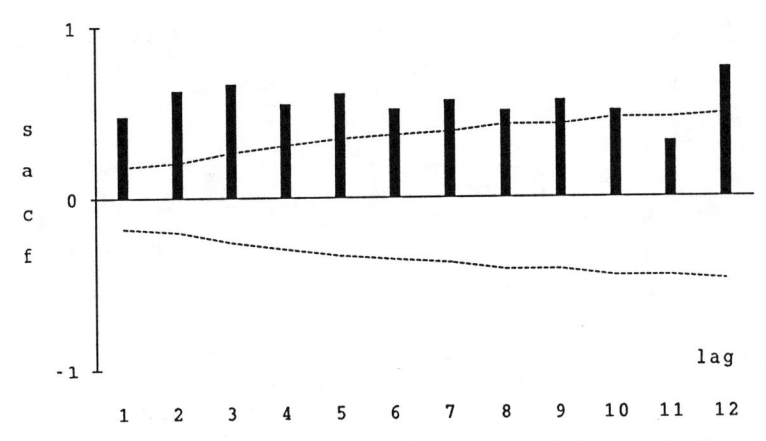

Figure C1.1 SACF for series z'_t (logs of U.S. budget receipts).

data in Figure 2.4 have a strong seasonal variation. This calls for seasonal differencing ($D = 1$) in addition to nonseasonal differencing.

Figure 2.5 is a plot of the log data after both nonseasonal and seasonal first differencing, $w_t = \nabla_{12} w_t^* = \nabla_{12} \nabla z'_t$. This series returns quickly to a constant overall mean (like the series in Figure 2.4), and the strong seasonal pattern is gone. The SACF for $w_t = \nabla_{12} \nabla z'_t$ appears in Figure C1.3; the SPACF appears in Figure C1.4. The SACF value at lag 12 (-0.10) is now smaller than its standard error (0.13); the other SACF seasonal lag coefficients (lags 24 and 36, not shown) are also small enough to ignore. We tentatively conclude that there is no seasonal pattern beyond that represented by seasonal differencing.

There are two defensible ways to account for the remaining nonseasonal pattern: We can include either an MA(l) or an AR(2) component. The MA(1) seems justified since the SACF in Figure C1.3 has a large spike at

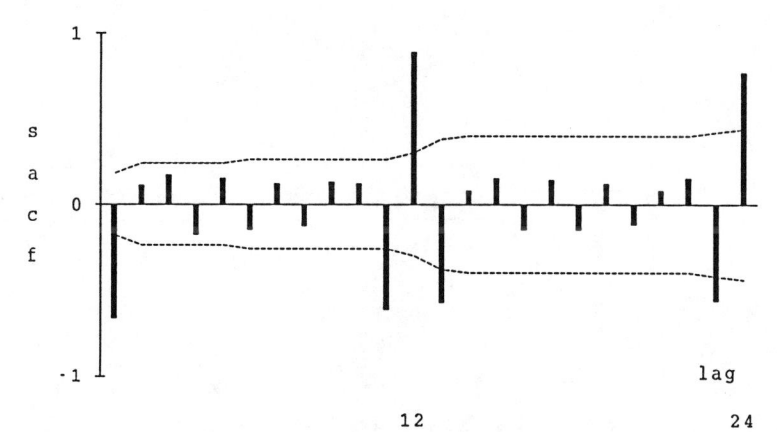

Figure C1.2 SACF for series $w_t^* = \nabla z'_t$.

Figure C1.3 SACF for series $w_t = \nabla_{12}\nabla z'_t$.

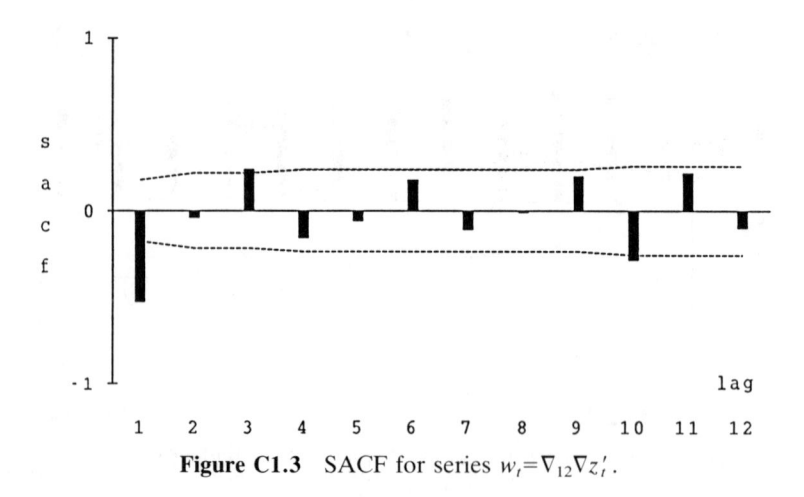

lag 1, followed by an irregular pattern that could be interpreted as a cut-off to zero (insignificant) values. Further, the SPACF in Figure C1.4 could be loosely interpreted as decaying on the negative side. This SACF and SPACF look somewhat like the theoretical MA(1) example in Figure 2.11(a). The MA(1) option is attractive since it is quite parsimonious: It calls for just one estimated coefficient after differencing. Further, choosing an MA(1) gives us an ARIMA(0, 1, 1) for the nonseasonal part of the model. This is attractive since, as discussed in Chapter 2, its forecasts are interpretable as an EWMA pattern. In addition, the ARIMA(0, 1, 1) occurs so often in practice that we might prefer it when it is justified.

A less parsimonious alternative for the nonseasonal part of the model is an AR(2). The irregular pattern in the SACF in Figure C1.3 could be interpreted as a rough decay, and the SPACF in Figure C1.4 has two spikes followed by a cut-off to insignificant values. Together the SACF and SPACF

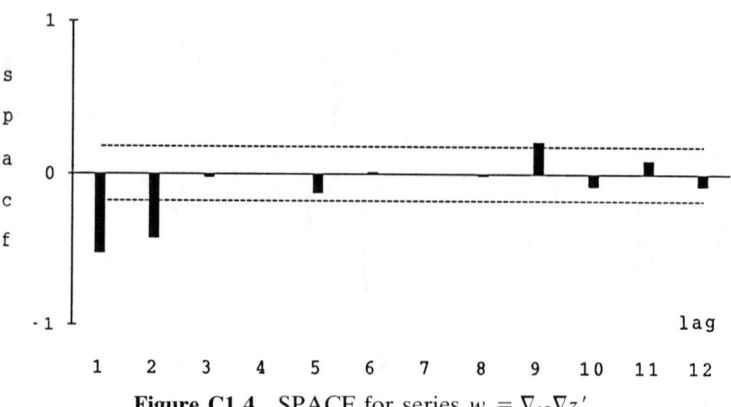

Figure C1.4 SPACF for series $w_t = \nabla_{12}\nabla z'_t$.

IDENTIFICATION

Table C1.1   Simplified ESACF for Series $\nabla_{12}z_t'$ [a]

$(q \rightarrow)$	0	1	2	3	4	5	6	7
$(p' = 0)$	X	X	X	X	X	X	0	0
$(p' = 1)$	X	0	X	0	0	0	0	0
$(p' = 2)$	X	0	X	0	0	0	0	0
$(p' = 3)$	0	0	0	0	0	0	0	0
$(p' = 4)$	X	0	0	0	0	0	0	0
$(p' = 5)$	X	X	0	0	0	0	0	0
$(p' = 6)$	X	X	0	0	0	0	0	0

[a] X = Significant at 5% level.

look somewhat like the theoretical AR(2) example in Figure 2.10(c). The AR(2) is less parsimonious than the MA(1), but it requires only one more estimated coefficient. To interpret the SACF as decaying (in support of an AR(2)), we must place more emphasis on the overall pattern and less emphasis on the significance of the individual coefficients.

The ESACF is especially useful for identifying mixed models. In this case we don't think a mixed model is called for. However, the ESACF can also be used to identify, or confirm the existence of, pure AR or pure MA models; thus it might help us to decide if the MA(1) or AR(2) is more appropriate. As discussed in Chapter 2, a useful feature of the ESACF is that we can use it without differencing the data. If the data are not differenced, the ESACF AR order (p') is equal to $p + d$. The evidence for a nonstationary mean would then appear at the estimation stage in the form of an AR component that does not meet the stationarity requirement(s). It is difficult to use the ESACF to identify seasonal orders. Since we are fairly sure that seasonal differencing is needed, we will examine the ESACF with $D = 1$, but with $d = 0$; that is, we study series $w_t = \nabla_{12}z_t'$.

This ESACF is shown in Table C1.1. One alternative is to locate a triangle (or, in this case, a rectangle) of zeros with its upper left vertex (or corner) in the $p' = 1$, $q = 3$ position; this is not a very parsimonious choice, and it is not entirely consistent with either the MA(1) or the AR(2) that we are considering based on the SACF and SPACF. However, the AR(1) part (due to $p' = 1$ in the ESACF) could be nonstationary; this would be consistent with our previous differencing choice, $d = 1$. And the MA(3) part (due to $q = 3$ in the ESACF) is somewhat consistent with our MA(1) option after regular differencing; but it requires that we increase the MA order to 3. (In fact, the SACF in Figure C1.3 does show a fairly large spike at lag 3.)

Another possibility is to place a triangle of zeros with its upper left vertex in the $p' = 3$, $q = 0$ position. This is consistent with an AR(2) *after regular first differencing*. That is, if we multiply an AR(2) operator by the regular differencing operator with $d = 1$, $(1 - \phi_1 B - \phi_2 B^2)\nabla$, we obtain a third-order AR operator; this is consistent with $p' = 3$ in the ESACF before regular differencing. We will proceed with the following tentative ARIMA(2, 1,0)(0, 1, 0)$_{12}$ model, without a constant:

$$(1 - \phi_1 B - \phi_2 B^2)\nabla_{12}\nabla z'_t = a_t \qquad (C1.1)$$

C1.2 ESTIMATION, CHECKING, AND REFORMULATION

Estimation of model (C1.1) gives the following results:

$$(1 + 0.7731B + 0.4496B^2)\nabla_{12}\nabla z'_t = \hat{a}_t$$
$$(9.26) \qquad (5.38)$$
$$\hat{\sigma}_a = 0.0620455 \qquad (C1.2)$$

The numbers in parentheses are the absolute t values of the coefficients (the ratios of the estimated coefficients to their respective standard errors). $\hat{\sigma}_a$ is the sample standard deviation of the residual series $\hat{a}_t$. The AR coefficients meet the three conditions for an AR(2) model to be stationary in the mean:

$$|\hat{\phi}_2| = 0.4496 < 1$$
$$\hat{\phi}_2 + \hat{\phi}_1 = -0.4496 - 0.7731 = -1.2227 < 1$$
$$\hat{\phi}_2 - \hat{\phi}_1 = -0.4496 + 0.7731 = 0.3235 < 1$$

The residual series $\hat{a}_t$ is computed from (C1.2) as

$$\hat{a}_t = w_t + 0.7731 w_{t-1} + 0.4496 w_{t-2}$$

where w_t is the differenced series $w_t = \nabla_{12}\nabla z'_t$. This series is computed and stored automatically by the SCA System using appropriate commands. The residual SACF for model (C1.2), shown in Figure C1.5, is satisfactory. (The r_k axis is scaled from 0.5 to -0.5 for more visual detail.) The residual autocorrelations are all fairly small relative to their standard errors, and the approximate χ^2 statistic (10.7) is insignificant at the 5% level for 10 degrees of freedom. However, the absolute value of the ratio of the residual autocorrelation at lag 12 (-0.14) to its standard error (0.10) is 1.4. As discussed in Chapter 2, when the absolute value of a *seasonal* autocorrelation divided by its standard error exceeds about 1.25, this is a warning that a coefficient at that lag may prove to be significant in an estimated model. Therefore we try a new model with a seasonal MA term at lag 12, an ARIMA(2, 1, 0)(0, 1, 1)$_{12}$:

$$(1 - \phi_1 B - \phi_2 B^2)\nabla_{12}\nabla z'_t = (1 - \theta_{12} B^{12})a_t \qquad (C1.3)$$

ESTIMATION, CHECKING, AND REFORMULATION

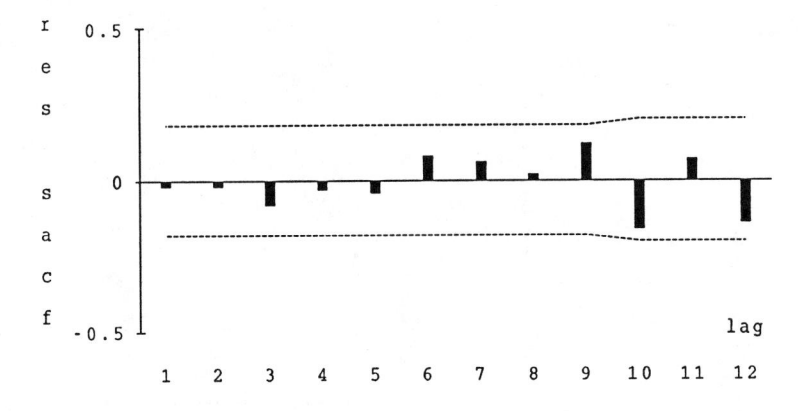

```
r    0.5
e
s

s    0
a
c
f   -0.5                                          lag
         1   2   3   4   5   6   7   8   9  10  11  12
```

Chi-squared (Q*) = 10.7 for df = 10

Figure C1.5 Residual SACF for model (C1.4).

The SCA System exact maximum-likelihood option estimation results for model (C1.3) are

$$(1 + 0.7747B + 0.4911B^2 \nabla_{12} \nabla z'_t = (1 - 0.2071B^{12})\hat{a}_t$$
$$(9.46) \quad (5.94) \quad\quad\quad (2.28)$$
$$\hat{\sigma}_a = 0.0609834 \tag{C1.4}$$

Each estimated coefficient is more then twice its standard error in absolute value. The residual SACF for model (C1.4), not shown, is satisfactory: r_{12} is 0.00 (rounded), and the approximate χ^2 statistic (8.0) for 9 degrees of freedom is insignificant at the 5% level. You should verify that (C1.4) is stationary (after differencing) and invertible. The histogram of the residuals in Figure C1.6 is roughly symmetrical, suggesting that the shocks may be normally distributed. The normal probability plot of the residuals in Figure

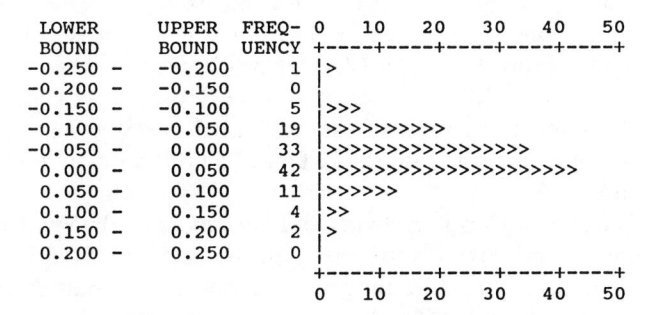

```
         LOWER    UPPER   FREQ-  0    10   20   30   40   50
         BOUND    BOUND   UENCY +----+----+----+----+----+
        -0.250 -  -0.200    1   |>
        -0.200 -  -0.150    0   |
        -0.150 -  -0.100    5   |>>>
        -0.100 -  -0.050   19   |>>>>>>>>>
        -0.050 -   0.000   33   |>>>>>>>>>>>>>>>>>
         0.000 -   0.050   42   |>>>>>>>>>>>>>>>>>>>>>
         0.050 -   0.100   11   |>>>>>>
         0.100 -   0.150    4   |>>
         0.150 -   0.200    2   |>
         0.200 -   0.250    0   |
                               +----+----+----+----+----+
                                0    10   20   30   40   50
```

Figure C1.6 Histogram of residuals from model (C1.4).

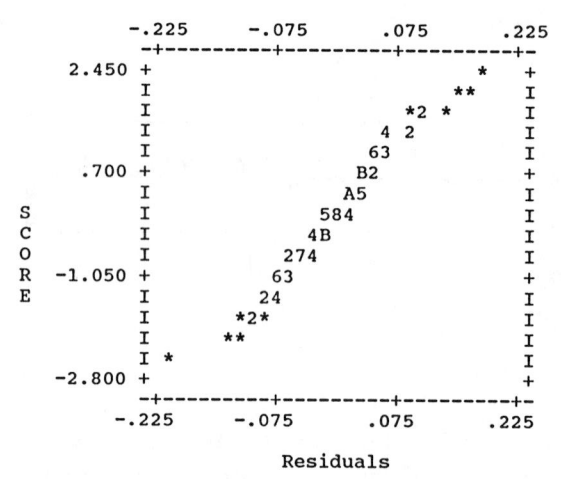

```
                  -.225       -.075        .075        .225
                 -+----------+----------+----------+-
          2.450 +                                *    +
                I                              **     I
                I                         *2 *        I
                I                        4 2          I
                I                        63           I
           .700 +                        B2           +
                I                        A5           I
          S     I                       584           I
          C     I                       4B            I
          O     I                       274           I
          R    -1.050 +                  63           +
          E     I                       24            I
                I                       *2*           I
                I                      **             I
                I  *                                  I
         -2.800 +                                     +
                 -+----------+----------+----------+-
                  -.225       -.075        .075        .225
                                Residuals
```

Figure C1.7 Normal probability plot of residuals from model (C1.4).

C1.7 does not deviate too badly from a straight line, again suggesting that the shocks are normal. The plot of the standardized residuals ($\hat{a}_t/\hat{\sigma}_a$) in Figure C1.8 has only one value that falls a little more than three standard deviations from zero. In a roughly normal distribution, this is not unusual. However, it might be worth while to do some research to find out if something unusual happened at $t = 53$ when this moderately large outlier occurs. (The formal outlier detection method discussed in Chapter 8 might be more revealing than our casual inspection.)

C1.3 MODEL INTERPRETATION

Model (C1.4) has an interesting interpretation. First, it has an ARIMA$(0, 1, 1)_{12}$ component. As discussed in Chapter 2, this introduces a seasonal EWMA component into the forecast pattern. Second, the nonseasonal AR(2) part of the model introduces a stochastic cycle into the forecasts. Using Equation (2.11.8) in Chapter 2, for this example we find $\hat{\phi}_1^2 + 4\hat{\phi}_2 = (-0.7747)^2 + 4(-0.4911) = -1.364 < 1$; therefore model (C1.4) includes a stochastic cycle. From Equation (2.11.9) in Chapter 2, the *average* period for this cycle is $p^* = 2.9$ time periods. Since the cycle is captured by an AR(2) in the *nonseasonal* part of the model, p^* is stated in the time units in which the data are measured; thus we have a cycle with an average period of 2.9 *months*.

When we find a stochastic cycle in *cumulated monthly* data with an average period of *about* 3 *months*, this suggests that the data may exhibit *trading day variation*. That is, they may be moving in response to the differing numbers of each day of the week (Monday, Tuesday, etc.) that occur in different

FORECASTING

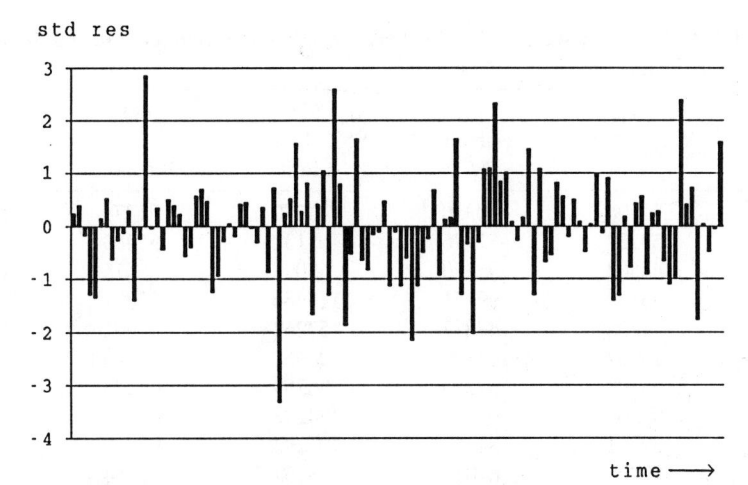

Figure C1.8 Standardized residuals from model (C1.4).

Table C1.2 Twelve Forecasts for Series z'_t from Model (C1.4), with Standard Errors

Time	Forecast	Standard Error
133	11.2939	0.0610
134	10.9591	0.0625
135	10.9209	0.0658
136	11.5044	0.0762
137	10.7973	0.0791
138	11.3362	0.0831
139	11.1158	0.0886
140	11.0306	0.0920
141	11.3440	0.0958
142	11.0668	0.0998
143	10.9623	0.1032
144	11.3259	0.1067

months. We revisit this issue in Case 2 where we build a DR model for the same data series.

C1.4 FORECASTING

Forecasts from model (C1.4) are shown in Table C1.2. These forecasts are in the log metric. Using the forecast standard errors (also shown in Table C1.2), we can construct a rough 95% confidence interval around each forecast as $f \pm 2s(f)$, where f is the forecast in the log metric and $s(f)$ is the forecast standard error for each forecast.

Table C1.3 Twelve Forecasts (F) for Series z_t (Original Metric) from Model (C1.4) from Time Origin $t = 132$

Time	(1) 95% Lower Limit	(2) Forecast (F)	(3) 95% Upper Limit	(4) Observed	(5) Error
133	71106	80330	90750	81771	1441
134	50722	57476	65131	55463	−2013
135	48503	55320	63095	56515	1195
136	85142	99151	115466	122897	23746
137	41738	48887	57262	47691	−1196
138	70967	83803	98961	82945	−858
139	56308	67225	80259	64223	−3002
140	51365	61736	74202	60213	−1523
141	69731	84457	102293	92410	7953
142	52425	64011	78157	62354	−1657
143	46910	57661	70875	56987	−674
144	67010	82943	102664	85525	2582
				RMSFE =	7405

We can state the forecasts in the original metric (F) by taking antilogs: $F = \exp(f)$. Approximate 95% lower and upper limits in the original metric are $\exp[f - 2s(f)]$, and $\exp[f + 2s(f)]$, respectively. (Retransforming forecasts to the original metric is potentially more complicated than is suggested here; this issue is discussed in more detail in Chapter 9.) The original metric forecasts and their rough 95% limits are shown in column (2) in Table C1.3. The future values for these periods, assumed to be unknown at the time the forecasts are made, are shown in column (4). All but one (at $t = 136$) of the 95% intervals contain the future observed values.

The root mean squared forecast error (RMSFE) can be used as an overall measure of accuracy for these 12 forecasts. The RMSFE is defined as

$$\text{RMSFE} = \sqrt{\frac{\sum_{t=n+1}^{n+m}(z_t - f_t)^2}{m}} \quad \text{(C1.5)}$$

Here we are considering m forecasts (f_t) for periods $n + 1$ through $n + m$, and m future observed values (z_t) that are unknown when we forecast from time origin $t = 132$; thus we can compute this RMSFE only after time period $t = 144$ has passed. We compute each forecast error ($z_t - f_t$), square it, compute the mean of the m squared errors, and return this figure to the units of the original data by taking the square root. The resulting RMSFE is interpreted as an average of the m forecast errors. The RMSFE of the 12 forecasts in Table C1.3, shown below the error column, is 7405.

C1.5 ALTERNATIVE MODELS

For exercise, try estimating and checking some alternative models. How could you justify each of the following models? What do you find are their advantages and disadvantages?

1. An ARIMA$(2, 1, 0)(0, 1, 1)_{12}$ with a constant term. How do you interpret the constant term? How is your interpretation affected by the fact that the data are modeled in the log metric? How do you determine if it is appropriate to include a constant term in the model?
2. An ARIMA$(0, 1, 1)(0, 1, 0)_{12}$. Is the residual SACF satisfactory? How might you modify this model if you want to retain the nonseasonal MA(1) component?
3. An ARIMA$(3, 1, 1)(0, 1, 0)_{12}$ with $\hat{\phi}_1$ and $\hat{\phi}_2$ excluded from the model (i.e., these coefficients are forced to be zero). Is this an acceptable revision of the result you obtained in alternative (2)? Is this model better than model (C1.4)?
4. An ARIMA$(0, 1, 3)(0, 1, 0)_{12}$ with $\hat{\theta}_2$ excluded from the model. Is this an acceptable revision of the model you found in alternative (2)? Is this model better than model (C1.4)?

CHAPTER 3

A Primer on Regression Models

This chapter is an overview of the main ideas about single-equation regression models. For a comprehensive treatment of these topics at an intermediate level, you may consult a textbook such as Gujarati (1988) or Kmenta (1986). The material in Sections 3.8.4 and 3.9 is especially important as background for the remainder of this book.

3.1 TWO TYPES OF DATA

The data listed in Table 3.1 occur in sets of three (triplets): variable $X_{1,i}$ is a father's height, variable $X_{2,i}$ is a mother's height, and variable Y_i is the height of a college-age son of that father and mother. (The data on $X_{1,i}$ and Y_i also appear in Table 1.5 and in Figure 1.10.) The i subscript identifies a particular triplet; for example, the fifth triplet ($i = 5$) is the set {$X_{1,5} = 72$, $X_{2,5} = 60$, $Y_5 = 69$}. These are *cross-section* data: The observations were recorded during a given time period (a single 5-minute time span), rather than over many time periods.

We will introduce the main ideas about regression using the heights data. However, in this book we focus on *time series* data rather than cross-section data. The main ideas about regression are the same for both cross-section and time series data. Examples of regression with time series data appear later in Section 3.9 and throughout the book. Recall that we use a t subscript instead of an i subscript to index time series observations. For simplicity we will sometimes drop the i or t subscript when it is not needed for clarity.

3.2 THE POPULATION REGRESSION FUNCTION (PRF) WITH ONE INPUT

The data in Table 3.1 are one *sample* drawn from the *population* of all possible triplets of father's height (X_1), mother's height (X_2), and son's

THE POPULATION REGRESSION FUNCTION (PRF) WITH ONE INPUT

Table 3.1 Sample Data on Sons', Fathers', and Mothers' Heights (inches); n = 25

i	$X_{1,i}$	$X_{2,i}$	Y_i
1	68	64	69
2	72	66	71
3	73	65	71
4	70	62	71
5	72	60	69
6	69	62	68
7	75	63	73
8	72	60	67
9	69	62	72
10	69	62	68
11	75	64	73
12	72	64	68
13	69	64	72
14	73	64	73
15	78	62	75
16	70	67	72
17	72	64	69
18	72	62	70
19	71	66	70
20	69	62	70
21	71	65	73
22	68	65	64
23	69	66	72
24	75	66	74
25	75	65	75

height (Y). For genetic reasons there may be a causal relationship in the population between X_1 and X_2, on the one hand, and Y on the other hand: Taller fathers and mothers are likely to have taller sons, and shorter fathers and mothers are likely to have shorter sons. Of course, other factors might also affect a son's height; for example, the quality of nutrition received during the growing years may affect a college-age son's height.

The inputs (X variables) may be either stochastic (having some random movements) or deterministic (with no random component). For example, in Chapter 1 the valve orders series is stochastic; but the indicator variable standing for the presence or absence of a tax rebate is deterministic. Examples of models with stochastic inputs appear in Cases 3, 4, and 5. Examples of models with deterministic inputs appear later in this chapter, in Chapters 7 and 8, and in Cases 2 and 6.

3.2.1 The PRF Assumptions

Linearity

In regression analysis we assume that the population output (Y) may be *linearly* related to the inputs (the X's) by means of *fixed coefficients*. By *linear* we mean *linear in the coefficients*. However, within this framework we can also handle relationships that are *nonlinear in the variables*; examples of nonlinear relationships are given in Section 3.8.5. The linearity assumption is a simplifying one; it permits us to develop (1) a relatively simple theory and (2) some useful statistics that are fairly easily obtained.

A Stochastic Relationship

We also assume that the relationship between the output and the inputs is *stochastic*: The inputs don't perfectly predict the output. This imperfection is represented by an additive stochastic disturbance term that represents *all variation in the output that is not associated with movement in the inputs.*

It is easier to illustrate graphically the ideas about regression if there is just one input. Therefore, to simplify the presentation, we start by considering only the relationship between father's height (X_1) and son's height (Y). In Section 3.7 we consider the case of several inputs (multiple regression). For the heights example the two assumptions about the population relationship between Y and X_1 stated thus far (that it is linear and stochastic) are written as

$$Y_i = C + b_1 X_{1,i} + a_i \qquad (3.2.1)$$

Equation (3.2.1) is the *population regression function (PRF)*. The output Y is often called the "dependent" variable, and the input X is called the "independent" variable. Coefficient b_1 in (3.2.1) is a fixed parameter to be estimated; it is the population *average change in Y per one unit change in X_1*. C is a constant to be estimated; it captures the effect of all excluded independent variables on the *overall level* of Y (as opposed to the variability of Y).

Disturbance Properties

The disturbance represents the effect of all excluded independent variables on the *variability* of Y (as opposed to the overall level of Y). The stochastic disturbance term a_i is usually assumed to be zero mean and normally distributed white noise with constant variance σ_a^2. The disturbance is also assumed to be independent of the X's: Knowing the values of the X's should not help us to predict the values of the disturbance. The normality assumption is sometimes dropped; but it permits us to do statistical inference using the sample estimates of the parameters C and b_1.

THE POPULATION REGRESSION FUNCTION (PRF) WITH ONE INPUT 85

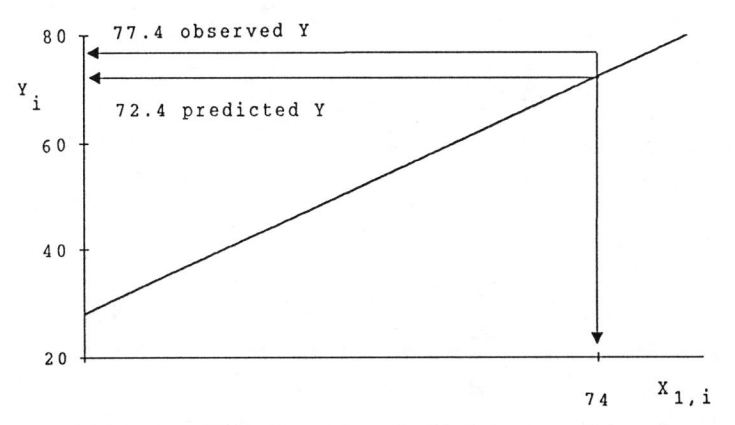

Figure 3.1 Illustration of predicting Y from PRF (3.2.1).

3.2.2 Prediction with a PRF (One Input)

If the parameters (C and b_1) in Equation (3.2.1) were known, we could use them to *predict* the value of Y for any *given* value of X_1. Of course, in practice we don't know the PRF parameters. But assuming that they are known for the moment will help us to picture the idea of a PRF.

In (3.2.1), Y_i on the left side of the equality is broken down into two additive parts on the right side of the equality. The first part is a set of *predictions* of Y (denoted $\hat{Y}$) based on the parameters C and b_1 and on the given values of X: $\hat{Y}_i = C + b_1 X_{1,i}$. The second part is the disturbance, which cannot be predicted from (3.2.1). For example, suppose $C = 28$ and $b_1 = 0.6$. Then given father's height $X_1 = 74$ inches, we can use (3.2.1) to predict son's height as $\hat{Y}_i = C + b_1 X_{1,i} = 28 + 0.6(74) = 72.4$ inches. In making this prediction, we assign a_i its expected value of zero since its value is not predicted by (3.2.1). If the son in question has an actual height that differs from the predicted 72.4 inches, the difference is equal to a_i. For example, if $a_i = 5$, then from (3.2.1) the son's *actual* height is $Y_i = \hat{Y}_i + a_i = 72.4 + 5 = 77.4$ inches.

The idea of predicting a son's height from PRF Equation (3.2.1) is illustrated in Figure 3.1. The PRF line for this example has Y-axis intercept $C = 28$ and slope $b_1 = 0.6$. Find $X_1 = 74$ inches along the horizonal axis. From that point, read up to the PRF prediction line, $\hat{Y}_i = 28 + 0.6X_{1,i}$. Now read from the PRF prediction line horizontally to the vertical axis. This gives the predicted value (72.4 inches) of the son's height given that the father's height is 74 inches. If the disturbance a_i for this son is 5, then we observe a Y value equal to 77.4 (at the intersection of the vertical arrow and the upper horizontal arrow in Figure 3.1); this lies 5 inches above the predicted value of 72.4 inches, given $X_1 = 74$ nches.

Figure 3.1 shows only one given value of X_1, along with its corresponding predicted Y and observed Y. Suppose we could observe the *population* of

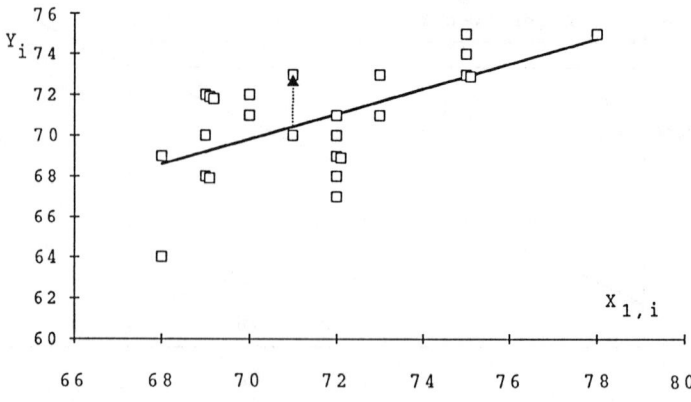

Figure 3.2 College-age son's height (Y_i) vs. father's height ($X_{1,i}$).

X_1 and Y values. Then we would have many possible given values of X_1 in Figure 3.1. For *each* given value of X_1 we would have *one* predicted Y from the PRF prediction line. But for each given X_1, there would be a *scatter of different possible observed Y's*, some falling *above* the predicted Y ($a_i > 0$) and some falling *below* the predicted Y ($a_i < 0$). For example, one father with height $X_1 = 74$ inches might have a son who is 78 inches tall, while another father with height also $X_1 = 74$ inches might have a son who is 69 inches tall.

The PRF in Equation (3.2.1) is a theoretical relationship in which the parameters typically are unknown. Next we consider how we can estimate those parameters, and the statistical properties of these estimates. Further, one or more of the assumptions underlying the PRF might not be satisfied. We must consider the effects on our estimates of violating these assumptions, ways of detecting violations, and methods of dealing with such violations.

3.3 THE SAMPLE REGRESSION FUNCTION (SRF) WITH ONE INPUT

The sample counterpart to the PRF in Equation (3.2.1) is the *sample regression function (SRF)*:

$$Y_i = \hat{C} + \hat{b}_1 X_{1,i} + \hat{a}_i \qquad (3.3.1)$$

$\hat{C}$ and $\hat{b}_1$ are sample estimates of the parameters C and b_1, respectively; they are statistics computed from the available sample data. The residual $\hat{a}_i$ has a circumflex (ˆ) to indicate that it is computed from sample estimated coefficients. [This is seen by solving (3.3.1) for $\hat{a}_i$ to get $\hat{a}_i = Y_i - \hat{C} - \hat{b}_1 X_{1,i}$; in practice $\hat{a}_i$ is computed from this expression.] We can picture the idea of the SRF for the heights data by imagining a "best-fit" line running through the sample data, plotted in Figure 1.10 and reproduced in Figure 3.2. This is a

THE SAMPLE REGRESSION FUNCTION (SRF) WITH ONE INPUT

plot of son's height (Y) against the corresponding father's height (X_1). The solid line is a possible best-fit line for these data; this line has Y-axis intercept $\hat{C}$ and slope $\hat{b}_1$.

3.3.1 Least-Squares Estimation

The sample regression line (the values of $\hat{C}$ and $\hat{b}_1$) can be chosen in many ways. It seems reasonable that we should choose a line that best fits the sample data according to some statistically appropriate definition of "best." Consider this: If we could find a single straight line that fit the data perfectly, then each observed Y would be *equal to* each predicted Y, and the *residual values* ($\hat{a}$) *would all be zero*. We can't expect to have a perfect fit, but this idea suggests how we can think about the matter of "best fit." It seems that a best-fit line would have residuals that are *as small as they can be*. The commonly used rule is: Choose the estimates $\hat{C}$ and $\hat{b}_1$ so that the *sum of the squared residuals* ($\Sigma_{i=1}^n \hat{a}_i^2$) is *minimized*. This is the so-called least-squares criterion. The summing takes into account all residuals and therefore all X, Y pairs in the sample. The squaring is done because the resulting estimator has desirable properties, discussed in Section 3.4.

The least-squares estimators are derived using calculus. See, for example, Gujarati (1988, Appendix 3A.1). For a regression with one independent variable, these estimators are

$$\hat{b}_1 = \frac{\sum_{i=1}^n (X_{1,i} - \overline{X}_1)(Y_i - \overline{Y})}{\sum_{i=1}^n (X_{1,i} - \overline{X}_1)^2} \tag{3.3.2}$$

$$\hat{C} = \overline{Y} - \hat{b}_1 \overline{X}_1 \tag{3.3.3}$$

where $\overline{Y}$ and $\overline{X}_1$ are the sample means of Y_i and $X_{1,i}$, respectively. These formulas involve computations applied to the n available sample observations. In this general sense they are just like any other formula for an estimator.

The idea of the least-squares rule is illustrated in Figure 3.2. The line through the data is the least-squares SRF prediction line, with intercept $\hat{C} = 26.72625$ and slope $\hat{b}_1 = 0.61568$. Consider the 21st X, Y pair in Table 3.1, $X_{1,21} = 71$, $Y_{21} = 73$. The arrow in Figure 3.2 points to the square symbol that represents this pair. *Given* $X_{1,21} = 71$, the Y value predicted from the SRF is $\hat{Y}_{21} = 26.72625 + 0.61568(71) = 70.43953$. The vertical distance between the observed $Y_{21} = 73$ and the predicted $\hat{Y}_{21} = 70.43953$, given $X_{1,21} = 71$, is $\hat{a}_{21} = Y_{21} - \hat{Y}_{21} = 73 - 70.43953 = 2.56047$. This value is represented in Figure 3.2 by the length of the dashed line arrow. The square of this value is $\hat{a}_{21}^2 = 2.56047^2 = 6.5560066$. Repeating this exercise for each of the other 24 X, Y pairs gives a total of 25 $\hat{a}_i^2$ values. The sum of these squared residual values for this example is 104.302.

Now imagine *any other straight line* through the sample scatterplot in Figure 3.2, with a different intercept and slope. This alternative line would have a different set of $\hat{Y}$ values and therefore a different set of $\hat{a}$ values. Would the sum of the squares of these $\hat{a}$ values be larger or smaller than 104.302? The answer must be "larger" since the line already shown in Figure 3.2 was chosen so that $\Sigma_{i=1}^n \hat{a}_i^2$ is minimized.

3.3.2 Residual Standard Deviation

It is useful to have an estimate ($\hat{\sigma}_a$) of the standard deviation of the random shocks (σ_a). This estimate measures how closely the observed Y's gather around the SRF, and it is used to compute confidence intervals for the forecasts from regression models. The estimate is usually computed as

$$\hat{\sigma}_a = \left[\frac{\sum_{i=1}^n (Y_i - \hat{Y}_i)^2}{n - k} \right]^{1/2} \qquad (3.3.4)$$

where k is the number of parameters estimated in the regression. For PRF (3.2.1) we estimate C and b_1, so $k = 2$. Some computer programs divide by n rather than $n - k$ in (3.3.4); with a large enough sample this makes little difference. Notice that $Y_i - \hat{Y}_i$ in the numerator of (3.3.4) is equal to $\hat{a}_i$, so $\Sigma(Y_i - \hat{Y}_i)^2$ is the sum of the squared residuals. For the heights data example, the sum of squared residuals is 104.302, so $\hat{\sigma}_a = [104.302/(25 - 2)]^{1/2} = 2.12952$.

3.4 PROPERTIES OF THE LEAST-SQUARES ESTIMATORS

Why do we choose the SRF using the least-squares rule? If our assumptions about the disturbance in the PRF are correct, and if the model is correctly specified, then the least squares estimators of the parameters are (1) unbiased and consistent and (2) have the least variance in repeated sampling (are "best") among all unbiased estimators whether linear or not (Rao, 1965).

Even without the normality assumption, the least squares estimators are still unbiased, and they are best among all linear unbiased estimators; see, for example, Gujarati (1988, Appendix 3A). Here "linear" refers to the fact that the estimates are a linear function of the Y observations.

GOODNESS OF FIT ($\overline{R}^2$)

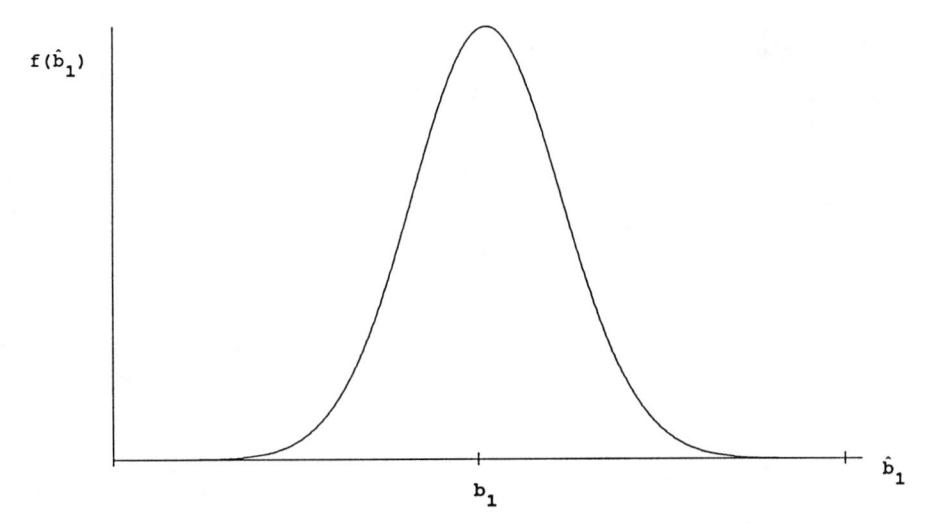

Figure 3.3 Repeated sampling distribution of statistic $\hat{b}_1$.

These ideas are illustrated in Figure 3.3. Consider parameter b_1. The estimated value $\hat{b}_1 = 0.61568$ for the heights data was found from just *one* sample of size $n = 25$. But imagine that we could draw *all possible samples* of size $n = 25$ from the population. Each sample of $n = 25$ would give a *different* SRF and therefore a *different* value for the statistic $\hat{b}_1$. If the disturbance a_i is normally distributed, then the *collection* of $\hat{b}_1$ values obtained by repeated sampling (see Figure 3.3) is t distributed, with $n - k$ degrees of freedom, where k is the number of parameters estimated in the regression. Under the stated PRF, the expected value of the repeated sampling distribution of the statistic $\hat{b}_1$ is the parameter b_1, $E(\hat{b}_1) = b_1$; that is, $\hat{b}_1$ is an unbiased estimator. This is illustrated in Figure 3.3, with the repeated sampling distribution centered on b_1. The variance of this sampling distribution is estimated by

$$s^2(\hat{b}_1) = \frac{\sum_{i=1}^{n} \hat{a}_i^2 / (n - k)}{(n - 1) s_{x_1}^2} \qquad (3.4.1)$$

where $s_{x_1}^2$ is the estimated variance of input X_1.

3.5 GOODNESS OF FIT ($\overline{R}^2$)

In this section we discuss the adjusted R^2 statistic, denoted $\overline{R}^2$. The $\overline{R}^2$ is a measure of how useful the inputs (the X's) are in explaining the movements of the output (Y). For this reason $\overline{R}^2$ is a measure of "goodness of fit." We

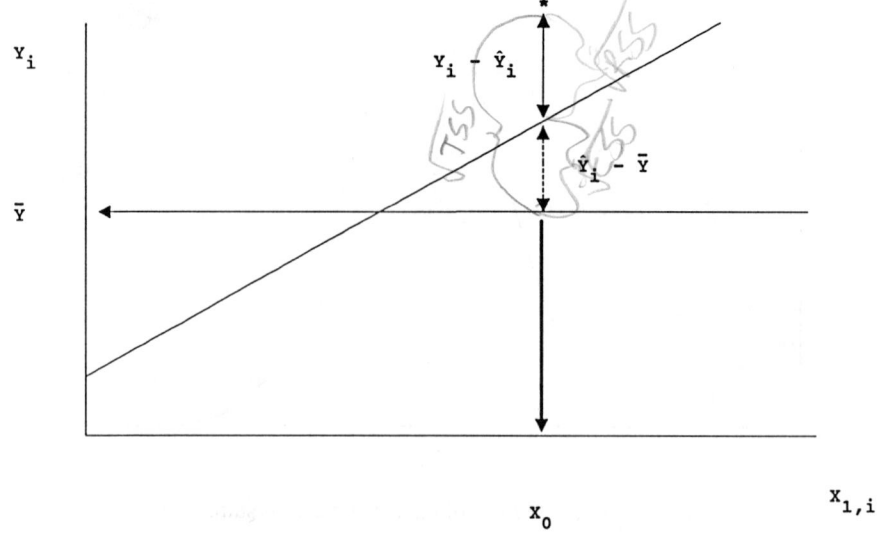

Figure 3.4 Illustration of Equation (3.5.1).

will not make much use of the $\overline{R}^2$ statistic in this text; but we present the basic ideas since $\overline{R}^2$ is a standard measure of goodness of fit in ordinary regression models. A dynamic regression (DR) model often includes an autoregressive integrated moving average (ARIMA) model for the disturbance. When it does, the $\overline{R}^2$ statistic loses its standard interpretation. Pierce (1979) presents an alternative $\overline{R}^2$ measure that is more useful for time series models. In this book we will emphasize forecast standard errors and confidence intervals (discussed in Chapter 9), rather than the $\overline{R}^2$, to measure the usefulness of a DR model.

3.5.1 Sums of Squares

To understand $\overline{R}^2$, we must first establish certain other ideas. Consider Figure 3.4. $\overline{Y}$ is the sample mean of the Y observations. The asterisk is one X, Y pair that has been used to obtain the SRF. We may think of getting to this observed Y value in three steps: (1) Given value $X_{1,i} = X_0$, go to the mean of the Y's, $\overline{Y}$. (2) Next, at the given value $X_{1,i} = X_0$, go from $\overline{Y}$ to the SRF predicted value $\hat{Y}_i$ on the SRF prediction line; that is, add the amount $\hat{Y}_i - \overline{Y}$ to $\overline{Y}$. The amount $\hat{Y}_i - \overline{Y}$ is represented by the dashed vertical arrow in Figure 3.4. (3) Finally, go from the SRF predicted value to the observed value Y_i (the asterisk); that is, add the amount $Y_i - \hat{Y}_i$ to $\hat{Y}_i$. The amount $Y_i - \hat{Y}_i$ is represented by the upper solid vertical arrow in Figure 3.4. Thus we may think of any observed Y_i as being composed of the following additive parts:

GOODNESS OF FIT ($\bar{R}^2$)

$$Y_i = \bar{Y} + (\hat{Y}_i - \bar{Y}) + (Y_i - \hat{Y}_i) \qquad (3.5.1)$$

Figure 3.4 illustrates this idea for just one observed Y_i value; in practice we have n observed Y_i values in a sample, with each observation composed of the three additive parts shown in (3.5.1).

If we now subtract $\bar{Y}$ from both sides of (3.5.1), square the result, and sum across all n observations, we obtain

$$\Sigma(Y_i - \bar{Y})^2 = \Sigma(\hat{Y}_i - \bar{Y})^2 + \Sigma(Y_i - \hat{Y}_i)^2$$
$$+ 2\Sigma(\hat{Y}_i - \bar{Y})(Y_i - \hat{Y}_i) \qquad (3.5.2)$$

But note that $(\hat{Y}_i - \bar{Y}) = (\hat{C} + \hat{b}_1 X_i) - (\hat{C} + \hat{b}_1 \bar{X}) = \hat{b}_1(X_i - \bar{X})$, and that $(Y_i - \hat{Y}_i) = \hat{a}_i$. Therefore the last term in (3.5.2) is $2\hat{b}_1 \Sigma(X_i - \bar{X})\hat{a}_i$, which is zero since X_i and $\hat{a}_i$ are independent by assumption. Therefore (3.5.2) reduces to

$$\Sigma(Y_i - \bar{Y})^2 = \Sigma(\hat{Y}_i - \bar{Y})^2 + \Sigma(Y_i - \hat{Y}_i)^2 \qquad (3.5.3)$$

which may also be written as

$$\text{TSS} = \text{ESS} + \text{RSS} \qquad (3.5.4)$$

where TSS $= (Y_i - \bar{Y})^2$, ESS $= (\hat{Y}_i - \bar{Y})^2$, and RSS $= \Sigma(Y_i - \hat{Y}_i)^2$. Here TSS stands for "total sum of squares," which is the total variation of the Y values around their own mean. We may think of $\bar{Y}$ as a *baseline predictor* of Y; that is, *we could use $\bar{Y}$ to predict Y even if we had no knowledge of the X's.* Therefore TSS represents the total variation of the Y's (around their own mean) that the X's could possibly "explain" (or be associated with). Then ESS stands for "explained (or regression) sum of squares," which is a measure of the extent to which the X's help to explain the deviation of the observed Y's from their own mean. And RSS is the "residual (or error, or unexplained) sum of squares," which is the remaining variation in the observed Y's that is not associated with movements in the X's; this is also the sum of squared residuals. For the heights data example, TSS = 166.56; ESS = 62.258; and RSS = 104.302.

3.5.2 Unadjusted and Adjusted R^2

We are now in a position to understand $\bar{R}^2$. The unadjusted R^2 (without the bar) is a raw measure of the *fraction of the total sum of squares that is explained by (associated with) the X's;* it is defined as $R^2 = \text{ESS/TSS}$. An alternative expression is found by dividing both sides of (3.5.4) by TSS, and subtracting RSS/TSS from both sides of the result to get

$$R^2 = \text{ESS/TSS} = 1 - \text{RSS/TSS} \tag{3.5.5}$$

Of course, a squared value is nonnegative. In fact, R^2 falls in the (inclusive) range between 0 and 1.0: (1) If the X's explain none of the movement in the Y's around their own mean, then we must have $\hat{b}_1 = 0$, and the SRF is the horizontal line at the level of $\overline{Y}$ in Figure 3.4. Then ESS = 0, and the residual variation (RSS) is equal to the total variation in the Y's (TSS). In this case, from (3.5.5) we see that $R^2 = 1 - \text{RSS/TSS} = 1 - \text{TSS/TSS} = 1 - 1 = 0$. (2) On the other hand, if the X's explain all of the movement of the Y's around their own mean, then all observed Y's are *on* the SRF, in which case ESS = TSS and RSS = 0. Then from (3.5.5) we see that $R^2 = 1 - \text{RSS/TSS} = 1 - 0/\text{TSS} = 1 - 0 = 1$. Usually R^2 falls somewhere between these two extremes. For the heights example, $R^2 = 1 - \text{RSS/TSS} = 1 - 104.302/166.56 = 1 - 0.62621 = 0.37379$. That is, about 37% of the variation in sons' heights (around the mean $\overline{Y} = 70.76$) is associated with movement in fathers' heights.

The sample R^2 is a biased estimate of the population R^2. This bias is reduced (but not eliminated) through an adjustment for degrees of freedom. The adjusted R^2 (denoted $\overline{R}^2$) is computed as

$$\overline{R}^2 = 1 - (1 - R^2)(n-1)/(n-k) \tag{3.5.6}$$

For the heights example, $\overline{R}^2 = 1 - (1 - 0.37379)(25 - 1)/(25 - 2) = 0.34656$. Therefore, adjusted for degrees of freedom, variation in fathers' heights explains almost 35% of the variation in sons' heights around their own mean value in this sample.

3.6 STATISTICAL INFERENCE

3.6.1 Testing a Hypothesis about b_1

Since we don't know the value of parameter b_1, we don't know where the sample value $\hat{b}_1 = 0.61568$ falls relative to b_1 on the horizontal axis of the sampling distribution in Figure 3.3. However, we can *test a hypothesis* about b_1 using the sample value $\hat{b}_1$ and its standard error, which is the square root of the variance in Equation (3.4.1). The standard null hypothesis in regression analysis is that the independent variable plays no role in explaining the movements in Y, that is, $H_0 : b_1 = 0$. The sample t statistic for this null hypothesis for the heights data SRF is $t = (\text{statistic} - \text{parameter})/(\text{standard error}) = (0.61568 - 0)/0.16617 = 3.70523$. Thus the statistic $\hat{b}_1 = 0.61568$ falls more than 3.7 standard errors above zero, which is greater than the critical t value of 2.069 for a 5% t test with 23 degrees of freedom. Therefore we reject H_0.

MULTIPLE REGRESSION

3.6.2 Testing a Hypothesis about a Set of Coefficients (*F* test)

In Section 3.7 we consider regression with several inputs, each with a coefficient b_i. For a regression with $k - 1$ independent variables, consider the null hypothesis, $H_0: b_1 = b_2 = \cdots = b_{k-1} = 0$. If H_0 is true, if the model includes a constant, and if the disturbance is not autocorrelated, then in repeated sampling the following statistic has the *F* distribution, with $k - 1$ and $n - k$ degrees of freedom: *# of independent variables X*

$$F = \frac{\text{ESS}/(k - 1)}{\text{RSS}/(n - k)} \qquad (3.6.1)$$

For the heights example there is only one coefficient (b_1), so the null simplifies to $H_0: b_1 = 0$. In this case the test statistic is $F = [62.258/1]/[104.302/23] = 13.7287$. From the table of *F* values (Table C in Appendix at end of book), this is significant at the 5% level; that is, the test statistic exceeds the critical value (between 4.35 and 4.17), which would be exceeded only 5% of the time in repeated sampling if H_0 were true. Since the sample *F* exceeds this critical value, we reject H_0.

Relationship between F and t
Note that the null hypothesis for this example is the same as the one we tested earlier using a *t* test. In fact the *t* and *F* statistics have a simple relationship when the SRF has just one independent variable: $F = t^2$. For the heights example $13.7287 = 3.70523^2$. When only one *X* variable is present, either the *F* test or the *t* test is redundant. This is not true, however, with a multiple regression (one with more than one independent variable). In a multiple regression, *t* statistics are used to test the significance of individual coefficients, while the *F* test is used to test the significance of groups of coefficients.

3.7 MULTIPLE REGRESSION

A multiple regression has more than one independent variable. For the heights example, consider a PRF with $X_{2,i}$ = mother's height, in addition to $X_{1,i}$ = father's height, as an explanatory variable for Y_i = son's height:

$$Y_i = C + b_1 X_{1,i} + b_2 X_{2,i} + a_i \qquad (3.7.1)$$

Coefficient b_1 in (3.7.1) has almost the same interpretation as it has in (3.2.1): It is the population average change in Y_i, per one unit change in $X_{1,i}$; but now we must add the phrase *taking into account the role* of $X_{2,i}$. Sometimes this last phrase is stated as "holding $X_{2,i}$ constant." Likewise, b_2 is the population average change in Y_i, per one unit change in $X_{2,i}$, taking into account the role of $X_{1,i}$. As before, C captures the effect of any excluded variables on the *overall level* of Y_i, and a_i captures the effect of any excluded variables on the *variability* of Y_i. We retain all of the assumptions made for the simple bivariate PRF (3.2.1), and we add one more: There is *no exact linear relationship among any of the X variables*. In (3.7.1) this means that there must not be perfect correlation between $X_{1,i}$ and $X_{2,i}$.

3.7.1 Prediction with a PRF (Several Inputs)

Just as with (3.2.1), we can use PRF (3.7.1) to predict values of Y_i for known parameters, given values for $X_{1,i}$ and $X_{2,i}$. For example, suppose it is known that $C = 27$, $b_1 = 0.5$, and $b_2 = 0.2$; and suppose $X_{1,i} = 70$ and $X_{2,i} = 65$. Then our prediction is $\hat{Y}_i = C + b_1 X_{1,i} + b_2 X_{2,i} = 27 + 0.5(70) + 0.2(65) = 75$ inches. Again, a_i is assigned its expected value of zero since it cannot be predicted from (3.7.1). If the son in question has an actual height of 72 inches, then the disturbance is $a_i = Y_i - \hat{Y}_i = 72 - 75 = -3$ inches.

We pictured the idea of predicting Y_i for PRF (3.2.1) in Figure 3.1; with just one independent variable, the PRF prediction equation $\hat{Y}_i = C + b_1 X_{1,i}$ is a straight line in a two-dimensional diagram. It is more difficult to picture the idea of predicting Y_i for a multiple regression. For a PRF with two independent variables, like (3.7.1), we need a three-dimensional diagram: $X_{1,i}$ is on one axis, $X_{2,i}$ is on a second axis, and Y_i is on a third axis. The PRF prediction equation $\hat{Y}_i = C + b_1 X_{1,i} + b_2 X_{2,i}$ is then represented as a plane. With more than two independent variables, the prediction equation is represented as a hyperplane, which is yet more difficult to picture.

3.7.2 Sample Regression Function (Several Inputs)

The SRF for a multiple regression is found using the least-squares rule. For PRF (3.7.1), the SRF is

$$Y_i = \hat{C} + \hat{b}_1 X_{1,i} + \hat{b}_2 X_{2,i} + \hat{a}_i \qquad (3.7.2)$$

We choose $\hat{C}$, $\hat{b}_1$, and $\hat{b}_2$ so as to minimize the sum of squared residuals, $\Sigma \hat{a}_i^2$, where each $\hat{a}_i$ is found by solving (3.7.2) for $\hat{a}$ to get

$$\hat{a}_i = Y_i - \hat{C} - \hat{b}_1 X_{1,i} - \hat{b}_2 X_{2,i}$$

and substituting sample observed values for each triplet $\{Y_i, X_{1,i}, X_{2,i}\}$.

The SRF for the heights example, with absolute t values in parentheses, is

$$Y_i = 2.00556 + 0.61531 X_{1,i} + 0.38863 X_{2,i} + \hat{a}_i$$
$$(0.11) \qquad (3.88) \qquad (1.79) \qquad (3.7.3)$$

MULTIPLE REGRESSION

The formulas for the estimated coefficients in a multiple regression are somewhat complicated in their ordinary algebraic form and are not presented here; see Gujarati (1988, p. 172). In Section 3.7.4 we will show the general matrix form solution for estimated coefficients and their standard errors in a multiple regression. The standard deviation of the disturbance for a multiple regression is estimated using (3.3.4). For the heights example, $\hat{\sigma}_a = [91.0959/(25 - 3)]^{1/2} = 2.03488$.

3.7.3 Testing a Hypothesis about One Coefficient

We can test an individual coefficient in a multiple regression for significance using a t test. For example, for b_2 we write the null hypothesis $H_0: b_2 = 0$, with alternate $H_a: b_2 \neq 0$. Then the sample t statistic is $t = (\hat{b}_2 - 0)/s(\hat{b}_2)$. For our example, we have $t = (0.38863 - 0)/0.21761 = 1.78587$.

The TSS for our example (166.56) is the same as it was when the only independent variable was father's height. (Why is this true?) However, to the extent that mother's height helps to explain the variation in son's height, we expect the ESS to be larger, and the RSS to be smaller, when $X_{2,i}$ is included. In fact, for (3.7.3) ESS = 75.4641, which is larger, and RSS = 91.0959, which is smaller.

The R^2 for a multiple regression is found using (3.5.5). In this example, $R^2 = 1 - (91.0959/166.56) = 0.453$. Again we focus on $\overline{R}^2$, which is found from (3.5.6). For the heights example $\overline{R}^2 = 1 - (1 - 0.453)(25 - 1)/(25 - 3) = 0.403$. That is, about 40% of the variation of son's height Y_i around its own mean is associated with movement in father's height and mother's height ($X_{1,i}$ and $X_{2,i}$).

We can test the joint null $H_0: b_1 = b_2 = 0$ with the F statistic given in (3.6.1). For this example $F = [75.4641/(3 - 1)]/[91.0959/(25 - 3)] = 9.11244$. The 5% critical F for $(k - 1, n - k) = (3 - 1, 25 - 3) = (2, 22)$ degrees of freedom is between 3.49 and 3.32. Since the sample F statistic exceeds this critical level, we reject H_0.

3.7.4 Matrix Notation

Many of the ideas in regression analysis are conveniently presented in matrix form. An SRF with $k - 1$ independent variables and n observations in the sample is written as

$$\mathbf{Y} = \mathbf{X}\hat{\mathbf{B}} + \hat{\mathbf{a}} \qquad (3.7.4)$$

where $\mathbf{Y}$ is an $n \times 1$ vector of n observations on the dependent variable; $\mathbf{X}$ is a $n \times k$ matrix of n observations on $k - 1$ inputs, plus a column of 1's to represent the constant; $\hat{\mathbf{B}}$ is a $k \times 1$ vector of estimated coefficients; and $\hat{\mathbf{a}}$ is an $n \times 1$ vector of residuals:

$$Y = \begin{bmatrix} Y_1 \\ Y_2 \\ \vdots \\ Y_n \end{bmatrix} \quad X = \begin{bmatrix} 1 & X_{1,1} & \cdots & X_{k-1,1} \\ 1 & X_{1,2} & \cdots & X_{k-1,2} \\ \vdots & \vdots & & \vdots \\ 1 & X_{1,n} & \cdots & X_{k-1,n} \end{bmatrix}$$

$$\hat{B} = \begin{bmatrix} \hat{C} \\ \hat{b}_1 \\ \hat{b}_2 \\ \vdots \\ \hat{b}_{k-1} \end{bmatrix} \quad \hat{a} = \begin{bmatrix} \hat{a}_1 \\ \hat{a}_2 \\ \vdots \\ \hat{a}_n \end{bmatrix}$$

Notice that the estimated constant term is composed of two parts: a set of 1's that are "observations" on the constant and a coefficient $\hat{C}$. The observations on the constant could be any fixed real value but using 1's is convenient since $\hat{C}$ times 1 equals $\hat{C}$.

The least-squares estimates in matrix form are obtained using calculus. For example, see Gujarati (1988, Appendix 9). The result is

$$\hat{B} = (X'X)^{-1}X'Y \qquad (3.7.6)$$

with variance-covariance matrix

$$\text{var-cov}(\hat{B}) = \hat{\sigma}_a^2 (X'X)^{-1} \qquad (3.7.7)$$

where the prime denotes matrix transposition.

3.8 SELECTED ISSUES IN REGRESSION

In this section we discuss several issues in regression. Most of these topics are dealt with in much more detail in econometrics textbooks. Since this chapter is only a primer on regression methods, we will discuss them but briefly. Perhaps the most important of these issues, because of its implications for our work in the remainder of this book, is autocorrelation in the disturbance discussed in Section 3.8.4.

3.8.1 Causation

The relationship between the Y and X variables may be a causal one, in the sense that changes in the X variables physically or logically cause changes in the Y variable. To argue for the existence of such causation requires an underlying theory, such as an economic, psychological, or physical theory. For the heights example there is support from genetic theory for the idea that son's height depends in a causal way on father's height and mother's height. The data in a sample might refute, or fail to refute, such a theory at a chosen significance level; but the data cannot prove the theory in the sense of establishing that the theory is true beyond any doubt. An SRF could show a statistical relationship between Y and the X's when there is no true causal connection at all.

3.8.2 Colinearity

Colinearity refers to *correlation among the independent variables*. An assumption in multiple regression is that there is *no perfect* colinearity among the inputs. If there is perfect colinearity, then at least one input is exactly a linear combination of one or more of the other inputs. For example, suppose father's height $(X_{1,i})$ could always be stated exactly as some linear function of mother's height $(X_{2,i})$: $X_{1,i} = c_0 + c_1 X_{2,i}$, where c_0 and c_1 are constants; then there would be perfect colinearity between $X_{1,i}$ and $X_{2,i}$. Or, suppose we plan to regress company president's salary (Y_i) on total assets $(X_{1,i})$, total liabilities $(X_{2,i})$, and owners' equity $(X_{3,i})$ for a sample of business firms. Again we have a perfect colinearity problem since owners' equity is defined as $X_{3,i} = X_{1,i} - X_{2,i}$; that is, X_3 is always an exact linear combination of X_1 and X_2.

If any perfect correlation exists among the X's, then the least-squares estimates cannot be found. This may be understood formally in a matrix context as follows: If there is any perfect colinearity among the X's, then the matrix $\mathbf{X'X}$ is singular; therefore its inverse does not exist, and solution (3.7.6) cannot be computed. Intuitively, the idea can be put this way: If two or more of the X's are *statistically identical* (perfectly correlated) in a given sample, then the least-squares method cannot estimate for us their *separate effects*, in the form of their separate b_i coefficients, based on that sample.

So far we have discussed perfect colinearity. But colinearity is a matter of *degree*, not a matter of its presence or absence. If the X's are correlated with each other *at all*, then there is *some* colinearity. It may be possible to create a set of uncorrelated X's, if their values are subject to our control. However, often we must use X values that have been generated by historical experience not subject to our control; in these cases it is nearly impossible to avoid at least some degree of colinearity.

The greater the degree of colinearity, the larger are the standard errors of the estimated coefficients, and the smaller are their absolute t values, other things equal. Thus colinearity may sometimes make it difficult to determine the importance of individual independent variables. For example, because of colinearity we could have an SRF in which all of the absolute t values are quite small despite the fact that $\overline{R}^2$ is quite large. In other words, all of the X's together could explain a lot of Y's movement, but we might not be able to tell which input, or set of inputs, is responsible for this explanatory power, or to what degree. High colinearity does not always destroy our ability to discern clearly the separate roles of the X's, since the coefficient standard errors (and t values) depend not only on the degree of colinearity but also on $\hat{\sigma}_a$ and the variability of the X's. But colinearity is never helpful.

3.8.3 Heteroskedastic Disturbance

If σ_a^2 is constant (as assumed), the disturbance is said to be *homoskedastic*. If σ_a^2 is not constant, the disturbance is said to be *heteroskedastic*. A heteroskedastic disturbance leads to several unhappy results; in particular, the least-squares estimator is no longer efficient (not least variance) in repeated sampling among all unbiased estimators. There are several methods for detecting and treating heteroskedastic disturbances. These methods are largely beyond the scope of our discussion. Gujarati (1988, Chapter 11) gives an introductory discussion in an econometrics context; Tsay (1988) gives a procedure in a time series (ARIMA) context.

As discussed in Chapter 2, we can often achieve a sufficiently constant variance in a time series (and therefore in the regression disturbance series) through transformation of the data. This is the extent of our discussion and treatment of heteroskedastic disturbances in this book.

3.8.4 Autocorrelated Disturbance

So far we have assumed that the disturbance in a PRF is a set of mutually independent values (not autocorrelated). However, the disturbance is often autocorrelated when the regression involves time series data. Let N_t represent the disturbance in a PRF involving time series data. Then (3.2.1), for example, is written as

$$Y_t = C + b_1 X_{1,t} + N_t \tag{3.8.1}$$

Suppose that N_t is autocorrelated and that its behavior is described by an ARIMA process of the type discussed in Chapter 2. For example, suppose N_t follows an ARIMA(1, 0, 1) process:

$$N_t = \phi_1 N_{t-1} - \theta_1 a_{t-1} + a_t \tag{3.8.2}$$

or, in backshift form,

$$(1 - \phi_1 B)N_t = (1 - \theta_1 B)a_t \tag{3.8.3}$$

If we apply ordinary least squares to estimate the coefficients in an equation whose disturbance is autocorrelated, without correcting for this autocor-

SELECTED ISSUES IN REGRESSION

relation, there are several unhappy consequences: (1) The estimator of each parameter is no longer best (not minimum variance); (2) the residual variance ($\hat{\sigma}_N^2$) is likely to underestimate the true σ_N^2; (3) the $\overline{R}^2$ is likely to be overestimated; (4) the usual t and F statistics are invalid; and (5) forecasts from the equation will be less accurate than they could be since the autocorrelation pattern in the residuals (the estimated disturbance, $\hat{N}_t$) contains useful information that we are ignoring.

In this book our approach to this problem is as follows. Suppose we are working with a PRF like (3.8.1), with an autocorrelated disturbance N_t whose autocorrelation pattern is described, for example, by the ARIMA(1, 0, 1) in (3.8.3). Then we can *solve* (3.8.3) *for* N_t,

$$N_t = \frac{1 - \theta_1 B}{1 - \phi_1 B} a_t$$

and *substitute the result into* (3.8.1) *to obtain the dynamic regression*

$$Y_t = C + b_1 X_{1,t} + \frac{1 - \theta_1 B}{1 - \phi_1 B} a_t \qquad (3.8.4)$$

At this point you need not be concerned with the mathematical meaning of the ratio $(1 - \theta_1 B)/(1 - \phi_1 B)$. (You may think of B as an ordinary algebraic variable. Thus, we are dividing one polynomial in B by another polynomial in B.) The main point is that the last term in (3.8.4) is a convenient way of saying "the regression disturbance N_t has a time pattern described by an ARIMA(1, 0, 1) process." An alternative form of (3.8.4) is obtained by substituting (3.8.2) into (3.8.1):

$$Y_t = C + b_1 X_{1,t} + \phi_1 N_{t-1} - \theta_1 a_{t-1} + a_t \qquad (3.8.5)$$

If an ARIMA model for N_t is included in a complete DR model, as in (3.8.4) or (3.8.5), then the DR model residuals are $\hat{a}_t$ rather than $\hat{N}_t$. For example, solving (3.8.1) for N_t, we have

$$N_t = Y_t - C - b_1 X_{1,t}$$

That is, N_t ($\hat{N}_t$ in a sample) is the difference between the observed Y_t and the regression predicted value ($C + b_1 X_{1,t}$). Thus the disturbance $\hat{N}_t$ may be thought of as *the set of "residuals" if we predict Y_t using only the "regression part" of model* (3.8.1). On the other hand, solving (3.8.5) for a_t, we have

$$a_t = Y_t - C - b_1 X_{1,t} - \phi_1 N_{t-1} + \theta_1 a_{t-1}$$

That is, a_t ($\hat{a}_t$ in a sample) is the difference between the observed Y_t and the

DR predicted value $(C + b_1 X_{1,t} + \phi_1 N_{t-1} - \theta_1 a_{t-1})$. Thus $\hat{a}_t$ is *the set of residuals if we predict Y_t using the entire DR model* (3.8.4) or (3.8.5).

We can obtain expressions like (3.8.4) or (3.8.5) for any DR model regardless of the ARIMA process for N_t. Suppose the time structure of N_t is described by a *general* ARIMA$(p, d, q)(P, D, Q)_s$ process. Using the notation developed in Chapter 2, we write this as

$$\phi(B^s)\phi(B)\nabla_s^D \nabla^d N_t = \theta(B^s)\theta(B) a_t \qquad (3.8.6)$$

Solve (3.8.6) for N_t:

$$N_t = \frac{\theta(B^s)\theta(B)}{\phi(B^s)\phi(B)\nabla_s^D \nabla^d} a_t$$

Substitute this result into (3.8.1) to get

$$Y_t = C + b_1 X_{1,t} + \frac{\theta(B^s)\theta(B)}{\phi(B^s)\phi(B)\nabla_s^D \nabla^d} a_t \qquad (3.8.7)$$

The last term in (3.8.7) merely says that the regression disturbance N_t has an ARIMA$(p, d, q)(P, D, Q)_s$ time pattern.

In general, the coefficients in a DR equation like (3.8.4), (3.8.5), or (3.8.7) can't be estimated directly with ordinary least squares regression methods. As discussed in Chapter 9, we can estimate the coefficients in equations like this using maximum-likelihood methods. An application is shown in Section 3.9.

Durbin–Watson Statistic

A commonly used statistic to test for the presence of autocorrelation in regression residuals is the *Durbin–Watson* (D–W) statistic. In using this statistic, it is assumed that a constant term is present in the model, that the X's are fixed in repeated sampling, that the disturbance series follows an AR(1) process, that the regression does not include any time-lagged Y's, and that there are no missing observations. The null hypothesis is $H_0: \rho_1 = 0$ where ρ_1 is the first-order (lag $k = 1$) autocorrelation coefficient. The D–W statistic is computed as

$$d = \frac{\sum_{t=2}^{n} (\hat{e}_t - \hat{e}_{t-1})^2}{\sum_{t=1}^{n} \hat{e}_t^2} \qquad (3.8.8)$$

where $\hat{e}_t$ is a regression model sample residual. If the model is estimated

SELECTED ISSUES IN REGRESSION

without the ARIMA model for N_t, then $\hat{e}_t = \hat{N}_t$; if the ARIMA model for N_t is included, or if $N_t = a_t$, then $\hat{e}_t = \hat{a}_t$. If the sample size (n) is not too small, then *approximately*

$$d = 2[1 - r_1(\hat{e})] \qquad (3.8.9)$$

where $r_1(\hat{e})$ is the lag $k = 1$ sample autocorrelation coefficient for the residual series $\hat{e}_t$. Thus the D–W statistic provides limited information about the autocorrelation structure of the disturbance series. The sample autocorrelation function (SACF), the sample partial autocorrelation function (SPACF), and the extended SACF (ESACF) (discussed in Chapter 2) provide more information since they cover more time lags and help us to identify a parsimonious form for the ARIMA structure of the disturbance.

3.8.5 Functional Form

We pointed out in Section 3.2 that a PRF is assumed to be linear in the coefficients; that is, it has the form of Equation (3.2.1) (and appears as a straight line in a two-dimensional graph) if there is one independent variable; it has the form of Equation (3.7.1) (and appears as a plane in a three-dimensional graph) if there are two independent variables. By contrast, the PRF equation $Y_t = C X_{1,t}^{b_1} X_{2,t}^{b_2} a_t$ is nonlinear in the coefficients; in a three-dimensional graph, a plot of Y versus X_1 and X_2 would not appear as a plane.

Presumably there are many nonlinear relationships in the world; restricting the PRF to relationships that are linear in the coefficients may seem to impose severe limits on the usefulness of regression models. However, linear regression can be *nonlinear in the variables*. Thus, by transforming the data (the variables), we might obtain a model that is linear in the coefficients. For example, by taking natural logarithms of the PRF $Y_t = C X_{1,t}^{b_1} X_{2,t}^{b_2} a_t$, we obtain $Y_t' = C' + b_1 X_{1,t}' + b_2 X_{2,t}' + a_t'$, where the prime denotes natural logarithms. This latter PRF is linear in the coefficients and can be estimated using ordinary least squares.

As another example, consider the data plotted in Figure 3.5. Y_i is crop output, and $X_{1,i}$ is fertilizer input; the data were generated by changing fertilizer input, holding other inputs (seed, water, etc.) constant, and observing how crop output changed. The curvilinear nature of the data plot reflects the law of diminishing marginal productivity; this law states that, as one input to a production process is increased while other inputs are held constant, eventually the output from the process will increase *at a decreasing rate*.

It seems fairly clear that a straight-line regression equation will not adequately describe this relationship since it will not capture its curvilinear nature. However, the data look as if the relationship might be described fairly well by the following equation:

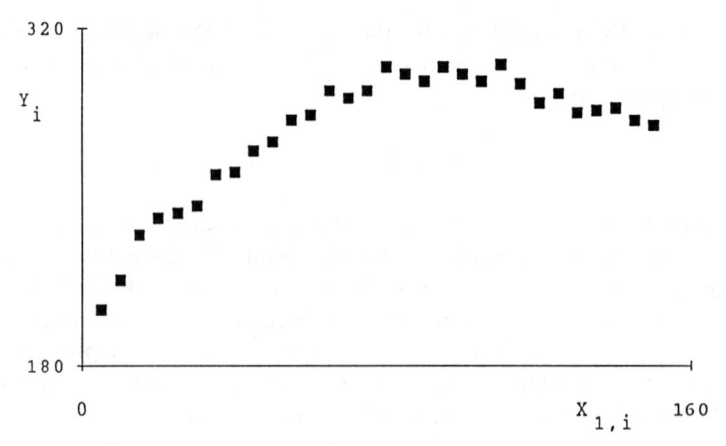

[Figure 3.5: Fertilizer input (X_i) and crop output (Y_i). Scatter plot with Y_i on vertical axis ranging from 180 to 320, and $X_{1,i}$ on horizontal axis from 0 to 160.]

Figure 3.5 Fertilizer input (X_i) and crop output (Y_i).

$$Y_i = C + b_1 X_{1,i} + b_2 X_{1,i}^2 + a_i \qquad (3.8.10)$$

This equation is a second-degree polynomial in $X_{1,i}$; as such it can describe a relationship in two dimensions that contains one "turn" in the data. While this equation is nonlinear in the variables (it includes $X_{1,i}^2$), it is linear in the coefficients, as required for ordinary linear least-squares estimation. This may be seen by writing $X_{2,i} = X_{1,i}^2$, and substituting this into (3.8.10) to get

$$Y_i = C + b_1 X_{1,i} + b_2 X_{2,i} + a_i \qquad (3.8.11)$$

which is identical to (3.7.1).

3.8.6 Excluded Variables

In Chapter 1 we pointed out that using a simple bivariate regression could lead to poor estimation results if some relevant independent variables are excluded. This is a difficult problem since we usually don't know the PRF with certainty; rather, we try to develop a hypothetical PRF using some underlying theory for guidance.

In this book we focus on the situation where Y_t may be related not only to X_t, but also to X_{t-1}, X_{t-2}, and so on. We will attempt to include all relevant lagged time periods of X. If we exclude relevant lagged time periods, and if the values of X for the excluded time periods are correlated with those included in our model, then several undesirable results arise: (1) The least-squares estimates are biased and inconsistent; that is, the expected values of the estimated coefficients are not equal to the corresponding parameters in repeated sampling, and the bias does not tend to disappear as the sample size grows; (2) the residual standard deviation is larger than need be, since we have left out useful information; thus (3) forecasts from the model are less accurate than they could be; and (4) the usual tests of significance may be misleading.

Starting in Chapter 5 we present methods designed to ensure that we have included all relevant *lagged time periods* of any X variable included in our regression models. However, these methods cannot ensure that we have included all relevant *variables*, or that these excluded variables are uncorrelated with the ones that we have included. The implication of this problem is less severe if our main purpose is to forecast. It is more serious if our purpose is to obtain estimates of the "true" effects of an X variable for evaluating hypotheses or forming policy conclusions.

3.8.7 Simultaneous Equations

Using *single-equation* regression models is appropriate only when Y depends logically on the X's, but not vice versa. Yet many situations arise where Y and the X's may be interdependent. When this happens, a *set* of equations is needed to properly specify these relationships, and generally methods other than ordinary single-equation least squares must be used to estimate the coefficients in this *system* of equations.

There is a well-developed econometrics literature for dealing with multiple equations. For an introduction see Gujarati (1988, Part IV). In this book we focus on single-equation methods for forecasting time series data. We touch on the matter of multiple equations in a time series context in Chapter 10.

3.9 APPLICATION TO TIME SERIES DATA

In this section we apply the ideas presented in this chapter to develop a multiple regression forecasting model for quarterly housing starts in the United States (Y_t). Because we include an ARIMA model for the disturbance, this is an example of a DR model. The observations (shown in the Data Appendix as Series 11) are from the first quarter of 1965 through the fourth quarter of 1975. A partial listing of these data is shown in Table 3.2. The plot of the data in Figure 3.6 suggests that the variance of the series is fairly constant, so no transformation is needed to stabilize the variance. Inspection of the plot shows that the data have a seasonal pattern, with first- and fourth-quarter figures generally lower than second- and third-quarter figures.

3.9.1 Seasonal Dummy Variables

We will use a set of quarterly seasonal dummy (binary) independent variables to represent the seasonal pattern. These variables are defined as

Table 3.2 Partial Listing of Quarterly Housing Starts and Quarterly Seasonal Dummy Variables

Date		Housing Starts	$X_{1,t}$	$X_{2,t}$	$X_{3,t}$
1965	1	181.504	1	0	0
	2	296.690	0	1	0
	3	266.195	0	0	1
	4	219.299	0	0	0
1966	1	180.158	1	0	0
	2	258.919	0	1	0
	3	197.834	0	0	1
	4	141.714	0	0	0
	⋮	⋮	⋮	⋮	⋮

$X_{1,t} = 1$ when t is a first quarter
$\phantom{X_{1,t}} = 0$ otherwise

$X_{2,t} = 1$ when t is a second quarter
$\phantom{X_{2,t}} = 0$ otherwise

$X_{3,t} = 1$ when t is a third quarter
$\phantom{X_{3,t}} = 0$ otherwise

A partial listing of the X's is shown in Table 3.2.
The PRF we are considering is

$$Y_t = C + b_1 X_{1,t} + b_2 X_{2,t} + b_3 X_{3,t} + N_t \qquad (3.9.1)$$

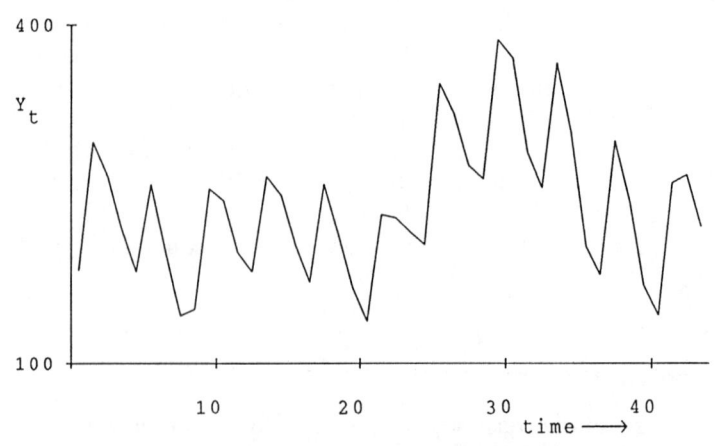

Figure 3.6 Quarterly housing starts, 1965 I–1975 IV.

APPLICATION TO TIME SERIES DATA

where N_t is a disturbance series that may be autocorrelated. From the definitions of the X's, it appears that the fourth quarter is not represented in Equation (3.9.1). However, consider the PRF predicted value for Y_t when t is a first quarter:

$$\hat{Y}_t = C + b_1 X_{1,t} + b_2 X_{2,t} + b_3 X_{3,t}$$
$$= C + b_1(1) + b_2(0) + b_3(0)$$
$$= C + b_1$$

where N_t is assigned its expected value of zero. (We are ignoring for simplicity the possible autocorrelation pattern of N_t.) Similarly, we may obtain PRF predicted values of Y_t when t is a second, third, or fourth quarter. The PRF predicted values of Y_t for the four quarters are:

First quarter: $\hat{Y}_t = C + b_1$
Second quarter: $\hat{Y}_t = C + b_2$
Third quarter: $\hat{Y}_t = C + b_3$
Fourth quarter: $\hat{Y}_t = C$ (3.9.2)

Thus, C is the PRF prediction of the level of Y during fourth quarters; b_1 is the amount that must be added to the fourth-quarter prediction (C) to obtain the PRF prediction of the first-quarter level of Y; b_2 is the amount that must be added to the fourth-quarter prediction (C) to obtain the PRF prediction of the second-quarter level of Y; and b_3 is the amount that must be added to the fourth quarter prediction (C) to obtain the PRF prediction of the third-quarter level of Y. By excluding the fourth quarter dummy from (3.9.1), we make the fourth quarter the base quarter, captured by C. The b coefficients then capture all other quarterly effects relative to the fourth quarter.

We have seen that including a fourth quarter-dummy in (3.9.1) is not necessary; in fact, doing so would introduce perfect colinearity. This may be seen by recalling that the constant term in a regression is actually the "observed" value of the constant (1.0) *times the coefficient C*. We adopt the common practice of referring to C as "the constant," though C is actually the *coefficient* of the constant (1.0). Suppose we add another column to Table 3.2: $X_{4,1} = 1$ if t is a fourth quarter and $X_{4,t} = 0$ otherwise. Then *for each t the sum of the variables $X_{1,t}$ through $X_{4,t}$ is 1*, which is exactly equal to the constant. Thus, the constant (1.0) would be a perfect linear combination of the four seasonal dummies, and we would have perfect colinearity. Dropping $X_{4,t}$ (or any other quarterly dummy) solves this problem. However, we don't lose information about the level of Y during fourth quarters, as shown in (3.9.2): The value of C represents the average level of Y during fourth quarters.

Table 3.3 Ordinary Least-Squares Estimation Results for Model (3.9.1)

Parameter	Estimate	Standard Error	t Value
C	209.2191	13.5631	15.43
b_1	-23.6836	19.1812	-1.23
b_2	84.2744	19.1812	4.39
b_3	55.4067	19.1812	2.89

$\hat{\sigma}_N = 44.9839$
Total $n = 44$ $R^2 = 0.477$ $\bar{R}^2 = 0.438$ $F_{3,40} = 12.14$
Total sum of squares = 170097
Residual sum of squares = 89036.3
Explained sum of squares = 81060.7

Correlation Matrix of Parameter Estimates

	1	2	3	4
1	1.00			
2	-0.71^a	1.00		
3	-0.71^a	0.50^a	1.00	
4	-0.71^a	0.50^a	0.50^a	1.00

[a] Significant at 5% level.

Models like (3.9.1) that capture seasonality with dummy variables assume that the seasonal elements are *globally constant*. That is, from (3.9.2) we see that each second quarter level of Y, *regardless of the year*, is predicted to be b_2 units above (if $b_2 > 0$) or below (if $b_2 < 0$) the fourth-quarter level C. This is in contrast to seasonal ARIMA models (discussed in Chapter 2) where seasonal patterns are treated as stochastic.

3.9.2 Model Estimation

Estimating (3.9.1) (assuming $N_t = a_t$) using the least-squares method gives the results shown in Table 3.3. The negative sign on $\hat{b}_1$ suggests that first quarters are, on average, below the fourth-quarter average by 23.6836 units; the positive signs on $\hat{b}_2$ and $\hat{b}_3$ suggest that the second and third quarters are, on average, above the fourth-quarter average by 84.2744 and 55.4067 units, respectively. It is tempting to use the t values of the estimated coefficients to test hypotheses about the b coefficients. However, three points must be borne in mind:

1. The presence or absence of seasonality involves all three X variables; we should consider a joint test involving all three coefficients, rather than just testing one coefficient at a time.
2. The correlations among the estimated b coefficients could make individual t tests somewhat misleading; again, a joint test is indicated.

APPLICATION TO TIME SERIES DATA

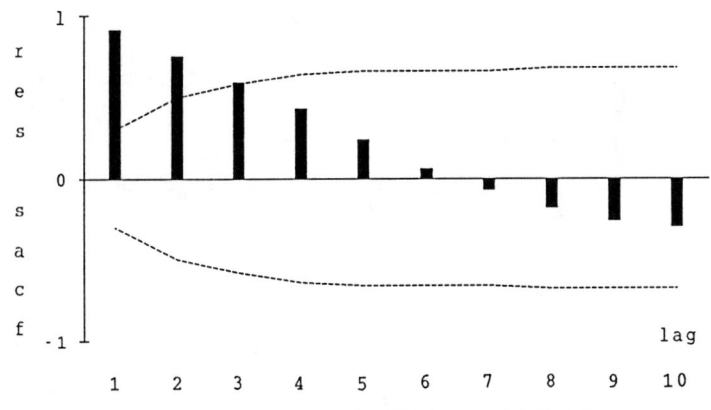

Figure 3.7 Residual SACF for model (3.9.1).

3. Tests on the *b* coefficients at this point are invalid if the disturbance in (3.9.1) is autocorrelated. We should first examine the autocorrelation pattern of the sample disturbance from our estimated model. (There is another reason to examine the sample disturbance for autocorrelation: If it is autocorrelated, this represents useful information that could help us to produce better forecasts.)

3.9.3 Model Checking and Reformulation

The sample disturbance series in model (3.9.1) is calculated as

$$\hat{N}_t = Y_t - \hat{C} - \hat{b}_1 X_{1,t} - \hat{b}_2 X_{2,t} - \hat{b}_3 X_{3,t}$$

That is, solve (3.9.1) for N_t and replace parameters with their estimates. The sample disturbance series is computed automatically by the SCA System using appropriate commands. The SACF and SPACF for $\hat{N}_t$ are shown in Figures 3.7 and 3.8, respectively. The first three coefficients in the SACF are more than twice their standard errors; this suggests that the disturbance series in (3.9.1) is autocorrelated, so that the *t* tests and *F* test in Table 3.3 are invalid. We will attempt to identify an ARIMA model for the disturbance series and incorporate that model into (3.9.1).

The SACF in Figure 3.7 moves to insignificant values fairly quickly, which implies that the disturbance series is stationary in the mean. The SACF seems to decay, suggesting an autoregressive (AR) or mixed [autoregressive moving average (ARMA)] model. The SPACF in Figure 3.8 could be interpreted as having two spikes; along with the decay in the SACF, this suggests an AR(2). Alternatively, the SPACF could be interpreted as decaying; along with the decay in the SACF, this suggests a mixed model. Since a mixed model is a good candidate, the ESACF might be especially helpful. In fact, the ESACF in Table 3.4 suggests an ARMA(1, 1); that is, we can

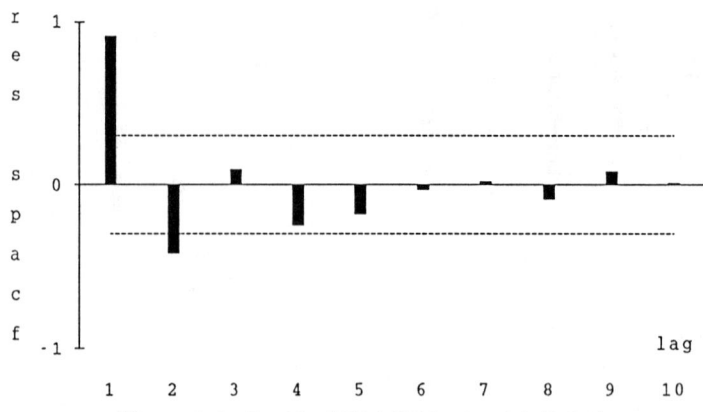

Figure 3.8 Residual SPACF for model (3.9.1).

place a triangle of zeros with its upper left vertex in the ($p' = 1, q = 1$) position. With $d = 0$, $p' = p = 1$. All things considered, an ARMA(1, 1) model seems like the best candidate. Thus our model for N_t is

$$(1 - \phi_1 B) N_t = (1 - \theta_1 B) a_t \quad (3.9.3)$$

Solving (3.9.3) for N_t and substituting the result into (3.9.1) gives this DR model:

$$Y_t = C + b_1 X_{1,t} + b_2 X_{2,t} + b_3 X_{3,t} + \frac{1 - \theta_1 B}{1 - \phi_1 B} a_t \quad (3.9.4)$$

3.9.4 Estimation and Checking (Final Model)

We will estimate (3.9.4) using the method of maximum likelihood (ML). Ordinary least squares is not applicable when the disturbance series has an ARIMA pattern; a nonlinear estimation routine is needed. A more detailed

Table 3.4 Residual ESACF for Model (3.9.1)[a]

($q \to$)	0	1	2	3	4	5	6
($p' = 0$)	X	X	0	0	0	0	0
($p' = 1$)	X	0	0	0	0	0	0
($p' = 2$)	0	0	0	0	0	0	0
($p' = 3$)	0	X	0	0	0	0	0
($p' = 4$)	X	0	0	0	0	0	0
($p' = 5$)	0	0	0	0	0	0	0
($p' = 6$)	0	0	0	0	0	0	0

[a] X = significant at 5% level.

APPLICATION TO TIME SERIES DATA   109

Table 3.5 ML Estimation Results for Model (3.9.4)

Parameter	Estimate	Standard Error	t Value
C	211.1774	24.3323	8.68
b_1	−22.3190	4.7966	−4.65
b_2	85.1021	6.3082	13.49
b_3	55.7842	4.6625	11.96
θ_1	−0.4991	0.1387	−3.60
ϕ_1	0.8384	0.0880	9.53

$\hat{\sigma}_a = 17.164$

Correlation Matrix of Parameter Estimates

	1	2	3	4	5	6
1	1.00					
2	−0.07	1.00				
3	−0.12	0.68[a]	1.00			
4	−0.09	0.11[a]	0.68[a]	1.00		
5	0.03	0.04	0.02	0.01	1.00	
6	0.07	0.01	0.01	0.01	0.33[a]	1.00

[a] Significant at 5% level.

discussion of this topic is deferred until Chapter 9. When nonlinear methods are used, the usual statistical tests are approximate.

We can use the SCA System to obtain ML estimates of our DR model coefficients. The estimation results for model (3.9.4) are shown in Table 3.5. The estimated moving average (MA) coefficient is significant, and it satisfies the relevant invertibility condition. The estimated AR coefficient is also significant, and it satisfies the relevant stationarity condition. The SACF in Figure 3.9 for the residuals ($\hat{a}_t$) of model (3.9.4) indicates that there is no remaining significant residual autocorrelation.

We test the overall equation by testing the null $H_0: b_1 = b_2 = b_3 = \phi_1 = \theta_1 = 0$. The test statistic is the F statistic with $(5, 37)$ degrees of freedom:

$$F = \frac{\text{ESS}/(k-1)}{\text{RSS}/(n-k)} = \frac{157429.1/(6-1)}{12667.9/(43-6)} = 91.96 \qquad (3.9.5)$$

From Table C (see Appendix at end of book), the critical value for $(5, 37)$ degrees of freedom is about 2.5; the sample F exceeds this critical value by a wide margin, and we reject H_0.

The histogram of the standardized residuals ($\hat{a}_t/\hat{\sigma}_a$) in Figure 3.10 is skewed slightly to the positive side, and the distribution looks a little too flat to be normal. But the normal probability plot of the residuals in Figure 3.11 is a fairly straight line, except for two values in the lower left corner; this suggests that the shocks are fairly close to a normal distribution.

110   A PRIMER ON REGRESSION MODELS

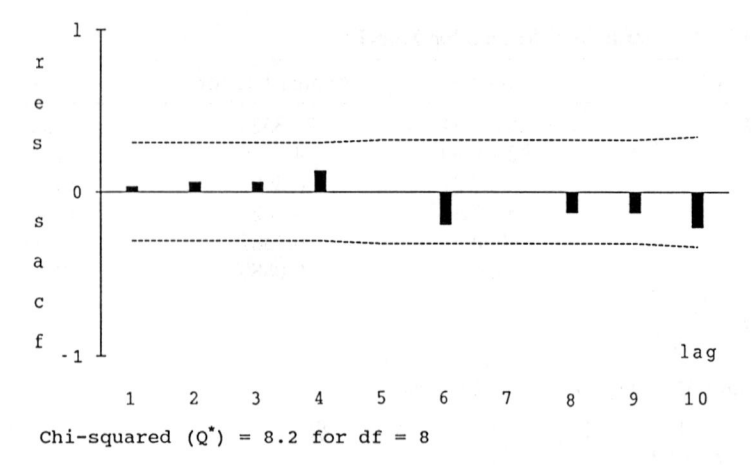

```
  1
r
e
s
  0
s
a
c
f
 -1                                                                    lag

      1    2    3    4    5    6    7    8    9   10
Chi-squared (Q*) = 8.2 for df = 8
```

Figure 3.9 Residual SACF for residuals from model (3.9.4).

```
    LOWER      UPPER    FREQ-  0           5          10
    BOUND      BOUND    UENCY  +----+----+----+----+
   -3.000  -  -2.400      0    |
   -2.400  -  -1.800      0    |
   -1.800  -  -1.200      6    |>>>>>>>>>>>
   -1.200  -  -0.600      8    |>>>>>>>>>>>>>>>
   -0.600  -   0.000      8    |>>>>>>>>>>>>>>>
    0.000  -   0.600      9    |>>>>>>>>>>>>>>>>>
    0.600  -   1.200      6    |>>>>>>>>>>>
    1.200  -   1.800      5    |>>>>>>>>>
    1.800  -   2.400      1    |>>
    2.400  -   3.000      0    |
                                +----+----+----+----+
                                0           5          10
```

Figure 3.10 Histogram of standardized residuals from model (3.9.4).

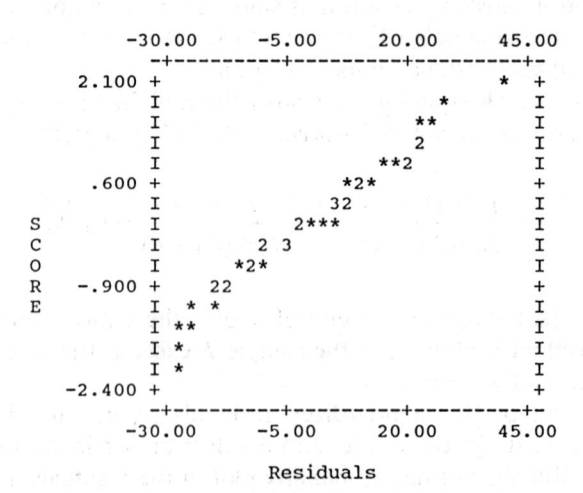

```
                  -30.00    -5.00     20.00     45.00
                -+---------+---------+---------+-
         2.100 +                         *      +
               I                              * I
               I                         *     I
               I                        **     I
               I                         2     I
               I                       **2     I
          .600 +                    *2*        +
               I                      32       I
         S     I                   2***        I
         C     I                  2 3          I
         O     I                  *2*          I
         R    -.900 +         22              +
         E     I        * *                    I
               I       **                      I
               I      *                        I
               I      *                        I
         -2.400 +                               +
                -+---------+---------+---------+-
                  -30.00    -5.00     20.00     45.00
                              Residuals
```

Figure 3.11 Normal probability plot of residuals from model (3.9.4).

APPLICATION TO TIME SERIES DATA 111

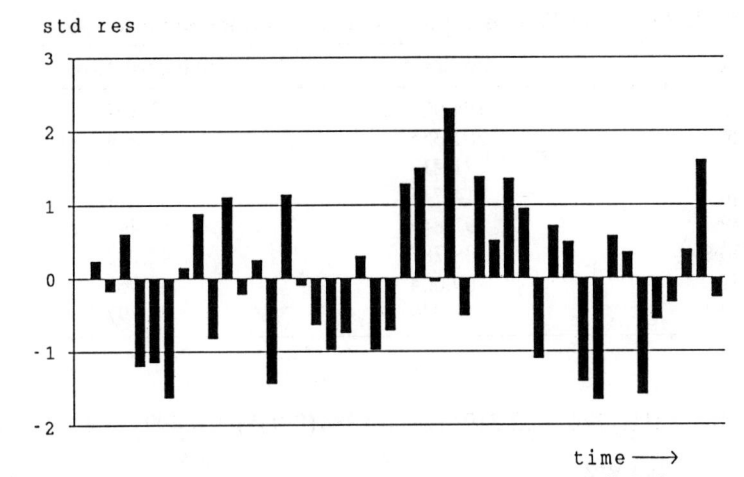

std res

Figure 3.12 Standardized residuals from model (3.9.4).

Figure 3.12 shows the standardized residuals. Only one value falls beyond the two standard deviation limits, suggesting that there are no important outliers. The formal outlier detection procedure discussed in Chapter 8 might give different results.

3.9.5 Forecasting

The matter of producing forecasts from DR models is discussed in detail in Chapter 9. In practice, DR forecasts are produced by the same software used to estimate the DR model. However, to illustrate the idea, consider how we can produce a one-step ahead forecast from (3.9.4). First, rewrite (3.9.4) as

$$Y_t = \hat{C} + \hat{b}_1 X_{1,t} + \hat{b}_2 X_{2,t} + \hat{b}_3 X_{3,t} + \hat{\phi}_1 \hat{N}_{t-1} - \hat{\theta}_1 \hat{a}_{t-1} + \hat{a}_t \quad (3.9.6)$$

Since the last observation in the sample is 1975 IV (period $t = 44$), a one-step ahead forecast from origin $t = 44$ is for 1976 I (period $t = 45$), which is a first quarter. Therefore $X_{1,45} = 1$, while $X_{2,45} = X_{3,45} = 0$. The random shock a_{45} is unknown, and it cannot be estimated by $\hat{a}_{45}$ since we have estimated the model only over periods $t = 1, 2, \ldots, 44$. Therefore we assign $\hat{a}_{45}$ its expected value of zero. From (3.9.6) we see that we need $\hat{N}_{44}$ and $\hat{a}_{44}$ to forecast Y_{45}. The SCA System estimation routine provides the needed values as a by-product of estimating model (3.9.4); these values are $\hat{N}_{44} = 9.945$ and $\hat{a}_{44} = -4.578$. Using this information, and the estimated coefficients from Table 3.5, Equation (3.9.6) for $t = 45$ is

Table 3.6 Eight Forecasts from Origin $t = 44$ from Model (3.9.4)

Time	Forecast	Standard Error
45	194.912	17.164
46	301.355	28.664
47	271.217	34.526
48	214.745	38.111
49	191.849	40.441
50	298.787	42.002
51	269.064	43.065
52	212.940	43.797

$$\hat{Y}_{45} = 211.1774 - 22.319(1) + 0.8384(9.945) + 0.4991(-4.578)$$
$$= 194.912 \qquad (3.9.7)$$

This forecast, as well as seven additional ones from origin $t = 44$ are shown in Table 3.6. The additional forecasts were computed by the SCA System in a fashion similar to the procedure used here. When a needed value for $\hat{N}_t$ or $\hat{a}_t$ is not available, that variable is assigned its expected value of zero. Procedures for finding the forecast standard errors shown in Table 3.6 are given in Chapter 9.

3.9.6 Joint Test of Significance for Seasonal Dummy Coefficients

We may test the significance of the set of three seasonal dummy coefficients ($H_0: b_1 = b_2 = b_3 = 0$) with the following approximate χ^2 statistic, with 3 degrees of freedom:

$$Q = \hat{\mathbf{B}}' \mathbf{V}^{-1} \hat{\mathbf{B}} \qquad (3.9.8)$$

where $\hat{\mathbf{B}} = (\hat{b}_1, \hat{b}_2, \hat{b}_3)'$, and $\mathbf{V}$ is the variance-covariance matrix of the elements in $\hat{\mathbf{B}}$. The diagonal elements of $\mathbf{V}$ are the variances of the b's, which are the squares of the coefficient standard errors given in Table 3.5. Each off-diagonal element of $\mathbf{V}$ (v_{ij}, $i \neq j$) is the covariance of $\hat{b}_i$ and $\hat{b}_j$. These covariances are found by multiplying the correlation between any two coefficients by their standard errors; these items are found in Table 3.5. For example, the covariance between $\hat{b}_1$ and $\hat{b}_2$ is $(0.68)(4.7966)(6.3082) = 20.5754$. The resulting test statistic from (3.9.8) is $Q = 665.95$, which exceeds the critical χ^2 (7.81) for 3 degrees of freedom. We reject H_0 and conclude that the housing starts series exhibits a seasonal pattern.

QUESTIONS AND PROBLEMS

1. You have a sample of size $n = 2$ (two X observations and two Y observations) to estimate this PRF: $Y_i = C + b_1 X_i + e_i$.
   a. Illustrate this situation using an X, Y graph that shows the two pairs of observations and the least-squares SRF.
   b. What is the numerical value of the unadjusted sample R^2? Explain briefly.
   c. What can you say about the adjusted sample R^2 in this situation? Start from general formulas. Show all work.

2. You have this SRF: $Y_i = \hat{C} + \hat{b}_1 X_i + \hat{a}_i$, with $\hat{C} = 100$, and $\hat{b}_1 = -4$.
   a. Find the predicted value of Y for $X = 3$.
   b. For $X = 3$, $Y = 84$. Find $\hat{a}$.
   c. Draw an X, Y graph to illustrate parts (a) and (b).

3. An SRF has this F statistic:

$$F = \frac{1139/(3-1)}{624/(33-3)}$$

   a. Find the adjusted R^2. Interpret this number.
   b. Find the residual standard deviation. Interpret this number.

4. Consider the following SRF results, where Y_t = annual performance of an investment manager (% return) and X_t = annual performance of overall stock market (% return), based on $n = 16$ observations.

Parameter	Estimate	Standard Error
C	9.88	2.74
b_1	0.53	0.19

   a. Suppose the overall stock market returns 15% in a certain year. What is your best prediction of the investment manager's performance in that year?
   b. Is the estimated constant term significantly different from zero at the 5% level? Explain. Start from general formulas. Illustrate with a t-distribution graph.
   c. Is the estimated slope significantly different from 1.0 at the 5% level? Explain. Start from general formulas. Illustrate with a t-distribution graph.
   d. Is the estimated slope significantly different from zero at the 5% level?

Find the answer using a t statistic. Find the answer using an F statistic. Explain.

5. Using the data in Question 4 at the end of Chapter 1, estimate the multiple regression equation $Y_t = C + v_0 X_t + v_1 X_{t-1} + v_2 X_{t-2} + N_t$. Do the results contradict your answer to Question 4 in Chapter 1?
   a. Do the estimated regression disturbance values ($\hat{N}_t$) that you obtained contradict the null hypothesis that the population disturbance sequence (N_t) is not autocorrelated? Answer by computing the sample autocorrelation function for the estimated disturbance series and the Durbin–Watson statistic. (How is the Durbin–Watson statistic related to the sample autocorrelation function?) If you reject the null hypothesis of no autocorrelation in N_t, what are the consequences for your estimated regression?
   b. Can you tell if there is feedback from Y_t to X_t? How? With graphs? With regression? (We will revisit this issue in Chapters 5 and 10.) If there is feedback, what are the consequences for your regression results?
   c. What problems arise in trying to assess the relationship between Y_t and X_t using a series of two-dimensional graphs, as suggested in Question 4 in Chapter 1? (Hint: Consider the number of dimensions of the graphs and the phrase "other things held constant.")

6. Build an ARIMA model for Series 9. Does your ARIMA model fit the data better than DR model (3.9.4)? Does it seem preferable to model the seasonality with seasonal dummy variables?

7. Consider the data in Case 1 following Chapter 2 (Series 7 in the Data Appendix). Build a regression model where Y_t' is the natural logs of the government receipts data and where the inputs are monthly seasonal dummy variables, using the first $n = 132$ observations.
   a. Does the SRF disturbance series show evidence of autocorrelation? If so, modify the model to take this pattern into account.
   b. In your final model do you find that the coefficients of the seasonal dummy variables are significant as a set? Do the seasonal dummies account for all of the seasonality in the data? (Is there evidence of seasonality in the residual SACF?)
   c. Use your final model to forecast Y_t' for $t = 133$.

CASE 2

Federal Government Receipts (Dynamic Regression)

In Case 1 we built an ARIMA model for monthly federal government receipts (series z_t) shown in Figure 2.2. We applied a log transformation to stabilize the variance, creating series $z_t' = \ln(z_t)$ shown in Figure 2.3. We used both nonseasonal and seasonal first differencing to induce a constant mean. The final model for series z_t' was an ARIMA(2, 1, 0) (0, 1, 1)$_{12}$ shown in Equation (C1.4). We saw that one component of the model, the nonseasonal AR(2) part, represents a *stochastic cycle* with an *average period* of *about 3 months*. When a stochastic cycle with an average period of about 3 months occurs in *cumulated monthly* data, this is a clue that the series may contain a *trading day* pattern. Discussions of modeling data with trading day patterns are found in Hillmer (1982), Bell and Hillmer (1983), and Liu (1986a), for example. In the present case we try to account for the trading day pattern in the government receipts data using a DR model.

C2.1 REPRESENTING TRADING DAY VARIATION IN A REGRESSION MODEL

A trading day pattern means that the data vary according to the number of times that each day of the week occurs in each month. Each day (Monday, Tuesday, etc.) occurs either four or five times in each month. For the government receipts data, we might expect that a month with five Sundays will have a smaller volume of receipts since receipts will not be recorded on Sundays. On the other hand, a public park might have more visitors on Sundays than on other days; thus the monthly total of visitors will tend to be higher in a month with five Sundays.

In this case study we build a regression model with dependent variable Y_t' (the log of government receipts). The independent variables are the numbers of times that each day of the week occurs in each month. If Y_t' varies according to a trading day pattern, then the coefficients of these

Table C2.1 Partial Listing of Series Y_t' and Trading Day Variables

Date		Y_t'	$d_{1,t}$	$d_{2,t}$	$d_{3,t}$	$d_{4,t}$	$d_{5,t}$	$d_{6,t}$	$d_{7,t}$
1976	1	10.152	4	4	4	5	5	5	4
	2	9.945	4	4	4	4	4	4	5
	3	9.925	5	5	5	4	4	4	4
	4	10.415	4	4	4	5	5	4	4
	5	10.029	5	4	4	4	4	5	5
	6	10.513	4	5	5	4	4	4	4
	7	10.025	4	4	4	5	5	5	4
	8	10.216	5	5	4	4	4	4	5
	9	10.366	4	4	5	5	4	4	4
	10	9.953	4	4	4	4	5	5	5
	11	10.154	5	5	4	4	4	4	4
	12	10.291	4	4	5	5	5	4	4
1977	1	10.307	5	4	4	4	4	5	5
	2	10.093	4	4	4	4	4	4	4
	3	10.119	4	5	5	5	4	4	4
					etc.				

independent variables should be significant as a group. Table C2.1 illustrates the trading day variables. It shows the first 15 observations of the dependent variable (Y_t') and each independent variable ($d_{i,t}$). Variables $d_{1,t}$ through $d_{7,t}$ are the number of Mondays through Sundays per month. For example, January 1976 (the first set of observations) has four Mondays ($d_{1,t}$), Tuesdays ($d_{2,t}$), Wednesdays ($d_{3,t}$), and Sundays ($d_{7,t}$); but it has five Thursdays ($d_{4,t}$), Fridays ($d_{5,y}$), and Saturdays ($d_{6,t}$).

It remains to be seen how Y_t' responds to changes in these independent variables. But to illustrate the idea, suppose that receipts generally are higher on Thursdays and lower on Mondays. Then $d_{4,t}$ (number of Thursdays) might have a large positive coefficient and variable $d_{1,t}$ (number of Mondays) might have a smaller positive or even negative coefficient. And we would expect receipts for January 1976, for example, to be somewhat higher, other things equal, since there are five Thursdays (when higher receipts occur) but only four Mondays (when lower receipts occur) in that month. On the other hand we would expect receipts for March 1976 to be somewhat lower, other things equal, since there are only four Thursdays but five Mondays in that month.

We are considering the following regression model:

$$Y_t' = C + \alpha_1 d_{1,t} + \alpha_2 d_{2,t} + \alpha_3 d_{3,t} + \alpha_4 d_{4,t} + \alpha_5 d_{5,t}$$
$$+ \alpha_6 d_{6,t} + \alpha_7 d_{7,t} + N_t \quad (C2.1)$$

where the α's are the PRF coefficients to be estimated, and N_t is a disturbance that may be autocorrelated.

A DATA TRANSFORMATION

Table C2.2 Partial Listing of Series Y_t' and Transformed Trading Day Variables

Date		Y_t'	$X_{1,t}$	$X_{2,t}$	$X_{3,t}$	$X_{4,t}$	$X_{5,t}$	$X_{6,t}$	$X_{7,t}$
1976	1	10.152	0	0	0	1	1	1	31
	2	9.945	−1	−1	−1	−1	−1	−1	29
	3	9.925	1	1	1	0	0	0	31
	4	10.415	0	0	0	1	1	0	30
	5	10.029	0	−1	−1	−1	−1	0	31
	6	10.513	0	1	1	0	0	0	30
	7	10.025	0	0	0	1	1	1	31
	8	10.216	0	0	−1	−1	−1	−1	31
	9	10.366	0	0	1	1	0	0	30
	10	9.953	−1	−1	−1	−1	0	0	31
	11	10.154	1	1	0	0	0	0	30
	12	10.291	0	0	1	1	1	0	31
1977	1	10.307	0	−1	−1	−1	−1	0	31
	2	10.093	0	0	0	0	0	0	28
	3	10.119	0	1	1	1	0	0	31
					etc.

C2.2 A DATA TRANSFORMATION

Hillmer (1982) and Bell and Hillmer (1983) point out that variables $d_{1,t}$ through $d_{7,t}$ tend to be highly correlated with each other. This colinearity can make it difficult to obtain reliable estimates from model (C2.1). (Recall our discussion of colinearity in Chapter 3.) These authors suggest a transformation of the data that tends to reduce the colinearity problem. They suggest that we define the following variables:

$$X_{i,t} = d_{i,t} - d_{7,t}, \quad i = 1, 2, \ldots, 6$$

$$X_{7,t} = \sum_{i=1}^{7} d_{i,t} \quad \text{(C2.2)}$$

In other words, $X_{1,t}$ is the number of Mondays in month t minus the number of Sundays in that month; $X_{2,t}$ is the number of Tuesdays in month t minus the number of Sundays in that month; and so forth. $X_{7,t}$ is the total number of days in month t. Table C2.2 shows the first 15 observations on Y_t' and the X's. For example, for January 1976, $X_{1,t} = d_{1,t} - d_{7,t} = 4 - 4 = 0$; $X_{6,t} = 5 - 4 = 1$; and $X_{7,t} = 4 + 4 + 4 + 5 + 5 + 5 + 4 = 31$ (January has 31 days). With these new variables we now consider the following regression:

$$Y'_t = C + b_1 X_{1,t} + b_2 X_{2,t} + b_3 X_{3,t} + b_4 X_{4,t} + b_5 X_{5,t}$$
$$+ b_6 X_{6,t} + b_7 X_{7,t} + N_t \qquad (C2.3)$$

This equation may be written more compactly as

$$Y'_t = C + \sum_{i=1}^{7} b_i X_{i,t} + N_t \qquad (C2.4)$$

Clearly each $X_{i,t}$ variable is different from each $d_{i,t}$ variable. Therefore the coefficients in model (C2.4) are different from those in model (C2.1). However, model (C2.4) is not fundamentally different from model (C2.1). In fact, it can be shown (see the exercise at the end of this case study) that each α coefficient in (C2.1) is a linear combination of the b coefficients in (C2.4):

$$\alpha_i = b_i + b_7, \qquad i = 1, 2, \ldots, 6$$
$$\alpha_7 = b_7 - \sum_{i=1}^{6} b_i \qquad (C2.5)$$

Having estimated the b coefficients in (C2.4), with less colinearity, we can then find the estimated α coefficients using (C2.5). While the two sets of coefficients are different, and the t values in (C2.4) should be more informative, the two regression results will not differ in any other respect.

C2.3 MODEL ESTIMATION, CHECKING, AND REFORMULATION

C2.3.1 Tentative Model 1

Table C2.3 shows the results of estimating model (C2.4) as an ordinary regression (without an ARIMA model for N_t). None of the estimated coefficients is significant since none of the t values is large in absolute value (except for the constant). However, assessing these t values is premature since the sample disturbance ($\hat{N}_t$) for (C2.4) is highly autocorrelated. As discussed in Chapters 1 and 3, when the disturbance in a regression is autocorrelated, the usual t tests are invalid. Although we cannot observe the PRF disturbance N_t, we can study an estimate of it in the form of the SRF residual series $\hat{N}_t$. Solving (C2.4) for N_t and substituting estimated values for parameters, $\hat{N}_t$ is computed as

$$\hat{N}_t = Y'_t - \hat{C} - \sum_{i=1}^{7} \hat{b}_i X_{i,t}$$

MODEL ESTIMATION, CHECKING, AND REFORMULATION 119

Table C2.3 Estimation Results for Model (C2.4)

Parameter	Estimate	Standard Error	t Value
C	11.50662	1.15406	9.97
b_1	0.03107	0.08519	0.36
b_2	−0.00701	0.08530	−0.08
b_3	0.04019	0.08300	0.48
b_4	−0.01551	0.08411	−0.18
b_5	0.01184	0.08413	0.14
b_6	−0.02721	0.08407	−0.32
b_7	−0.02669	0.03790	−0.70
$\hat{\sigma}_N = 0.3507$			

The autocorrelation in this sample disturbance is clear from its SACF (Figure C2.1). This SACF falls toward zero very slowly, if at all, suggesting that the disturbance in (C2.4) is nonstationary in the mean.

We apply regular first differencing ($d = 1$) in an effort to create a disturbance with a constant overall mean. Thus we have a simple ARIMA(0, 1, 0) model for N_t:

$$\nabla N_t = a_t \qquad (C2.6)$$

In other words the first differenced series $\nabla N_t = (1 - B)N_t$ may now have a constant mean; in fact, after differencing this series may be zero-mean white noise (a_t). Of course, we don't know if first differencing produces an uncorrelated series; but we can proceed as if that were true for the time being.

We must now consider an important point that will recur throughout this book: when N_t is differenced, the *same differencing* applies to every other variable in the regression equation. To see this, write (C2.6) as $N_t = a_t/\nabla$;

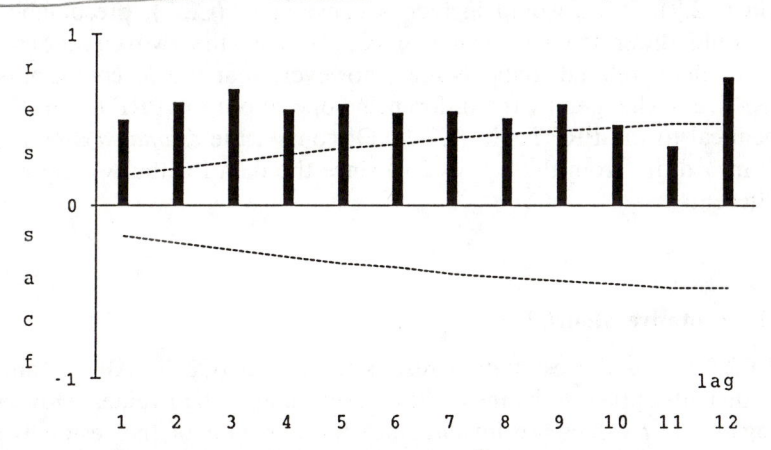

Figure C2.1 Residual SACF for $\hat{N}_t$ in model (C2.4).

Table C2.4 Estimation Results for Model (C2.7)

Parameter	Estimate	Standard Error	t Value
b_1	0.0381	0.0484	0.79
b_2	−0.0099	0.0485	−0.20
b_3	0.0212	0.0468	0.45
b_4	0.0032	0.0478	0.07
b_5	0.0126	0.0471	0.27
b_6	−0.0278	0.0477	−0.58
b_7	−0.0731	0.0206	−3.54

$\hat{\sigma}_a = 0.327235$

substitute this expression for N_t into (C2.4) and multiply through by ∇ to get

$$\nabla Y'_t = \sum_{i=1}^{I} b_i \, \nabla X_{i,t} + a_t \tag{C2.7}$$

In other words we first compute the regular first differences of Y'_t ($\nabla Y'_t$) and each $X_{i,t}$ ($\nabla X_{i,t}$), and then we use these new variables as the dependent variable and independent variables in the new regression model (C2.7).

When multiplying C in (C2.4) by ∇ we get $\nabla C = C - BC = C - C = 0$, so we could write (C2.7) without a constant. However, we may reinsert a constant term in (C2.7). After differencing, a nonzero constant represents a deterministic time trend, rather than an element helping to fix the overall constant level of Y'_t. In (C.2.7) we choose to let the differencing capture any trend elements in a stochastic fashion, so we do not insert a constant term in (C2.7). If we would include a constant in (C2.7), presumably its value would differ from its value in (C2.4) since the two equations use different, albeit related, data. Notice, however, that the b_i coefficients in the PRF are unchanged by the differencing operation: the PRF b's in (C2.7) are identical to the PRF b's in (C2.4). Of course, the *estimated* SRF $\hat{b}$'s in (C2.7) may differ from those in (C2.4) since the data for the two equations are different.

C2.3.2 Tentative Model 2

Table C2.4 shows the estimation results for model (C2.7). All but one of the coefficients appear to be insignificant according to the t values. However, once again the t values are invalid since the behavior of the residuals ($\hat{a}_t$) suggests that the shocks in (C2.7) are autocorrelated. This is seen from the

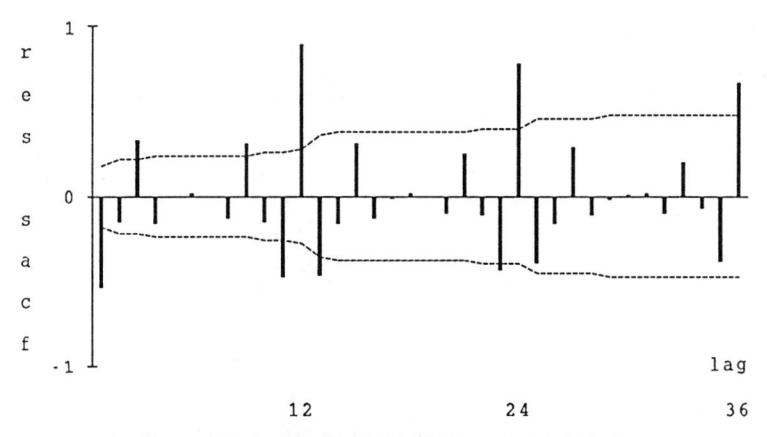

MODEL ESTIMATION, CHECKING, AND REFORMULATION 121

Figure C2.2 Residual SACF for model (C2.7).

residual SACF in Figure C2.2. The outstanding feature of this SACF is the slow decay at the seasonal lags (12, 24, 36). This implies that the data have a nonstationary seasonal pattern so that seasonal first differencing ($D = 1$) is called for. Many of the other large autocorrelations (e.g., lags 9, 11, 13, and 15) are probably a reflection of the strong seasonality, and they may become smaller after seasonal differencing. There also seems to be a nonseasonal autocorrelation pattern (at lags 1, 2, 3) in this SACF. But it is wise to wait until we have done seasonal differencing to identify any possible nonseasonal pattern.

We have now identified a tentative ARIMA(0, 1, 0)(0, 1, 0)$_{12}$ model for N_t:

$$\nabla_{12} \nabla N_t = a_t \tag{C2.8}$$

Again, the differencing applied to N_t also applies to the other variables in (C2.4). That is, rewrite (C2.8) as $N_t = a_t/\nabla_{12}\nabla$, substitute this expression for N_t in (C2.4), and multiply through by $\nabla_{12}\nabla$ to get

$$\nabla_{12} \nabla Y'_t = \sum_{i=1}^{7} b_i \nabla_{12} \nabla X_{i,t} + a_t \tag{C2.9}$$

According to (C2.9), we first compute the regular and seasonal first differences of Y'_t and each $X_{i,t}$. Then we use these new variables as the dependent variable and independent variables in the new regression model (C2.9).

C2.3.3 Tentative Model 3

Table C2.5 shows the estimation results for model (C2.9). Three of the coefficients now have absolute t values larger than 2.0, suggesting that there

Table C2.5 Estimation Results for Model (C2.9)

Parameter	Estimate	Standard Error	t Value
b_1	0.0208	0.0096	2.17
b_2	−0.0024	0.0058	−0.42
b_3	−0.0009	0.0055	−0.17
b_4	0.0240	0.0064	3.77
b_5	−0.0150	0.0080	−1.88
b_6	−0.0073	0.0068	−1.08
b_7	0.0892	0.0219	4.08
$\hat{\sigma}_a = 0.0678775$			

may, indeed, be a trading day pattern in the government receipts data. Since we have removed much of the autocorrelation pattern in the random shocks via the differencing operations, these t values should be more reliable than those from our earlier regressions. However, the residual SACF and SPACF for model (C2.9), shown in Figures C2.3 and C2.4, indicate that there is still some remaining autocorrelation pattern in the random shocks in (C2.9). The significant spike at lag 1 in the SACF and the decay pattern in the SPACF suggest a possible MA(1) for the differenced disturbance. Notice that this SACF and SPACF look much like the theoretical pair for an MA(1) in Figure 2.11(a).

We have now tentatively identified an ARIMA$(0, 1, 1)(0, 1, 0)_{12}$ model for N_t:

$$\nabla_{12}\nabla N_t = (1 - \theta_1 B)a_t \qquad (C2.10)$$

Rewrite (C2.10) as $N_t = (1 - \theta_1 B)a_t/\nabla_{12}\nabla$. Substitute this expression into

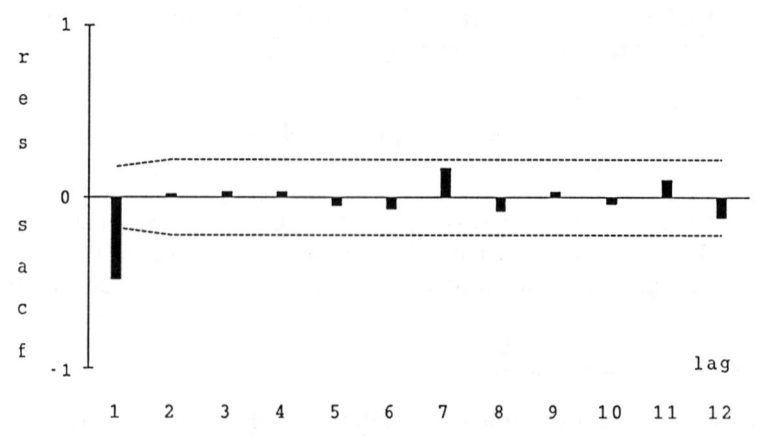

Figure C2.3 Residual SACF model (C2.9).

MODEL ESTIMATION, CHECKING, AND REFORMULATION

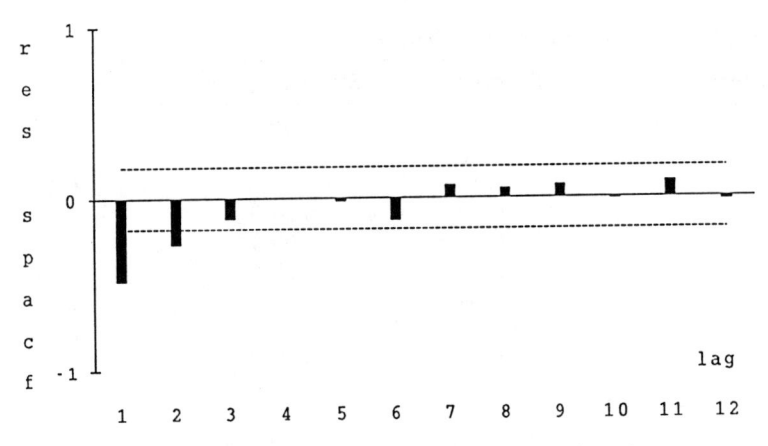

Figure C2.4 Residual SPACF for model (C2.9).

(C2.4), and multiply through by $\nabla_{12}\nabla$ to get the following tentative DR model:

$$\nabla_{12}\nabla Y'_t = \sum_{i=1}^{7} b_i \nabla_{12}\nabla X_{i,t} + (1 - \theta_1 B) a_t \qquad (C2.11)$$

C2.3.4 Tentative Model 4

With a moving average term in (C2.11), we can't estimate the model with ordinary linear regression methods. Instead we must use a nonlinear ML estimation routine of the type discussed in Chapter 9. The estimation results shown in Table C2.6 were obtained using the SCA System exact ML estimation option. Two of the trading day variables now have coefficients with absolute t values larger than 2.0, and a third has an absolute t value close to 2.0. The MA(1) coefficient is significant, and it satisfies the invertibility condition $|\hat{\theta}_1| < 1$.

Table C2.6 Estimation Results for Model (C2.11)

Parameter	Estimate	Standard Error	t Value
b_1	0.0178	0.0104	1.72
b_2	−0.0034	0.0063	−0.54
b_3	−0.0013	0.0063	−0.21
b_4	0.0246	0.0073	3.38
b_5	−0.0104	0.0094	−1.11
b_6	−0.0093	0.0080	−1.16
b_7	0.0656	0.0233	2.81
θ_1	0.5988	0.0744	8.05

$\hat{\sigma}_a = 0.0568593$

Table C2.7 Estimation Results for Model (C2.12)

Parameter	Estimate	Standard Error	t Value
b_1	0.0239	0.0107	2.24
b_2	−0.0040	0.0064	−0.63
b_3	0.0010	0.0066	0.16
b_4	0.0222	0.0079	2.80
b_5	−0.0131	0.0099	−1.33
b_6	−0.0099	0.0088	−1.13
b_7	0.0589	0.0250	2.35
θ_{12}	0.2013	0.0915	2.20
θ_1	0.6217	0.0727	8.56

$\hat{\sigma}_a = 0.0557972$

Correlation Matrix of Parameter Estimates

	1	2	3	4	5	6	7	8	9
1	1.00								
2	−0.45[a]	1.00							
3	−0.17	0.19[a]	1.00						
4	−0.01	−0.06	−0.58[a]	1.00					
5	−0.46[a]	−0.15	0.04	−0.38[a]	1.00				
6	0.27[a]	−0.44[a]	−0.12	0.01	−0.32[a]	1.00			
7	−0.01	−0.08	−0.13	0.16	−0.06	0.13	1.00		
8	0.15	0.03	0.04	0.01	−0.14	0.02	−0.07	1.00	
9	−0.03	−0.06	−0.02	−0.01	0.09	0.00	0.02	−0.04	1.00

[a] Significant at 5% level.

The residual SACF for model (C2.11), not shown, is satisfactory; however, the residual autocorrelation coefficient at lag 12 is −0.16, which is 1.6 times its standard error (0.10) in absolute value. As a practical rule, when a seasonal r_k is more than about 1.25 times its standard error in absolute value, an estimated model coefficient at that lag is likely to be significant. Therefore we introduce an $MA(1)_{12}$ component into our model for the shocks. This gives the following tentative DR model:

$$\nabla_{12}\nabla Y'_t = \sum_{i=1}^{7} b_i \nabla_{12}\nabla X_{i,t} + (1 - \theta_1 B)(1 - \theta_{12} B^{12}) a_t \quad (C2.12)$$

C2.4 ESTIMATION AND CHECKING (FINAL MODEL)

Table C2.7 shows the estimation results for model (C2.12), along with the correlation matrix of the estimated coefficients; Figure C2.5 shows its residual SACF. Three of the $\hat{b}$ coefficients have absolute t values larger than 2.0; in

ESTIMATION AND CHECKING (FINAL MODEL) 125

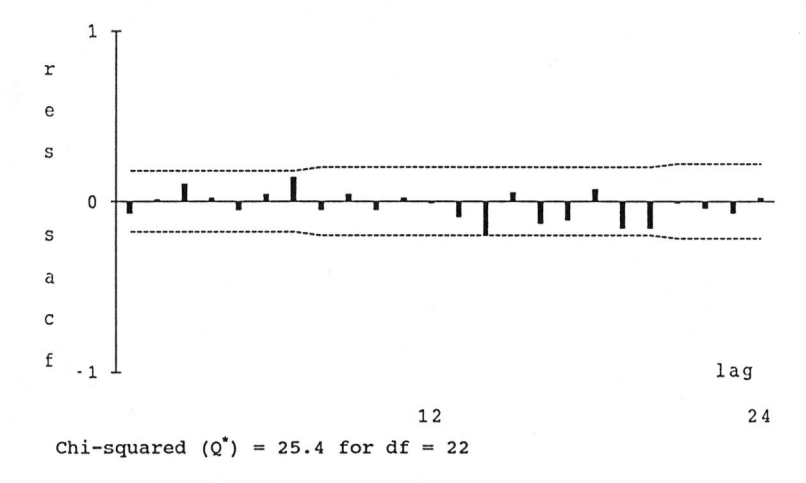

```
r
e
s

s        0
a
c
f       -1
                              12                    24
                                                         lag
Chi-squared (Q*) = 25.4 for df = 22
```

Figure C2.5 Residual SACF for model (C2.12).

the following we discuss how we can test the significance of the trading day coefficients as a set. Both MA coefficients are significant, and each satisfies its invertibility condition. (What are these conditions?) The residual standard error is slightly smaller than the one for model (C2.11), which has no seasonal MA term. The residual SACF in Figure C2.5 is satisfactory, with no large autocorrelations at important lags; its approximate χ^2 statistic (Q^*) is insignificant at the 5% level.

The histogram of the standardized residuals ($\hat{a}_t/\hat{\sigma}_a$) in Figure C2.6 is roughly symmetrical, and the normal probability plot of the standardized residuals in Figure C2.7 is roughly a straight line; this suggests that the unobserved random shocks of Equation (C2.12) are approximately normally distributed. The plots of the standardized residuals in Figure C2.8 shows only one value slightly beyond the three standard deviation limits, so outliers do not seem to be a problem. The formal outlier detection procedure discussed in Chapter 8 might be more revealing.

```
    LOWER      UPPER    FREQ-  0    10    20    30    40    50
    BOUND      BOUND    UENCY +----+----+----+----+----+
   -3.500  -  -2.800      1   |>
   -2.800  -  -2.100      0   |
   -2.100  -  -1.400     10   |>>>>>
   -1.400  -  -0.700     15   |>>>>>>>>
   -0.700  -   0.000     33   |>>>>>>>>>>>>>>>>>
    0.000  -   0.700     40   |>>>>>>>>>>>>>>>>>>>>
    0.700  -   1.400     11   |>>>>>>
    1.400  -   2.100      6   |>>>
    2.100  -   2.800      2   |>
    2.800  -   3.500      1   |>
                              +----+----+----+----+----+
                               0    10    20    30    40    50
```

Figure C2.6 Histogram of standardized residuals from model (C2.12).

```
                    -.180      -.060       .060       .180
                   -+----------+----------+----------+-
           2.450 +                                   * +
                 I                              2      I
                 I                          3 *        I
                 I                        4**          I
                 I                       342*          I
            .700 +                        292         +
                 I                         *E          I
         S       I                        2F           I
         C       I                        A5           I
         O       I                       373           I
         R     -1.050 +                   73          +
         E       I                       42            I
                 I                   *3                I
                 I                    2                I
                 I *                                   I
          -2.800 +                                    +
                   -+----------+----------+----------+-
                    -.180      -.060       .060       .180
```

Standardized Residuals

Figure C2.7 Normal probability plot of standardized residuals from model (C2.12).

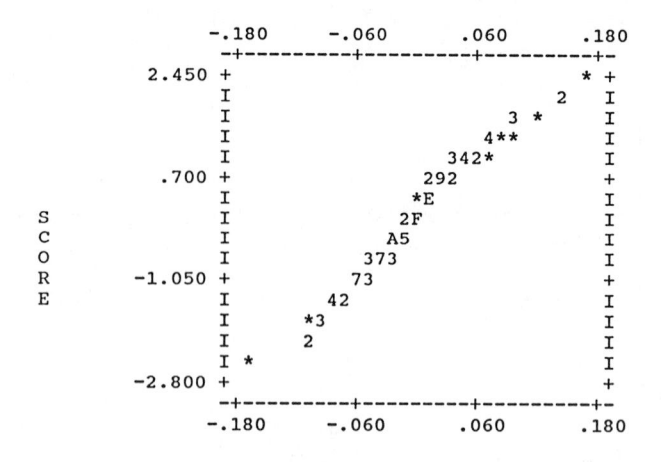

The correlation matrix of the estimated coefficients at the bottom of Table C2.7 shows that coefficients $\hat{b}_1$ through $\hat{b}_6$ (corresponding to rows and columns 1 through 6, respectively, in this matrix) are fairly highly correlated. Eight of the 15 correlation coefficients for the $\hat{b}$'s are significant at the 5% level, with absolute correlations ranging between about 0.3 and 0.6. This degree of correlation among the estimated coefficients is not fatal, but it does make it more difficult to assess the relative importance of the individual trading day variables using their t statistics.

To deal with this problem, Hillmer (1982) suggests that we test the following joint null hypothesis:

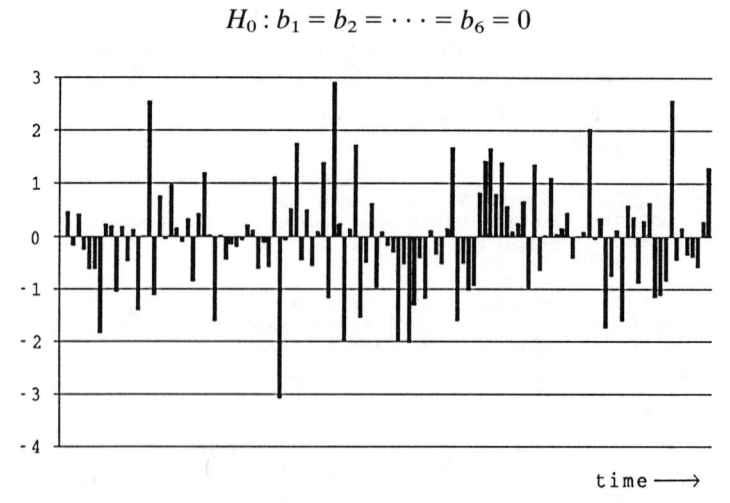

$$H_0 : b_1 = b_2 = \cdots = b_6 = 0$$

Figure C2.8 Standardized residuals from model (C2.12).

FORECASTS

against the alternate hypothesis, H_a: not all b's are zero. If H_0 is true, then in repeated sampling the following statistic is approximately χ^2 distributed, with 6 degrees of freedom:

$$H = \hat{\mathbf{B}}' \mathbf{V}^{-1} \hat{\mathbf{B}} \qquad (C2.13)$$

where $\hat{\mathbf{B}} = (\hat{b}_1, \hat{b}_2, \ldots, \hat{b}_6)'$ is a column vector of sample b coefficients, and $\mathbf{V}$ is the variance–covariance matrix of the sample b coefficients.

The elements in $\mathbf{V}$ can be found from information in Table C2.7. Each diagonal element v_{ii} (variance of $\hat{b}_i$) for $i = 1, 2, \ldots, 6$, is the square of the standard error of the estimated coefficient $\hat{b}_i$, found at the top of Table C2.7. Each off-diagonal element (covariance of $\hat{b}_i$ and $\hat{b}_j$) v_{ij}, $i, j = 1, 2, \ldots, 6$, $i \neq j$, is found by multiplying the coefficient correlation r_{ij} by the standard errors of the estimated coefficients $\hat{b}_i$ and $\hat{b}_j$. Inserting the appropriate values from model (C2.12) into (C2.13) gives $H = 29.8$, which exceeds the 5% critical χ^2 value of 12.6 for 6 degrees of freedom. We therefore reject H_0 (the null hypothesis that there is no trading day pattern in the government receipts data).

Equation (C2.5) gives the following estimated α coefficients for the original Equation (C2.1):

$$\hat{\alpha}_1 = 0.0828$$
$$\hat{\alpha}_2 = 0.0549$$
$$\hat{\alpha}_3 = 0.0599$$
$$\hat{\alpha}_4 = 0.0811$$
$$\hat{\alpha}_5 = 0.0458$$
$$\hat{\alpha}_6 = 0.0490$$
$$\hat{\alpha}_7 = 0.0388$$

The smallest of these estimated coefficients is $\hat{\alpha}_7$, which is associated with the number of Sundays in each month. It is not surprising that Sundays are associated with less federal tax remittance or recording activity than any other day of the week.

C2.5 FORECASTS

The forecasts from model (C2.12) have an interesting interpretation. Recall from Chapter 2 that an ARIMA(0, 1, 1) model is an exponentially weighted moving average (EWMA) model, and an ARIMA(0, 1, 1)$_{12}$ model is a seasonal EWMA. Our model for N_t is an ARIMA(0, 1, 1)(0, 1, 1)$_{12}$. Thus, in addition to the trading day pattern, the forecast includes a combination of

Table C2.8 Log Metric Forecasts and Standard Errors from Model (C2.12) from Origin $t = 132$

Time	Forecast	Standard Error
133	11.3208	0.0558
134	10.9469	0.0597
135	10.9067	0.0633
136	11.5333	0.0667
137	10.7521	0.0700
138	11.3475	0.0731
139	11.1280	0.0761
140	11.0153	0.0789
141	11.3394	0.0817
142	11.0521	0.0844
143	10.9696	0.0870
144	11.3449	0.0895

two EWMAs. One is a seasonal EWMA, with exponentially declining weights applied to the values of series $\hat{N}_t$ that occur 1 year prior to the forecast period, 2 years prior to the forecast period, and so forth. The other is a nonseasonal EWMA, with exponentially declining weights applied to the values of series $\hat{N}_t$ that occur one month prior to the forecast period, two months prior to the forecast period, and so forth.

Forecasts and their associated standard errors from model (C2.12) in the log metric are shown in Table C2.8. Forecasts in the original metric along with approximate 95% lower and upper confidence limits are shown in Table C2.9. The original metric forecasts and confidence limits are found in the same way that we found these values in Case 1.

Table C2.9 also shows the "future" observed values and the forecast errors (observed minus forecast). As happened in Case 1 with the ARIMA model forecasts, the approximate 95% limits around the point forecasts contain all but one (at $t = 136$) of the future observed values. The forecasts from the DR model in this case study are generally more accurate than those from the ARIMA model in Case 1. This is summarized by the root mean squared forecast error (6657) at the bottom of Table C2.9, which is about 10% less than the comparable figure (7405) in Case 1. This is consistent with the fact that $\hat{\sigma}_a$ for the DR model (C2.12) (0.0557972) is smaller than $\hat{\sigma}_a$ for the ARIMA model (C1.4) in Case 1 (0.0609834). If the government receipts data do have a trading day pattern of the sort we have postulated in this case study, then a DR model should produce more accurate forecasts than an ARIMA model whose AR(2) stochastic cycle component captures the trading day pattern less closely.

COMMENTS 129

Table C2.9 Twelve Forecasts (Original Metric) from Model (C2.12), from Origin $t = 132$

Time	(1) 95% Lower Limit	(2) Forecast (F)	(3) 95% Upper Limit	(4) Observed	(5) Error
133	73806	82519	92261	81771	−748
134	50391	56777	63971	55463	−1314
135	48059	54543	61902	56515	1972
136	89315	102063	116631	122897	20834
137	40625	46727	53746	47691	964
138	73229	84754	98094	82945	−1809
139	58449	68053	79237	64223	−3830
140	51917	60797	71196	60213	−584
141	71392	84068	98995	92410	8342
142	53277	63074	74673	62354	−720
143	48803	58079	69118	56987	−1092
144	70671	84529	101104	85525	996

RMSFE = 6657

C2.6 COMMENTS

1. In Chapter 1 we said that DR models are built on the assumption that the data occur at equally spaced time intervals. Accumulated monthly data (like the government receipts series) do not quite satisfy this assumption since the total number of days per month varies from 28 to 31. Even if the *composition* of this total (number of Mondays, number of Tuesdays, etc.) did not affect the data, we expect to see more receipts for months with more days. Thus these data are likely to be affected by the unequal spacing of the observations through time. The trading day inputs in a DR model can account for this variation in total number of days per month (as well as for variation in the composition of this total). Thus, the trading day inputs permit us to adjust for unequally spaced observations in cumulated monthly data if the data vary according to the total number of days per month. For a more complete discussion of time series models for unequally spaced data, see Parzen (1984).

2. Sometimes cumulated monthly data are affected by the total number of "business days" per month but unaffected by the composition of those days (number of Mondays, Tuesdays, etc.). For example, in the absence of overtime work, the output level at a factory for a given month may be determined largely by the number of total days in the month, minus the number of Saturdays, Sundays, and holidays. One way to deal with this, without building a trading day DR model, is to divide each monthly observation by the number of business days for that month. This puts the data on

a "per business day" basis. This new series may then be forecast using an ARIMA model (for example). Each forecast is then multiplied by the number of business days for that month to create a forecast for the total monthly output.

3. Model (C2.12) is a DR model since it includes an ARIMA model for the disturbance. However, we have not introduced any explicit relationship between Y_t and *past values of the inputs*. A similar DR model appears in Case 3. In Chapters 4 and 5 we consider models for representing responses of Y_t to past values of the inputs. Cases 4, 5, and 6 involve a relationship between Y_t and past values of the inputs.

QUESTIONS AND PROBLEMS

1. Find the numerical values of the EWMA weights on past $\hat{N}_t$ values implied by the ARIMA(0, 1, 1) and ARIMA(0, 1, 1)$_{12}$ components of model (C2.12).

2. Show that the α weights are related to the b weights as shown in (C2.5).

CASE 3

Kilowatt-Hours Used

In this example we study how kilowatt-hours of electricity used (Y_t) in a certain geographic region responds to temperature changes. Temperature is measured by monthly heating degrees ($X_{1,t}$) and cooling degrees ($X_{2,t}$).

C3.1 THE DATA

Heating degrees measures our need to heat ourselves as the temperature falls. Heating degrees for one day is defined as 65° F minus the average daily temperature *when the daily average is below 65° F*; otherwise it is zero. Thus, if the average daily temperature is 60° F, we have 65 − 60°F = 5 heating degrees for that day. If the average daily temperature is 70° F, we have zero heating degrees since this figure is above 65° F. Variable $X_{1,t}$ in this case study is the monthly sum of daily heating degrees.

Cooling degrees measures our need to cool ourselves as the temperature rises. Cooling degrees for one day is defined as the average daily temperature minus 65° F *when the daily average is above 65° F*; otherwise it is zero. Thus, if the average daily temperature is 75° F, we have 75 − 65° F = 10 cooling degrees. If the average daily temperature is 60° F, we have zero cooling degrees since this figure is below 65° F. Variable $X_{2,t}$ in this case study is the monthly sum of daily cooling degrees.

In Case 2 the trading day input variables are *deterministic*: Their values are predetermined by the structure of the calendar. In the present case study we work with *stochastic* inputs: The values of X_1 and X_2 are not predetermined, but they vary in a probabilistic fashion.

Figure C3.1 is a plot of a monthly index of kilowatt-hours used, from January 1973 to December 1987, a total of 180 observations; we will use the first $n = 168$ observations to build a model, and use the last 12 to check the forecasts. Figure C3.2 is the heating degree days series from January 1973 to December 1986; Figure C3.3 is the cooling degree days series from January 1973 to December 1986. The observations for these three series are listed in the Data Appendix as Series 12, Series 13, and Series 14, respectively. Since

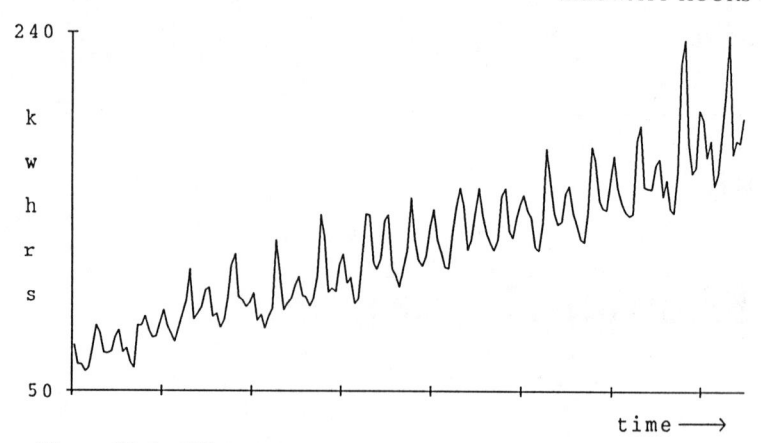

Figure C3.1   Kilowatt-hours used, January 1973–December 1987.

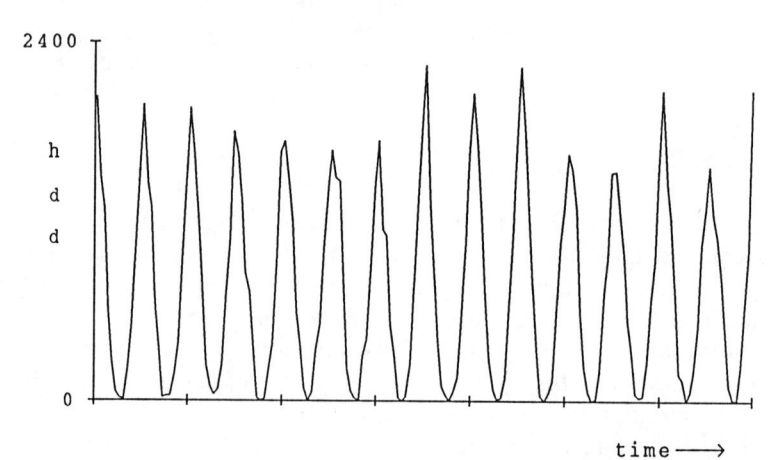

Figure C3.2   Heating degree days, January 1973–December 1986.

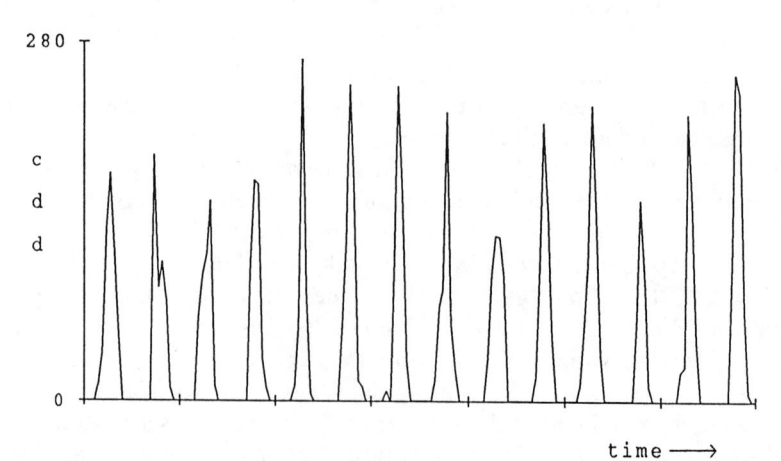

Figure C3.3   Cooling degree days, January 1973–December 1986.

PRELIMINARY STEPS   133

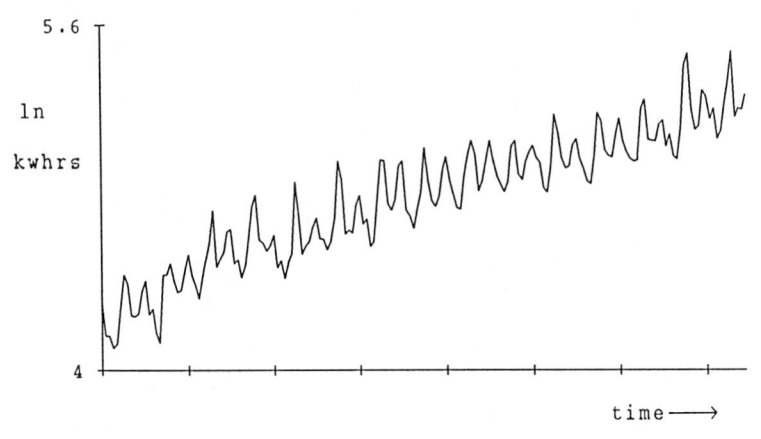

```
5.6

ln
kwhrs


  4
                                                              time ⟶
```

Figure C3.4 Log kilowatt hours used.

the data are monthly, the length of seasonality is $s = 12$. Inspection of the data plots reveals a strong seasonal pattern in all three series.

C3.2 PRELIMINARY STEPS

C3.2.1 Stabilizing the Variances

The variances of all three series are higher at higher levels of the series, so transformations to stabilize the variances may be used. Figure C3.4 is a plot of the natural logarithms ($Y'_t = \ln Y_t$) of the kilowatt-hours series; Figure C3.5 is a plot of the square roots ($X'_{1,t} = X^{1/2}_{1,t}$) of the heating degree days

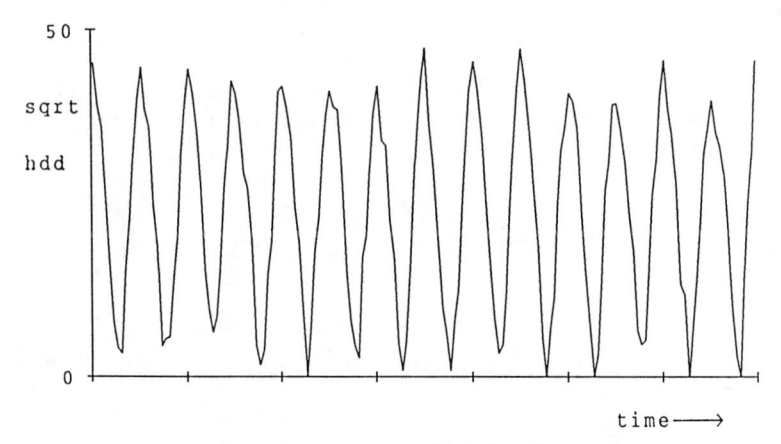

```
50

sqrt
hdd


  0
                                                              time ⟶
```

Figure C3.5 Square roots of heating degree days.

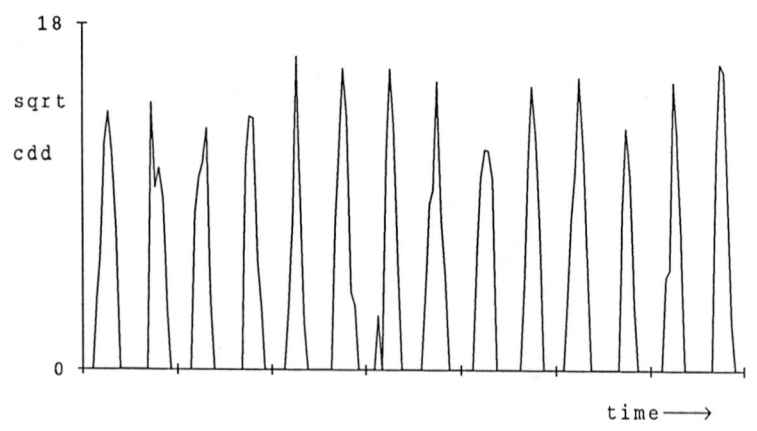

Figure C3.6 Square roots of cooling degree days.

series; Figure C3.6 is a plot of the square roots ($X'_{2,t} = X^{1/2}_{2,t}$) of the cooling degree days series. Each series now appears to have a more uniform variance.

C3.2.2 ARIMA Models

It is good practice to build an ARIMA model for each stochastic series in a dynamic regression. As discussed in Chapter 2, an ARIMA model for the output series gives us a baseline model: We can compare the fit and forecast accuracy of the DR model with those of the ARIMA model to see if it is worth the extra effort to use a DR model. Further, we need forecasts of the regression inputs to forecast the output; we can use the ARIMA models for the inputs to produce forecasts for those series. In addition, we may learn something useful about the series as we go through the ARIMA modeling steps. Finally, the ARIMA models for the inputs are needed to find the DR model forecast standard errors, as discussed in Chapter 9.

Following the procedures discussed in Chapter 2 leads to an ARIMA$(0, 1, 1)(0, 1, 1)_{12}$ model for Y'_t, an ARIMA$(0, 0, 2)(0, 1, 1)_{12}$ model for $X'_{1,t}$, and an ARIMA$(0, 0, 0)(1, 1, 1)_{12}$ model for $X'_{2,t}$. Estimation results (absolute t values are in parentheses) are:

$$\nabla_{12}\nabla Y'_t = (1 - 0.8105B^{12})(1 - 0.6974B)\hat{a}_t \quad \text{(C3.1)}$$
$$(15.69) \qquad (12.10)$$

$$\hat{\sigma}_a = 0.0512423$$

$$\nabla_{12}X'_{1,t} = (1 - 1.0067B^{12})(1 + 0.1847B + 0.241B^2)\hat{a}_t \quad \text{(C3.2)}$$
$$(22.77) \qquad (2.34) \quad (3.05)$$

$$\hat{\sigma}_a = 2.31795$$

MODEL ESTIMATION, CHECKING, AND REFORMULATION

$$(1 + 0.2897 B^{12}) \nabla_{12} X'_{2,t} = (1 - 0.676 B^{12}) \hat{a}_t \quad \text{(C3.3)}$$
$$(3.08) \qquad\qquad (9.91)$$
$$\hat{\sigma}_a = 1.82279$$

All three models have exponentially weighted moving average (EWMA) components. Recall from Chapter 2 that an ARIMA(0, 1, 1) is a (nonseasonal) EWMA model, and an ARIMA$(0, 1, 1)_s$ is a seasonal EWMA model. Model (C3.1) has two EWMA components, an ARIMA(0, 1, 1) and an ARIMA$(0, 1, 1)_{12}$. Model (C3.2) has one EWMA component, an ARIMA$(0, 1, 1)_{12}$. Model (C3.3) has an ARIMA$(0, 1, 1)_{12}$ component, along with an AR$(1)_{12}$ component.

The EWMA component of model (C3.2) has a further interpretation: Since the estimate of θ_{12} is virtually 1.0, the model implies that the data contain a *fixed (globally constant) seasonal* pattern. As discussed in Chapter 3, a fixed seasonal pattern can be modeled with, for example, seasonal dummy variables. This interpretation is discussed by Abraham and Ledolter (1983, Chapters 4 and 6). The details are beyond the scope of this text; but this interpretation of model (C3.2) can be understood intuitively as follows. In an ARIMA$(0, 1, 1)_s$ model the weights applied to observations for a given season in earlier years decline exponentially. The weights are given by $\theta_s^{k-1}(1 - \theta_s)$, where k is the number of years in the past. If $|\theta_s| < 1$, then as k increases the *decay factor* θ_s^{k-1} gets smaller for a given θ_s. The rate at which the weights decline as k increases is determined by the size of θ_s. For example, if $\theta_s = 0.5$, then the decay factor (rounded) applied to the first five observations ($k = 1, 2, \ldots, 5$) as we move back through preceding years is 1.0, 0.5, 0.25, 0.125, and 0.063. But if $\theta_s = 0.9$, then for $k = 1, 2, \ldots, 5$ the decay factor is: 1.0, 0.9, 0.81, 0.73, and 0.66. Thus, when θ_s is relatively large, the *decay rate is slower* than when θ_s is relatively small. For $\theta_s = 1.0$ the *implication* is that the decay factor is 1.0 throughout; that is, we give *equal weight* to all past observations that occur in a given season, which is equivalent to having a fixed seasonal effect. We will not pursue this further by constructing a regression model for $X_{1,t}$ with seasonal dummy variables. However, our discovery illustrates how building ARIMA models can lead to a better understanding of the data, and to further modeling.

C3.3 MODEL ESTIMATION, CHECKING, AND REFORMULATION

C3.3.1 Tentative Model 1

Now we are ready to build a regression model. The first model we entertain is

$$Y'_t = C + b_1 X'_{1,t} + b_2 X'_{2,t} + N_t \quad \text{(C3.4)}$$

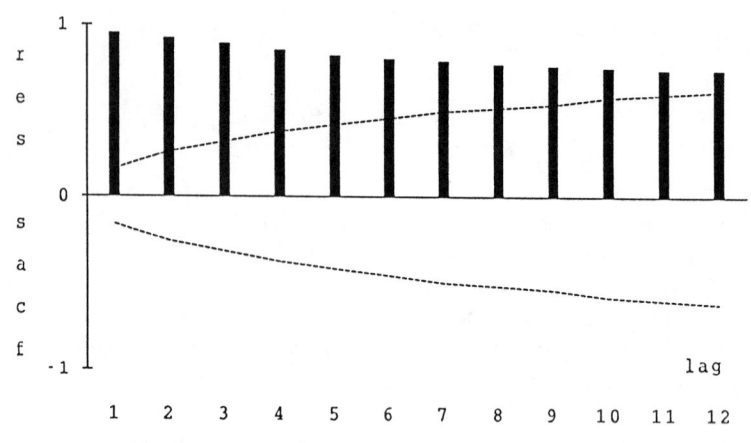

Figure C3.7 Residual SACF for model (C3.5).

We expect that both b_1 and b_2 are positive: as the temperature falls ($X_{1,t}$ rises) people should use more electricity to warm themselves, and as the temperature rises ($X_{2,t}$ rises) people should use more electricity to cool themselves. The estimation results for (C3.4), with absolute t values in parentheses, are

$$Y'_t = 4.4466 + 0.0081 X'_{1,t} + 0.0311 X'_{2,t} + \hat{N}_t$$
$$(38.78) \quad (2.40) \quad\quad (3.47)$$

$$\hat{\sigma}_N = 0.273512 \tag{C3.5}$$

Notice that $\hat{\sigma}_N$ is much larger for the regression model (C3.5) than $\hat{\sigma}_a$ for the ARIMA model (C3.1). Thus, it seems unlikely that the forecasts from regression model (C3.5) would be better than the ARIMA model forecasts.

As expected the estimated b coefficients in (C3.5) are positive. However, before we attempt to draw conclusions from this equation, we should examine the autocorrelation pattern of the estimated disturbance series $\hat{N}_t$. Rearranging model (C3.5), this series is computed as

$$\hat{N}_t = Y'_t - 4.4466 - 0.0081 X'_{1,t} - 0.0311 X'_{2,t}$$

The *slow decay* of the SACF for this series (Figure C3.7) shows that it is nonstationary in the mean. As discussed in Chapter 3, autocorrelation in a regression disturbance invalidates the t tests in the model. Before we can evaluate our regression model, we must modify it to reflect the autocorrelation pattern of the disturbance.

As discussed in Chapter 2, we use nonseasonal differencing to deal with a nonstationary overall mean. This gives us a *simple tentative ARIMA model for the disturbance* in Equation (C3.4): $\nabla N_t = a_t$. In this model we represent the shocks as zero-mean white noise (a_t). However, this is only a tentative

MODEL ESTIMATION, CHECKING, AND REFORMULATION

model; further ARIMA elements beyond differencing ($d = 1$) may be needed to explain the autocorrelation pattern of N_t. We can discover these elements with further analysis of the residuals after estimating a new regression model.

To see the new regression model, write the tentative model for N_t as $N_t = a_t/\nabla$, substitute this into (C3.4), and multiply through by ∇ to get

$$\nabla Y_t' = C + b_1 \nabla X_{1,t}' + b_2 \nabla X_{2,t}' + a_t \qquad (C3.6)$$

As discussed in Case 2, the *differencing applied to N_t is applied to all variables in the new regression model*. Note also that differencing introduces variables with *differing time subscripts* into the equation. For example, we now have $\nabla X_{1,t}' = X_{1,t}' - X_{1,t-1}'$ on the right side of (C3.6). Therefore we now have a *dynamic* regression (DR) model. It is understood that C in (C3.6) has a different value from C in (C3.4).

C3.3.2 Tentative Model 2

The estimation results for (C3.6), with absolute t values in parentheses, are:

$$\nabla Y_t' = 0.0059 + 0.0092 \, \nabla X_{1,t}' + 0.0323 \, \nabla X_{2,t}' + \hat{a}_t$$
$$\quad\;\; (1.13) \quad\; (9.33) \qquad\quad (16.56)$$
$$\hat{\sigma}_a = 0.0667875 \qquad\qquad\qquad\qquad\qquad\qquad (C3.7)$$

The residual standard deviation has fallen to less than 0.07 in (C3.7) from more than 0.27 in (C3.5). This implies that taking into account the nonstationary nature of the disturbance should add dramatically to our forecast accuracy. However, $\hat{\sigma}_a$ for (C3.7) is still larger than $\hat{\sigma}_a$ for the ARIMA model (C3.1).

We can compute the residual series $\hat{a}_t$ for (C3.7) in the same way that we computed the estimated disturbance series $\hat{N}_t$ for (C3.5). These computations are performed automatically by the SCA System using appropriate commands. The SACF for $\hat{a}_t$ is shown in Figure C3.8; the range of r_k values is from -0.5 to 0.5 to increase visual clarity. The key feature of this SACF is that it decays slowly at lags 12, 24, and 36. Apparently the shocks in (C3.6) are autocorrelated. In fact, the SACF pattern suggests a nonstationary seasonal behavior which calls for seasonal differencing ($D = 1$) of N_t.

Our *new tentative ARIMA model* for N_t is $\nabla_{12} \nabla N_t = a_t$. Solve this for N_t, substitute the result into (C3.4), and multiply through by $\nabla_{12} \nabla$ to get a new DR model:

$$\nabla_{12} \nabla Y_t' = C + b_1 \nabla_{12} \nabla X_{1,t}' + b_2 \nabla_{12} \nabla X_{2,t}' + a_t \qquad (C3.8)$$

Remember that C in (C3.8) is different in value from C in either of our two

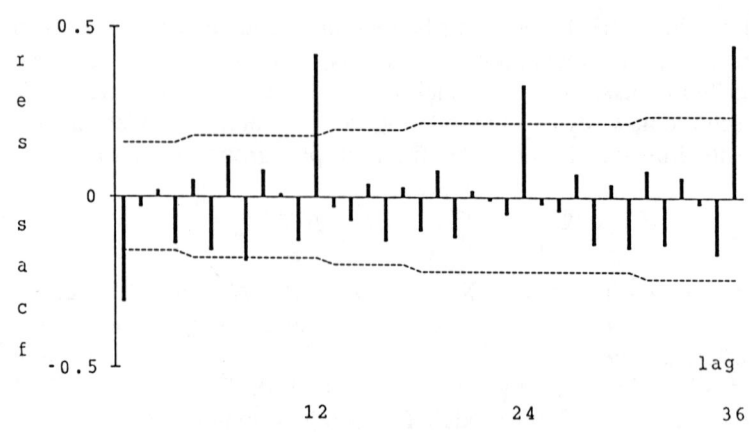

Figure C3.8 Residual SACF for model (C3.7).

earlier models. Note that *seasonal differencing applied to the disturbance* leads to seasonal differencing of *each variable* in the regression.

The estimation results for model (C3.8), with absolute t values in parentheses, are

$$\nabla_{12}\nabla Y'_t = -0.000085 + 0.0064\nabla_{12}\nabla X'_{1,t} + 0.0186\,\nabla_{12}\nabla X'_{2,t} + a_t$$
$$(0.02) \qquad (4.96) \qquad\qquad (11.20)$$

$$\hat{\sigma}_a = 0.052954 \qquad\qquad\qquad\qquad\qquad\qquad (C3.9)$$

The residual standard deviation for (C3.9) is smaller than the one for (C3.7); this suggests that seasonal differencing will tend to improve our forecast accuracy. But $\hat{\sigma}_a$ for (C3.9) is still larger than $\hat{\sigma}_a$ for the ARIMA model (C3.1). The residual SACF and SPACF for model (C3.9) in Figures C3.9

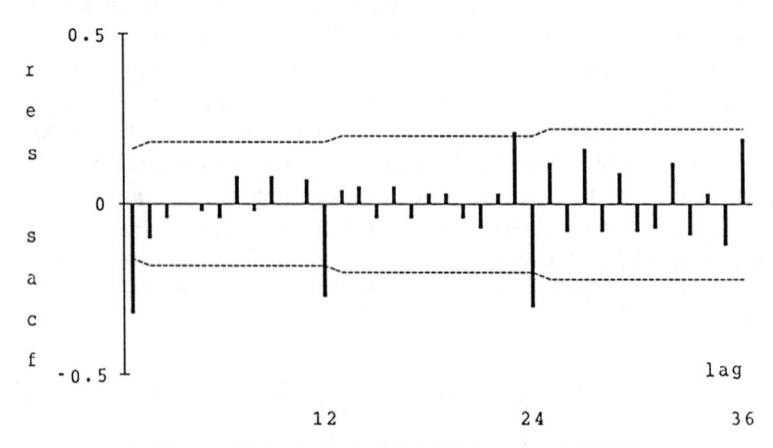

Figure C3.9 Residual SACF for model (C3.9).

MODEL ESTIMATION, CHECKING, AND REFORMULATION

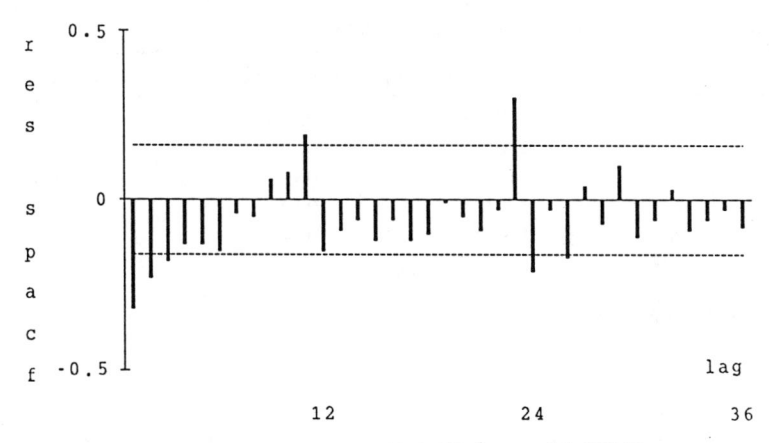

Figure C3.10 Residual SPACF for model (C3.9).

and C3.10 suggest that our ARIMA model for N_t is still incomplete. They indicate that the residual series $\hat{a}_t$ in (C3.9) has a nonseasonal MA(1) pattern, along with a seasonal pattern that seems more complicated.

After seasonal differencing, it is *most common* to see a negative SACF spike at lag s (=12 in this case), followed by a cutoff to insignificant values at lags $2s, 3s, \ldots$ (=24, 36, ... in this case). This leads to the ARIMA$(0, 1, 1)_s$, which is interpreted as a seasonal EWMA. This seasonal model occurs so often that we should be especially alert for it. However, in the present case the SACF in Figure C3.9 has significant coefficients at lags 12, 24, and 36; and the SPACF in Figure C3.10 has two significant spikes (at lags 12 and 24) followed by a cutoff to insignificance at lag 36. We don't have enough data to look at many seasonal lag coefficients; but the SACF and SPACF together look somewhat like the AR(2) theoretical pattern shown in Figure 2.10(c). Of course, in this case we are considering lags $s, 2s, 3s, \ldots$, rather than lags $1, 2, 3, \ldots$, as shown in Figure 2.10(c).

C3.3.4 A Final Model

We have tentatively identified an ARIMA$(0, 1, 1)(2, 1, 0)_{12}$ model for the disturbance N_t in Equation (C3.5); in backshift form this model is $(1 - \phi_{12}B^{12} - \phi_{24}B^{24})\nabla_{12}\nabla N_t = (1 - \theta_1 B)a_t$. Solve this expression for N_t, substitute the result into (C3.5), and multiply through that result by $\nabla_{12}\nabla$ to get

$$\nabla_{12}\nabla Y'_t = b_1 \nabla_{12}\nabla X'_{1,t} + b_2 \nabla_{12}\nabla X'_{2,t} + \frac{1 - \theta_1 B}{1 - \phi_{12}B^{12} - \phi_{24}B^{24}} a_t \quad (C3.10)$$

We have dropped the constant term from (C3.10) since it is insignificant in (C3.8).

Table C3.1   Estimation Results for Model (C3.10)

Parameter	Estimate	Standard Error	t Value
b_1	0.0077	0.0015	4.98
b_2	0.0208	0.0023	9.23
θ_1	0.5830	0.0720	8.10
ϕ_{12}	−0.5373	0.0856	−6.27
ϕ_{24}	−0.4667	0.0862	−5.41
$\hat{\sigma}_a = 0.0427775$			

Correlation Matrix of Parameter Estimates

	1	2	3	4	5
1	1.00				
2	0.64[a]	1.00			
3	0.03	0.08	1.00		
4	−0.16	−0.16	−0.02	1.00	
5	−0.15	−0.11	0.11	0.35[a]	1.00

[a] Significant at 5% level.

Estimation results for (C3.10) are shown in Table C3.1. All estimated coefficients are significant at about the 5% level since all absolute t values are greater then 2.0. (Do the estimates of θ_1, ϕ_{12}, amd ϕ_{24} satisfy the invertibility and stationarity conditions?) The residual SACF for model (C3.10), shown in Figure C3.11, does not contradict the hypothesis that the shocks (a_t) are uncorrelated. The histogram of the standardized residuals (Figure C3.12) is roughly normally distributed. The normal probability plot of the standardized residuals (Figure C3.13) is fairly close to a straight line; this provides additional evidence that the shocks are normally distributed. The standardized residuals (Figure C3.14) are all within three standard devia-

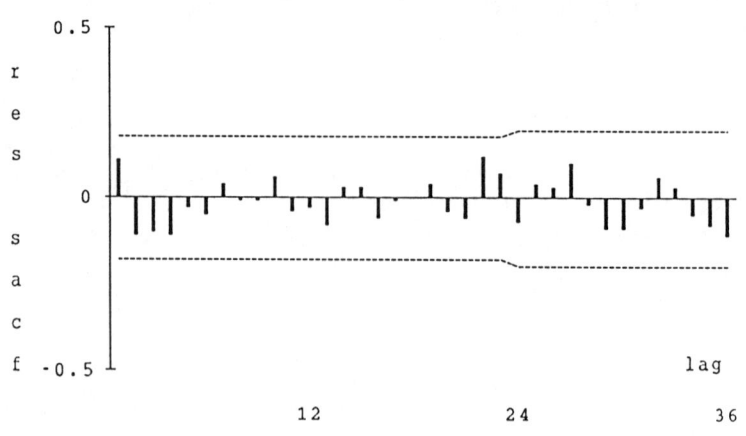

```
Chi-squared (Q*) = 24.1 for df = 33
```

Figure C3.11   Residual SACF for model (C3.10).

MODEL ESTIMATION, CHECKING, AND REFORMULATION

```
LOWER     UPPER    FREQ-  0     5    10   15   20   25   30   35   40
BOUND     BOUND    UENCY  +----+----+----+----+----+----+----+----+
-3.000  - -2.400      2   |>>
-2.400  - -1.800      3   |>>>
-1.800  - -1.200     12   |>>>>>>>>>>>
-1.200  - -0.600     19   |>>>>>>>>>>>>>>>>>>
-0.600  -  0.000     37   |>>>>>>>>>>>>>>>>>>>>>>>>>>>>>>>>>>>>
 0.000  -  0.600     26   |>>>>>>>>>>>>>>>>>>>>>>>>>
 0.600  -  1.200     18   |>>>>>>>>>>>>>>>>>
 1.200  -  1.800      9   |>>>>>>>>
 1.800  -  2.400      3   |>>>
 2.400  -  3.000      2   |>>
                          +----+----+----+----+----+----+----+----+
                          0     5    10   15   20   25   30   35   40
```

Figure C3.12 Histogram of the standardized residuals from model (C3.10).

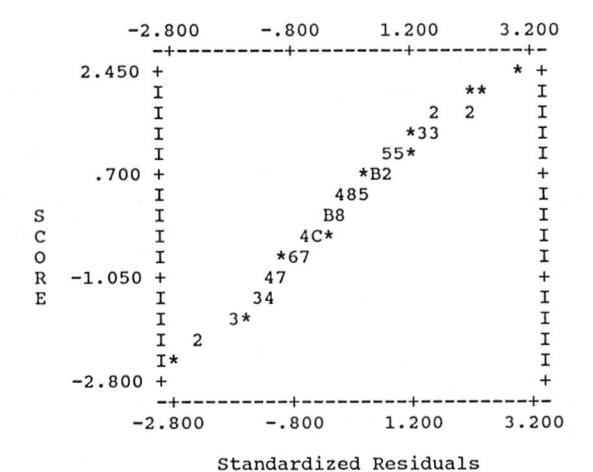

```
              -2.800       -.800       1.200       3.200
            -+----------+----------+----------+-
     2.450  +                                *   +
            I                             **      I
            I                        2   2        I
            I                      *33             I
            I                      55*             I
      .700  +                     *B2             +
            I                    485              I
       S    I                    B8               I
       C    I                    4C*              I
       O    I                  *67                I
       R  -1.050 +             47                 +
       E    I                  34                 I
            I              3*                     I
            I      2                              I
            I*                                    I
    -2.800  +                                     +
            -+----------+----------+----------+-
              -2.800       -.800       1.200       3.200

                      Standardized Residuals
```

Figure C3.13 Normal probability plot for the standardized residuals from model (C3.10).

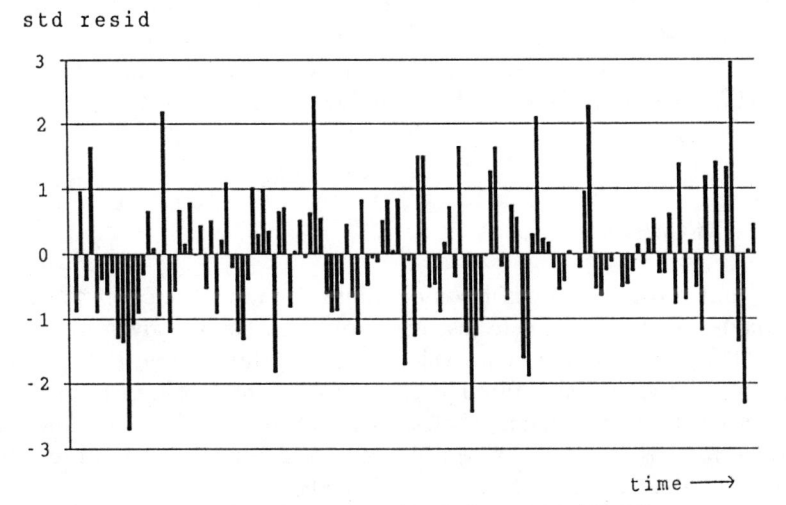

std resid

Figure C3.14 Standardized residuals for model (C3.10).

tions of zero, suggesting that no major outliers are present. The outlier detection procedure presented in Chapter 8 permits us to check for possible outliers in a more formal way.

Notice that $\hat{\sigma}_a$ for model (C3.10) is now somewhat smaller than $\hat{\sigma}_a$ for the ARIMA model (C3.1). Perhaps the regression model will produce more accurate forecasts than the ARIMA model.

C3.4 MODEL INTERPRETATION

1. As expected, the estimated coefficients of $\nabla_{12}\nabla X'_{1,t}$ and $\nabla_{12}\nabla X'_{2,t}$ are positive: Cooler temperatures apparently lead to more use of electricity for heating, and warmer temperatures apparently lead to more use of electricity for cooling.

2. As discussed in Section 3.8.4 in Chapter 3, we can use $X'_{1,t}$ and $X'_{2,t}$ in the "regression part" of the DR model (C3.9) to forecast Y'_t; but we then modify that forecast based on a weighted average of the one-step ahead forecast errors (disturbance values) that would occur if we used only $X'_{1,t}$ and $X'_{2,t}$ to forecast. That is, we use the ARIMA model for the disturbance series to help forecast Y'_t.

3. The ARIMA(0, 1, 1) component of the model for N_t is interpreted as an EWMA, as discussed in Chapter 2.

4. The AR(2)$_{12}$ component of the ARIMA model for N_t implies that the data contain a stochastic cycle of the type discussed in Chapter 2. A stochastic cycle is present in the model since $\hat{\phi}_{12}^2 + 4\hat{\phi}_{24} < 0$; that is, $-0.5373^2 + 4(-0.4667) = -1.578$. Using (2.11.9) in Chapter 2, the average period of this cycle is $p^* = 3.18$ years. If the AR(2) component were in the nonseasonal part of the model, the units of p^* would be in months; but since the cycle is represented in the seasonal part of the model, the units are in years. This cycle may correspond to a business cycle pattern.

C3.5 FORECASTS

We will not show the forecast computations for model (C3.10) since they are tedious. A more complete discussion of forecasting is given in Chapter 9. We will forecast the natural log values of Y_t (Y'_t) for $t = 169, 170, \ldots, 180$, using information available through time $t = 168$. Then we transform the log forecasts back to the original metric. The future values of $X'_{1,t}$ and $X'_{2,t}$ needed to forecast Y'_t are unknown. We replace these unknown future values with forecasts produced by models (C3.2) and (C3.3).

Twelve forecasts in the log metric from origin $t = 168$ are shown in Table

Table C3.2 Twelve Forecasts of Y_t' (Log Metric) from Model (C3.10) from Time Origin $t = 168$

Time	Forecast	Standard Error	Observed
169	5.2789	0.0599	5.2668
170	5.1708	0.0626	5.1596
171	5.1552	0.0652	5.2089
172	5.0787	0.0676	5.0676
173	5.1127	0.0699	5.1059
174	5.1501	0.0721	5.2300
175	5.4153	0.0743	5.3342
176	5.4298	0.0764	5.4756
177	5.2344	0.0785	5.1682
178	5.1548	0.0805	5.2095
179	5.1957	0.0824	5.2035
180	5.3002	0.0843	5.2715

C3.2, along with their standard errors and corresponding future observed values in the log metric. Procedures for finding the standard errors are discussed in Chapter 9. Approximate 95% confidence intervals for each log forecast (f) are found by computing $f \pm 2s(f)$, where $s(f)$ is the forecast standard error. We can transform each log metric forecast back to an original metric forecast (F) by computing $F = \exp(f)$. Retransformation of forecasts back to the original metric is discussed further in Chapter 9. To find approximate 95% limit values in the original metric, compute $\exp[f - 2s(f)]$ for the lower limit, and $\exp[f + 2s(f)]$ for the upper limit. The original metric forecasts and limits are shown in Table C3.3.

Table C3.4 shows the original metric forecasts for Y_t obtained from the ARIMA model (C3.1). Comparing these forecasts with those from the DR model (C3.10) in Table C3.3, we see that the ARIMA forecasts are not quite as accurate for the 12 forecasts considered here. The root mean squared forecast errors (RMSFE) for the DR forecasts and the ARIMA forecasts are 9.2 and 10.1 units, respectively. The corresponding means of the absolute percent errors are 3.8 and 4.3%, respectively.

C3.6 COMMENTS

1. In Chapters 1 and 3 we pointed out that single-equation regression models give inconsistent estimates of the coefficients when there is feedback from the output to the inputs. In the present case it is not reasonable to think that changes in kilowatt-hours used will lead to later changes in heating degree days or cooling degree days; it seems we should have no feedback problems. However, the data in this case study are temporal aggregates. That is, Y_t is the sum of all kilowatt-hours used during month t, $X_{1,t}$ is the

Table C3.3 Twelve Forecasts of Y_t (Original Metric) Based on DR Model (C3.10) from Time Origin $t = 168$

Time	(1) 95% Lower Limit	(2) Forecast	(3) 95% Upper Limit	(4) Observed	(5) Error	(6) % Error
169	174.0	196.2	221.1	193.8	−2.4	−1.2
170	155.4	176.1	199.5	174.1	−2.0	−1.1
171	152.1	173.3	197.5	182.9	9.6	5.2
172	140.3	160.6	183.8	158.8	−1.8	−1.1
173	144.4	166.1	191.0	165.0	−1.1	−0.7
174	149.3	172.4	199.2	186.8	14.4	7.7
175	193.8	224.8	260.8	207.3	−17.5	−8.4
176	195.8	228.1	265.8	238.8	10.7	4.5
177	160.4	187.6	219.5	175.6	−12.0	−6.8
178	147.5	173.3	203.5	183.0	9.7	5.3
179	153.1	180.5	212.8	181.9	1.4	0.8
180	169.3	200.4	237.2	194.7	−5.7	−2.9

Table C3.4 Twelve Forecasts of Y_t (Original Metric) Based on ARIMA Model (C3.1) from Time Origin $t = 168$

Time	(1) 95% Lower Limit	(2) Forecast	(3) 95% Upper Limit	(4) Observed	(5) Error	(6) % Error
169	182.5	202.2	224.0	193.8	−8.4	−4.3
170	162.2	180.5	200.9	174.1	−6.4	−3.7
171	159.0	177.9	198.8	182.9	5.0	2.8
172	147.4	165.5	185.8	158.8	−6.7	−4.2
173	147.0	165.7	186.8	165.0	−0.7	−0.5
174	162.5	183.9	208.2	186.8	2.9	1.5
175	200.9	228.2	259.2	207.3	−20.9	−10.1
176	199.8	227.8	259.7	238.8	11.0	4.6
177	167.6	191.8	219.5	175.6	−16.2	−9.2
178	160.1	183.9	211.2	183.0	−0.9	−0.5
179	162.6	187.3	215.9	181.9	−5.4	−3.0
180	180.2	208.3	240.9	194.7	−13.6	−7.0

monthly sum of all daily readings of heating degrees, and $X_{2,t}$ is the monthly sum of all daily readings of cooling degrees. As pointed out in Chapter 1, statistical feedback problems might arise, even when the underlying mechanism logically involves a one-way relationship, if the data are aggregated over time. In Chapters 5 and 10 we discuss feedback check procedures.

2. In this case study we assumed that Y responds immediately (during the same month) when X_1 and X_2 change during a given month. Further, we have assumed that Y does not respond in future months to any current change

in X_1 and X_2. Both of these points are seen in model (C3.4), where we relate Y' at time t to X_1' and X_2' at time t and *only* at time t. This seems like a reasonable assumption for this data set. For example, it seems reasonable that if a certain January is especially cold, people would use more electricity *during that January*, but that they would not continue to use more electricity in later months in response to the cold January. However, there may be other situations where the Y series may not respond immediately to changes in the X's. And after Y does begin to respond, it may continue to change over several, perhaps many, time periods in response to an earlier change in the X's. In Chapter 4 we discuss certain model types for dealing with such *delayed* and *distributed lag* responses.

(Although we have assumed that Y has no delayed or distributed lag reactions to changes in the X's in the present case, past X's do affect our forecasts of Y because of the differencing operations. This is different, however, from explicitly inserting past X's directly into the basic regression mechanism.)

3. We have shown that, for a single 12-month span, our DR model forecasts are more accurate than the ARIMA model forecasts for Y_t. This is consistent with the fact that $\hat{\sigma}_a$ for the DR model (C3.10) is smaller than for the ARIMA model (C3.1). However, we can't conclude that the DR model will definitely outperform the ARIMA model or any other forecast method. In particular, the accuracy of the DR model forecasts depends on the accuracy of the forecasts for the inputs. In practice we may continue to monitor the accuracy of the DR forecasts relative to other forecast methods.

Further, the DR forecasts are not drastically more accurate than the ARIMA model forecasts. Depending on the cost of forecast errors and the cost of model development, we might choose to work with a simpler model that is somewhat less accurate. Of course, in some situations small forecast errors are quite costly in terms of revenue foregone, reduced product quality, or deterioration of health or environmental conditions. In such cases even small gains in forecast accuracy might easily offset the cost of building and maintaining more complicated forecast models.

4. The $AR(2)_{12}$ component in the ARIMA model for the disturbance in the regression model (C3.10) is somewhat unusual. Fortunately we showed that it has a sensible interpretation, especially considering that we are working with economic data. We will see another similar example in Case 5.

QUESTIONS AND PROBLEMS

1. Replace the $AR(2)_{12}$ component in model (C3.10) with the more common $MA(1)_{12}$ component or with a higher-ordered seasonal MA. Do you obtain estimation and forecasting results that are superior to those for model (C3.10)?

2. ARIMA model (C3.2) for $X_{1,t}$ implies that this series has a fixed seasonal component. As an exercise, build a DR model for $X_{1,t}$ with seasonal dummy variables. Does this model fit the historical data better than (C3.2)? Does this model seem to forecast more accurately than (C3.2)? Does using the forecasts from this model seem to improve the forec· accuracy of model (C3.10) for Y_t?

CHAPTER 4

Rational Distributed Lag Models

In Chapter 1 we introduced the idea of a single-equation dynamic regression (DR) model where a time-ordered output variable (Y_t) is linearly related to one or more time-ordered input variables ($X_{1,t}, X_{2,t}, \ldots$), with observations at equally spaced time intervals. In a single-equation model there must be no feedback: Current values of the inputs must not be related to past values of the output ($Y_{t-1}, Y_{t-2}, \ldots$). If Y_t and X_t are contemporaneously related, we must be willing to attribute this relationship to dependence of Y_t on X_t, not vice versa, if a single-equation DR model is to be appropriate.

In Chapters 2 and 3 we discussed the main ideas about autoregressive integrating moving average (ARIMA) models and regression models, respectively. In Chapter 3 we discussed how we could incorporate an ARIMA model for the regression disturbance (N_t) into the regression; we illustrated this with an example involving seasonal dummy inputs. Cases 2 and 3 illustrate this idea further; Case 2 involves deterministic inputs (trading day variables), while Case 3 involves stochastic inputs (cooling and heating degrees variables).

A regression model with an ARIMA model for the disturbance is a *dynamic* regression, since Y_t depends partly on *past values of the disturbance*. But a *dynamic* regression might also involve *past values of the inputs*. The examples in Chapter 3, Case 2, and Case 3 did not include *time-lagged X's* in the regression. In this Chapter we begin to address this important aspect of DR models.

Regression models with time-lagged inputs are called *distributed lag* models. In this chapter we focus on a certain parsimonious form, called the *rational* distributed lag form, for writing distributed lag models. The word "rational" here refers to a mathematical *ratio*. Then in Chapters 5 and 6 we consider practical procedures for building DR models based on the available data.

Distributed lag relationships arise for a variety of technological, psychological, and institutional reasons. For example, if outdoor temperature (the input) drops suddenly, there will be only a gradual fall in indoor temperature as insulation slows heat transfer, and a gradual increase in the use of heating fuel (the output). Thus, the current value of the output is related to past values of the input. Or, if there is a decline in the cost of borrowed funds (the input), business managers may decide that it will be profitable to add to their stock of equipment. The managers must first recognize the change in the input, make decisions about desirable changes in equipment, appropriate funds, place orders or sign construction contracts, and wait for the production and delivery of the new goods. Thus, current investment expenditure by the managers (the output) responds only gradually to earlier changes in the input. See Nerlove (1958) for further discussion of the reasons for distributed lags in economic situations.

4.1 LINEAR DISTRIBUTED LAG TRANSFER FUNCTIONS

4.1.1 Transfer Functions

For simplicity we will discuss just one input. The ideas we develop here are easily extended to multiple inputs. If Y_t depends on X_t in some way, we may write this as

$$Y_t = f(X_t) \tag{4.1.1}$$

where $f(\cdot)$ is some mathematical function. The function $f(\cdot)$ is called a *transfer function*: The effect of a change in X_t is transferred to Y_t in some way specified by the function $f(\cdot)$. A simple example is $Y_t = 2X_t$: A one-unit change in X_t leads immediately (during the same time period) to a corresponding two-unit change in Y_t. We will focus on cases where $f(\cdot)$ is a *linear distributed lag* function: We assume that Y_t is a *linear combination* of *current* (X_t) and *past* ($X_{t-1}, X_{t-2}, \ldots$) values of the input. An important part of our modeling procedure will be to find a useful form of $f(\cdot)$ for a given data set.

Of course, (4.1.1) is too simple in its present form. It says that if the values of X_t are known, and if the function $f(\cdot)$ is known, then the values for Y_t are predicted without error from (4.1.1). In general, however, there are other factors causing variation in Y_t besides changes in the specified input(s). As discussed in Chapter 3, we capture those other factors with an additive stochastic disturbance (N_t) that may be autocorrelated. N_t represents the effects of all excluded inputs on the *variability* of Y_t. The input–output relationship may also have an additive constant term (C); this is a buffer term that captures the effect of excluded inputs on the *overall level* of Y_t. Thus we are considering models of the form

$$Y_t = C + f(X_t) + N_t \tag{4.1.2}$$

LINEAR DISTRIBUTED LAG TRANSFER FUNCTIONS

This is a regression-type model of the sort we considered in Chapter 3. For now we will concentrate on $f(X_t)$, the transfer function or "regression" component of (4.1.2).

4.1.2 Linear Distributed Lags

When X_t in (4.1.2) changes, there is no reason why the response in Y_t must all occur during one time period. Instead, Y_t might react to a change in X_t *with a time lag* that is *distributed across several time periods*. We assume that this distributed lag relationship is *linear*, so we can write the transfer function $f(X_t)$ as a linear combination of current and past X_t values:

$$Y_t = f(X_t)$$
$$= v_0 X_t + v_1 X_{t-1} + v_2 X_{t-2} + \cdots \qquad (4.1.3)$$

The true transfer function underlying a sample data set may not be linear. But we use the linearity assumption because (1) it simplifies the statistical analysis considerably; (2) despite their relative simplicity, linear models have proven to be useful in a wide variety of situations; and (3) a linear model is often a useful first step or approximation.

Coefficient v_0 is a weight that states how Y_t responds to a change in X_t (current time period change in X_t); coefficient v_1 states how Y_t responds to a change in X_{t-1} (one period earlier change in X_t); coefficient v_2 states how Y_t responds to a change in X_{t-2} (two periods earlier change in X_t): and so forth. In theory this distributed lag response could be of *infinite* length: the current level of Y_t may respond to changes in X_t that occurred infinitely far back into the past.

The v weights can be positive or negative. Of course, the larger the absolute value of any weight v_k, the larger the response of Y_t to a change in X_{t-k}. For example, an additional $1 million promotional expenditure by a business firm (change in $X_t = \$1$ million) might lead to no added sales during the current month ($v_0 = 0$), $2 million added sales during the first following month ($v_1 = 2$), and a loss of sales (due to customers having bought ahead) of $1 million during the second following month ($v_2 = -1$).

4.1.3 Dead Time

Y_t might not react immediately to a change in X_t: Some initial v weights may be zero. For example, suppose X_t is dosage of aspirin, and Y_t is headache pain level. Swallowing aspirin generally doesn't affect headache pain immediately. It may take up to an hour for the drug to be absorbed by the body and affect the pain level. If we are measuring Y_t every 5 minutes, we might find no change in pain at the time the aspirin is taken ($v_0 = 0$), no change in pain 5 minutes later ($v_1 = 0$), no change after 10 minutes ($v_2 = 0$), and so on. The *number of v weights equal to zero* (starting with v_0) is called *dead time*, denoted as b. In the promotional expenditure example in the preceding paragraph, starting with v_0 there is one v weight equal to zero ($v_0 = 0$), so $b = 1$. Alternatively, if $v_0 = v_1 = v_2 = 0$, and $v_3 \neq 0$, then $b = 3$.

4.1.4 Impulse Response Function

In Chapter 2 we discussed backshift notation for ARIMA models. We can also write a linear distributed lag transfer function in *backshift form* by defining $v(B)$ as

$$v(B) = v_0 + v_1 B + v_2 B^2 + v_3 B^3 + \cdots \qquad (4.1.4)$$

where B is the backshift operator defined such that $B^k X_t = X_{t-k}$. Using Equation (4.1.4), (4.1.3) may be rewritten as

$$Y_t = v(B) X_t \qquad (4.1.5)$$

To see that (4.1.5) is just a restatement of (4.1.3), substitute for $v(B)$ in (4.1.5) using (4.1.4), and use the definition of B^k to get (4.1.3). Equation (4.1.5) is a compact way of saying that there is a linear distributed lag relationship between changes in X_t and changes in Y_t.

The individual v weights in $v(B)$, ($v_0, v_1, v_2, \ldots$), are called the *impulse response weights*. The entire *set* of v weights is called the *impulse response function*. $v(B)$ tells us how Y_t reacts *through time* to a given change in X_t. The individual v weights state how Y_t reacts to a change in X_t at specific time lags. The signs and sizes of the v weights indicate whether Y_t increases (positive sign) or decreases (negative sign) by a lot (large v weight) or by a little (small v weight) when X_t changes. It is often helpful to present the impulse response function graphically, as shown in the next section.

4.2 A SPECIAL CASE: THE KOYCK MODEL

In this section we consider an important special case of the transfer function $v(B) X_t$, which arises often in practice. This section and the next one are especially important since they lay the foundation for our discussion of transfer functions and DR models throughout the rest of this book.

4.2.1 Parsimony

Consider that $v(B)$ in (4.1.5) can have a large number of v weights, possibly an infinite number. This can present serious estimation problems since our sample size is always limited. The main purpose of this section is to show that *if the v weights in v(B) follow a special pattern*, then we can *rewrite v(B)*

A SPECIAL CASE: THE KOYCK MODEL

in a *parsimonious form* that requires the estimation of *just a few* parameters. We first consider a special case, and then we generalize from that case in Section 4.3.

4.2.2 Exponential v-Weight Decay

The case we consider here was developed by Koyck (1954); we will call the result the *Koyck model*. Let δ_1 be a constant where $0 < |\delta_1| < 1$. Suppose dead time is zero ($b = 0$), and suppose we know that the v weights in $v(B)$ are related to each other as follows:

$$v_1 = \delta_1 v_0$$
$$v_2 = \delta_1 v_1$$
$$v_3 = \delta_1 v_2$$
$$\vdots$$
$$v_k = \delta_1 v_{k-1} \quad k = 1, 2, 3, \ldots \quad (4.2.1)$$

Starting from the initial response v_0, each succeeding (time-lagged) response of Y_t to a change in X_t, represented by the v weights $v_1, v_2, \ldots$, is a *constant fraction* (δ_1) *of the previous time period's response*. These v weights *decline exponentially*. For example, if $v_0 = 1$ and $\delta_1 = 0.5$, then we have this pattern:

$$v_1 = 0.5(1) = 0.5$$
$$v_2 = 0.5(0.5) = 0.25$$
$$v_3 = 0.5(0.25) = 0.125$$
$$\vdots$$

This impulse response function is graphed in Figure 4.1. The value k on the horizontal axis is the subscript on each v weight (v_k); k is therefore equal to the lag length associated with the input variable X_{t-k}. Notice that these v weights display an exponential decay pattern.

In the Koyck distributed lag pattern, each v_k (except v_0) is related in a known way to v_{k-1}. In fact, all the v weights are known if we know v_0 and δ_1. This is clear for v_1 in (4.2.1). Recursive substitution gives the following pattern for the remaining v weights:

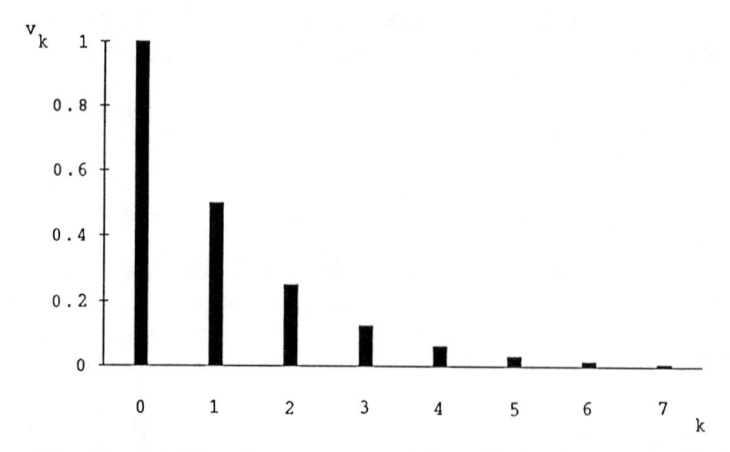

Figure 4.1 Graph of impulse response weights with simple exponential decay pattern, $v_0 = 1$, $\delta_1 = 0.5$.

$$v_2 = \delta_1 v_1 = \delta_1^2 v_0$$
$$v_3 = \delta_1 v_2 = \delta_1^3 v_0$$
$$v_4 = \delta_1 v_3 = \delta_1^4 v_0$$
$$\vdots$$

This pattern may be expressed in a compact way as

$$v_k = \delta_1^k v_0, \quad k \geq 0 \tag{4.2.2}$$

Notice that v_0 provides a *start-up value* for the later decay, and δ_1 dictates the *decay rate*. In practice, if we can estimate δ_1 and v_0, we will make good use of a limited sample. We can then find the other estimated v weights using (4.2.2).

Many different v-weight patterns are possible in practice. In a practical situation we won't know with certainty if pattern (4.2.2) is right for a given data set. As discussed in Chapters 5 and 6, using the Box–Jenkins modeling strategy we "let the data talk to us." That is, we let the data suggest whether the special pattern in (4.2.2) is appropriate for the data at hand. In fact we have a large *family* of $v(B)$ patterns from which to choose, and pattern (4.2.2) is just one member of that family. Fortunately, just a few of these patterns seem to occur often in practice. We consider the v-weight patterns of this family in more detail later in this chapter and in Chapter 5. Our purpose right now is to consider just the special case of pattern (4.2.2) and to ask: *If* the v weights really do follow the pattern in (4.2.2), can we state the transfer function (4.1.5) in a more parsimonious form? Can we find a

A SPECIAL CASE: THE KOYCK MODEL 153

statement of the transfer function (4.1.5) that has only δ_1 and v_0 as its parameters?

4.2.3 A Parsimonious Form of the Koyck Model

In this section we use Koyck's method to find a form of the Koyck transfer function that has only δ_1 and v_0 as its parameters. First, substitute for the v weights in (4.1.5) according to pattern (4.2.2) to get

$$Y_t = v_0 X_t + \delta_1 v_0 X_{t-1} + \delta_1^2 v_0 X_{t-2} + \cdots \quad (4.2.3)$$

Now consider how the response for Y_{t-1} is found. We find this from (4.2.3) in two steps. At the first step, replace t by $t-1$ throughout Equation (4.2.3) to obtain

$$Y_{t-1} = v_0 X_{t-1} + \delta_1 v_0 X_{t-2} + \delta_1^2 v_0 X_{t-3} + \cdots \quad (4.2.4)$$

At the second step, multiply both sides of (4.2.4) by δ_1:

$$\delta_1 Y_{t-1} = \delta_1 v_0 X_{t-1} + \delta_1^2 v_0 X_{t-2} + \delta_1^3 v_0 X_{t-3} + \cdots \quad (4.2.5)$$

If we now subtract (4.2.5) from (4.2.3), we obtain

$$Y_t - \delta_1 Y_{t-1} = v_0 X_t$$

or

$$Y_t = v_0 X_t + \delta_1 Y_{t-1} \quad (4.2.6)$$

Equation (4.2.6) is a transfer function with only v_0 and δ_1 as parameters to be estimated. Having estimated v_0 and δ_1, we can then estimate the remaining v weights using (4.2.2). The main point is this: *if* the v weights follow the special pattern stated in (4.2.2), then we can write the transfer function (4.1.5) with *just a few parameters*, as in (4.2.6). The form (4.2.6) does not contain less information than (4.1.5), *if the v weights follow pattern (4.2.2)*. But (4.2.6) has just two parameters (v_0 and δ_1), while (4.1.5) has an infinite number ($v_0, v_1, v_2, \ldots$). Having fewer parameters to estimate helps us to make more efficient use of a limited sample at the model estimation stage; and models with fewer parameters tend to produce more accurate forecasts.

4.2.4 Koyck Model with Dead Time $b > 0$

Let's consider the Koyck model when the dead time (b) is not zero. Suppose the dead time is $b = 3$, with the v weights following an exponential decay patterns after this. Then the v weights are

$$\begin{aligned}
v_0 &= 0 \\
v_1 &= 0 \\
v_2 &= 0 \\
v_3 &\neq 0 \\
v_4 &= \delta_1 v_3 \\
v_5 &= \delta_1^2 v_3 \\
&\vdots \\
v_k &= \delta_1^{k-b} v_3, \quad k > 3
\end{aligned} \qquad (4.2.7)$$

In (4.2.7), v_3 provides a *start-up value* for the later decay, while δ_1 dictates the decay rate. We can substitute for the v weights in (4.1.5) using pattern (4.2.7) to get the transfer function

$$Y_t = v_3 X_{t-3} + \delta_1 v_3 X_{t-4} + \delta_1^2 v_3 X_{t-5} + \cdots \qquad (4.2.8)$$

Lagging (4.2.8) by one time period and multiplying through the result by δ_1 yields

$$\delta_1 Y_{t-1} = \delta_1 v_3 X_{t-4} + \delta_1^2 v_3 X_{t-5} + \delta_1^3 v_3 X_{t-6} + \cdots \qquad (4.2.9)$$

Subtracting (4.2.9) from (4.2.8), we get

$$Y_t - \delta_1 Y_{t-1} = v_3 X_{t-3}$$

or

$$Y_t = v_3 X_{t-3} + \delta_1 Y_{t-1} \qquad (4.2.10)$$

The transfer functions in (4.2.6) (with $b = 0$) and (4.2.10) (with $b = 3$) have the same basic structure, but the *start-up coefficient* from which the decay proceeds is different: The start-up coefficient is v_0 in (4.2.6), while it is v_3 in (4.2.10).

4.2.5 Application 1

As we proceed we will use the sales and leading indicator data introduced in Chapter 1 (see Figures 1.4 and 1.5) to illustrate the ideas. It turns out that the transfer function for those data is described very well by the *Koyck model, with $b = 3$, after the data are differenced once nonseasonally ($d = 1$)*. (How we arrive at these conclusions will become more clear in the next two chapters.) The final model for these data has a transfer function, $v(B)\nabla X_t$, where the impulse response function $v(B)$ follows the pattern stated in (4.2.7).

Estimation using the first $n = 140$ observations and the differenced data

A SPECIAL CASE: THE KOYCK MODEL

$(d = 1)$ gives these results: $\hat{v}_3 = 4.82$, and $\hat{\delta}_1 = 0.72$. Ignoring the constant term and the disturbance for now, and stating the results in the form of (4.2.10), gives this transfer function:

$$\nabla Y_t = 4.82 \nabla X_{t-3} + 0.72 \nabla Y_{t-1}$$

From (4.2.7), the $\hat{v}$ weights for this transfer function are

$$\hat{v}_0 = 0$$
$$\hat{v}_1 = 0$$
$$\hat{v}_2 = 0$$
$$\hat{v}_3 = 4.82$$
$$\hat{v}_4 = (0.72)4.82 = 3.47$$
$$\hat{v}_5 = (0.72)^2 4.82 = 2.50$$
$$\vdots$$
$$\hat{v}_k = (0.72)^{k-3} 4.82 \qquad k > 3 \qquad (4.2.11)$$

So far we have considered only the transfer function or "regression" part of the DR model for the sales and leading indicator data. The complete DR model for these data has an autocorrelated disturbance and a nonzero constant, as we will see in Section 4.4.

4.2.6 Application 2

As we will see in Chapter 5, we take an essentially empirical approach to identifying the Koyck transfer function model for the sales and leading indicator data. At times, however, an underlying theory may lead to the use of the Koyck (or some other) distributed lag model.

For example, an economic theory of investment behavior suggests that there is some desired stock of capital goods (machinery, etc.) needed to produce a certain level of output, given the cost of borrowed funds, the state of technology, and so forth. For the sake of illustration, a simple form of this theory may be written as

$$Y_t^* = C^* + c_0 X_t \qquad (4.2.12)$$

where Y_t^* is the desired capital stock and X_t is output. When the desired stock Y_t^* exceeds the actual capital stock Y_t, businesses usually cannot adjust their Y_t immediately to the level of Y_t^* because it takes time to build and install new capital goods.

Nerlove (1958) hypothesizes a gradual adjustment mechanism whereby the *observed change* in Y_t is *proportional to the desired change* (the difference between the current desired capital stock and the previous actual stock):

$$Y_t - Y_{t-1} = \alpha(Y_t^* - Y_{t-1})$$

or

$$Y_t = \alpha Y_t^* + (1 - \alpha) Y_{t-1} \quad (4.2.13)$$

where $0 < \alpha < 1$ is the coefficient of adjustment (the fraction of desired change that can be achieved currently).

Substituting (4.2.12) into (4.2.13) gives

$$Y_t = \alpha C^* + \alpha c_0 X_t + (1 - \alpha) Y_{t-1} \quad (4.2.14)$$

which is the so-called *stock adjustment model*. This is the Koyck transfer function (with added constant) shown in (4.2.6), with $C = \alpha C^*$, $v_0 = \alpha c_0$, and $\delta_1 = 1 - \alpha$. In practice this relationship is also contaminated by a disturbance term N_t that may be autocorrelated.

4.3 RATIONAL DISTRIBUTED LAGS

The Koyck model is just one in a *family* of transfer functions we will consider in this book. This family is the set of *rational distributed lag models*. In this section we generalize from the Koyck model to consider this larger family. For more advanced discussions of rational distributed lags, see Jorgenson (1966), Baker (1975), and Dhrymes (1981).

4.3.1 Koyck Model in Rational Form

We can express transfer functions like (4.2.6) and (4.2.10) in backshift notation. For example, use the backshift operator to rewrite (4.2.6) as

$$(1 - \delta_1 B) Y_t = v_0 X_t \quad (4.3.1)$$

This gives the backshift form of (4.2.6). To get the rational form, divide (4.3.1) by $1 - \delta_1 B$ to obtain

$$Y_t = \frac{v_0}{1 - \delta_1 B} X_t \quad (4.3.2)$$

Comparing (4.1.5) with (4.3.2), we see that $v(B) = v_0/(1 - \delta_1 B)$ for the special case of the Koyck model with $b = 0$. Equation (4.3.2) is the same as Equations (4.2.6) and (4.3.1), but (4.3.2) is written in *rational (ratio) polynomial form*. That is, in (4.3.2) we have stated $v(B)$ as a *ratio* of two *finite-order polynomials in B*, where B is now treated as an ordinary algebraic variable

RATIONAL DISTRIBUTED LAGS

as well as the backshift operator. In (4.3.2), $v_0 = v_0 B^0$ is a zero-order polynomial in B, and $1 - \delta_1 B$ is a first-order polynomial in B. Importantly, we have replaced the infinite number of v weights in $v(B)$ with just two parameters, v_0 and δ_1, thanks to the special v-weight pattern in (4.2.2).

For the sales and leading indicator data, where the data are differenced and $b = 3$, we have $\hat{v}_3 = 4.82$ and $\hat{\delta}_1 = 0.72$. Using (4.2.10) this transfer function is written as

$$\nabla Y_t - 0.72 \nabla Y_{t-1} = 4.82 \nabla X_{t-3}$$

Rewriting this in backshift form gives

$$(1 - 0.72 B) \nabla Y_t = 4.82 B^3 \nabla X_t$$

Dividing both sides of this expression by $1 - 0.72B$ gives the following transfer function stated in rational polynomial form:

$$\nabla Y_t = \frac{4.82 B^3}{1 - 0.72 B} \nabla X_t$$

or

$$\nabla Y_t = \frac{4.82}{1 - 0.72 B} \nabla X_{t-3}$$

4.3.2 The Rational Distributed Lag Family

The Koyck impulse response function is just one member of the *family* of *rational polynomial distributed lag* models. This family is a set of impulse response functions $v(B)$ given by

$$v(B) = \frac{\omega(B) B^b}{\delta(B)} \qquad (4.3.3)$$

where

$$\omega(B) = \omega_0 + \omega_1 B + \omega_2 B^2 + \cdots + \omega_h B^h \qquad (4.3.4)$$

and

$$\delta(B) = 1 - \delta_1 B - \delta_2 B^2 - \cdots - \delta_r B^r \qquad (4.3.5)$$

and where B^b incorporates the dead time. Comparing (4.3.2) and (4.3.3), we see that, for the Koyck model with $b = 0$, we have $h = 0$, $r = 1$, and $\omega_0 = v_0$.

The notation for $\omega(B)$ in (4.3.4) is slightly different from that used by Box and Jenkins (1976). Box and Jenkins use negative signs in front of each term except ω_0, and they use s instead of h as the order of $\omega(B)$. The use of positive or negative signs does not matter if we are consistent. We will use positive signs so that the sign of any ω weight (plus or minus) corresponds to the direction of the effect on Y_t (up or down) resulting from a change in X_t. We use h (instead of s) for the order of $\omega(B)$ to avoid confusion since we use s as the length of seasonality in the ARIMA model for the disturbance series.

Rational lag models can capture a wide variety of impulse response patterns with just a few parameters. Although the rational lag family includes an infinite number of possible models, there are *just a few* that seem to occur often in practice. This makes it somewhat easier to identify a reasonable model for a given data set.

Figure 4.2(a) (the Koyck model) is one type of impulse response pattern that occurs often in practice. We will see examples of the Koyck model applied to real data in Chapters 5, 6, and 7 and in Cases 4 and 5. Examples (b) and (c) in Figure 4.2 show two other common impulse response weight patterns and the rational form transfer function associated with each pattern. Other impulse response function graphs and their rational form transfer functions are shown in Chapter 5 where we discuss model identification in detail.

The numerator $\omega(B)B^b$ and the denominator $\delta(B)$ in (4.3.3) play distinct roles in representing impulse response patterns such as those shown in Figure 4.2:

1. The numerator factor B^b captures dead time. In Figure 4.2(a), the dead time is $b = 0$; therefore there are no initial zero-valued v weights, and the first nonzero v weight is at lag 0. In Figure 4.2(b) and 4.2(c), the dead time is $b = 1$; therefore there is one zero-valued v weight in each case; and the first nonzero v weight is at lag 1.

2. The numerator factor $\omega(B)$ captures *unpatterned spikes* (not part of the decay pattern) in the impulse response graph, plus *decay start-up* values. In Figure 4.2(a) (the pure Koyck model, with $b = 0$), there are no unpatterned values. As we discussed with reference to Equation (4.2.2), in this case coefficient $v_0 = \omega_0$ is the decay start-up value. In Figure 4.2(b) there are two unpatterned spikes and no decay pattern. Coefficients ω_0 and ω_1 in the numerator of the rational form transfer function represent these two unpatterned spikes. In Figure 4.2(c) there are no unpatterned values. The first two v weights (lags 1 and 2) are start-up values for the subsequent decay; these are represented in the numerator of the rational form transfer function by coefficients ω_0 and ω_1.

(a) $\dfrac{\omega_0}{1-\delta_1 B}$, b = 0, h = 0, r = 1

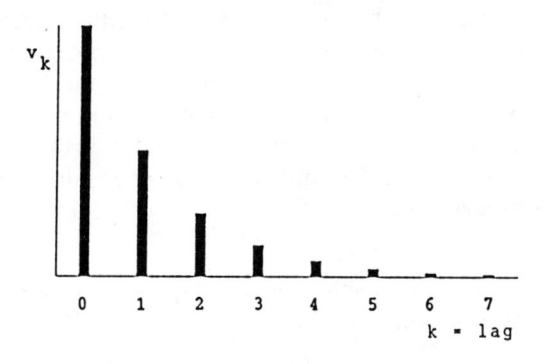

(b) $(\omega_0 + \omega_1 B) B$, b = 1, h = 1, r = 0

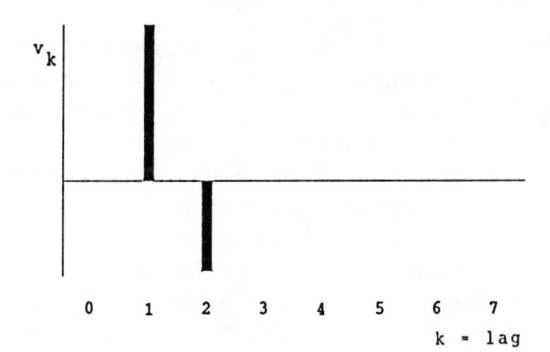

(c) $\dfrac{(\omega_0 + \omega_1 B) B}{1-\delta_1 B - \delta_2 B^2}$, b = 1, h = 1, r = 2

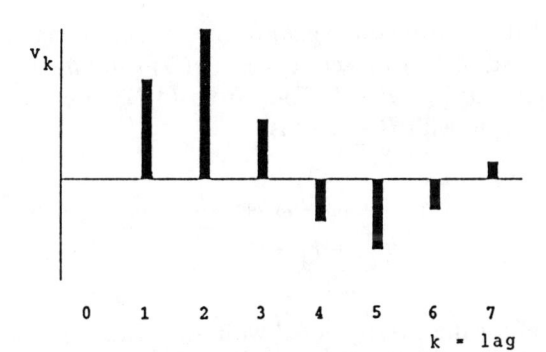

Figure 4.2 Examples of rational forms of $v(B)$ with corresponding impulse response function graphs. (a) $v(B) = \omega_0/(1 - \delta_1 B)$, $b = 0$, $h = 0$, $r = 1$. (b) $v(B) = (\omega_0 + \omega_1 B) B$, $b = 1$, $h = 1$, $r = 0$. (c) $v(B) = (\omega_0 + \omega_1 B) B/(1 - \delta_1 B - \delta_2 B^2)$, $b = 1$, $h = 1$, $r = 2$.

3. The denominator $\delta(B)$ in (4.3.3) represents the *decay pattern*. In Figure 4.2(a) we have $r = 1$; this is associated with a simple exponential decay in the impulse response weights. As we discussed with reference to Equations (4.2.2) and (4.2.7), in this case coefficient δ_1 dictates the decay rate. In Figure 4.2(b) we have $r = 0$. This corresponds to the fact that there is no decay in the v weights; thus the denominator term in (4.3.3) reduces to 1.0. In Figure 4.2(c) we have $r = 2$; this is associated with either a compound exponential decay or [as in Example (c)] a damped sine wave decay pattern in the remaining v weights.

For further discussion on the relationships among the v weights, the ω weights, and the δ weights, see Box and Jenkins (1976, Chapter 10).

4.3.3 Steady-State Gain

It is reasonable to consider only *stable* models where a finite, permanent change in X_t leads to a finite change in Y_t. That is, if X_t moves from one constant, *equilibrium* level to another, then Y_t should also move to a new *equilibrium* level, either immediately or eventually.

We are considering how much the equilibrium level of Y_t will eventually change in response to a one-unit change in X_t. Equation (4.1.3) says that Y_t will change by v_0 units immediately, by another v_1 units after one time period, by another v_2 units after two time periods, and so forth. Thus the *full (equilibrium) change over time* in Y_t in response to a one-unit change in X_t is

$$g = v_0 + v_1 + v_2 + \cdots \tag{4.3.6}$$

where g is called the *steady-state gain* of the transfer function. From inspection of (4.3.6) and (4.1.4) we see that $g = v(B)$ if we treat B as an ordinary algebraic variable and set $B = 1$. Then from (4.3.3) we may also write $g = \omega(B)B^b/\delta(B)$, again with $B = 1$; thus

$$g = \frac{\omega_0 + \omega_1 + \cdots + \omega_h}{1 - \delta_1 - \delta_2 - \cdots - \delta_r} \tag{4.3.7}$$

For example, for the Koyck model with $\omega_0 = 1$ and $\delta_1 = 0.5$, from (4.3.3) we find $g = \omega_0 B^b/(1 - \delta_1 B)$. Treating B as an ordinary algebraic variable, with $B = 1$, from (4.3.7) we have $g = \omega_0/(1 - \delta_1) = 1/(1 - 0.5)$ $1/0.5 = 2$. Thus, for this example, a one unit rise in X_t will lead to an eventual equilibrium rise in Y_t of two units.

RATIONAL DISTRIBUTED LAGS

4.3.4 Stability Conditions

From (4.3.6), we see that g is finite (and the transfer function relationship is stable) only if the series $v_0 + v_1 + v_2 + \cdots$ converges. More generally, g is finite if the series $v(B) = v_0 + v_1 B + v_2 B^2 + \cdots$ converges for $|B| \leq 1$, where B is again treated as an ordinary algebraic variable. Since $v(B) = \omega(B) B^b / \delta(B)$, stability requires that $\omega(B) B^b / \delta(B)$ must be finite for $|B| \leq 1$. This, in turn, requires that the δ coefficients in the denominator of (4.3.3) satisfy certain stability conditions.

To see these conditions, first write the *characteristic equation* $\delta(B) = 0$, where B is again treated as an ordinary algebraic variable. The *stability conditions* are that *all the roots* of this characteristic equation must lie *outside the unit circle* (in the complex plane). For the Koyck model, with $r = 1$, this condition is equivalent to

$$|\delta_1| < 1 \tag{4.3.8}$$

Condition (4.3.8) makes sense intuitively. After all, in (4.2.2) and (4.2.7) we see that δ_1 is the decay rate: Each period's response (except for v_b) is equal to δ_1 times the previous period's response. If (4.3.8) is satisfied, then δ_1 is a *fraction*: Each time period's new response (v_k) is *smaller in absolute value* than the previous period's response (v_{k-1}), and the *added* response of Y_t will approach zero as the time lag increases. But if $|\delta_1| \geq 1$, then the added responses of Y_t won't decay toward zero at longer lags.

For $r = 2$, the stability conditions (*all* must be met) are

$$|\delta_2| < 1$$
$$\delta_2 + \delta_1 < 1$$
$$\delta_2 - \delta_1 < 1 \tag{4.3.9}$$

For $r > 2$, which almost never occurs in practice, it is most convenient to use a computing program to find the roots of the characteristic equation $\delta(B) = 0$. We would use the values of the estimated δ weights to find the distances of the roots from the origin. For a transfer function response to be stable, all these distances must be larger than 1.0 in absolute value.

4.3.5 Finding v Weights from ω and δ Weights

Sometimes we will have the ω weights and δ weights for a model estimated in rational form, and we will want to find the corresponding v weights. We may do this by writing (4.3.3) as

$$\delta(B) v(B) = \omega(B) B^b \tag{4.3.10}$$

and then *equating coefficients of like powers of B on either side of* (4.3.10). This leads to the following general expressions:

$$v_j = 0, \qquad j < b$$

$$v_j = \sum_{i=1}^{r} \delta_i v_{j-i} + \omega_{j-b}, \qquad j \geq b \qquad (4.3.11)$$

where $\omega_{j-b} = 0$ if $j - b > h$.

To illustrate the derivation and use of (4.3.11) for a special case, let's find the $\hat{v}$ weights for the estimated sales and leading indicator model discussed earlier. The rational form model for these data is the Koyck model $\nabla Y_t = 4.82 \nabla X_{t-3} + 0.72 \nabla Y_{t-1}$, so that $b = 3$, $\hat{\omega}_0 = 4.82$, and $\hat{\delta}_1 = 0.72$. Ignoring circumflexes for simplicity, (4.3.10) for this example is

$$(1 - \delta_1 B)(v_0 + v_1 B + v_2 B^2 + v_3 B^3 + \cdots) = \omega_0 B^3 \qquad (4.3.12)$$

Expanding (4.3.12) and gathering terms,

$$v_0 + (v_1 - \delta_1 v_0) B + (v_2 - \delta_1 v_1) B^2 + (v_3 - \delta_1 v_2) B^3 + \cdots = \omega_0 B^3 \qquad (4.3.13)$$

Now equate coefficients of like powers of B on either side of (4.3.13). The coefficients of B^0 are v_0 on the left side and zero on the right side, so $v_0 = 0$. The coefficient of B^1 is $v_1 - \delta_1 v_0$ on the left side and zero on the right, so $v_1 - \delta_1 v_0 = 0$, or (substituting zero for v_0), $v_1 = 0$. Following the same procedure for B^2, B^3, and so forth, gives this set of equations:

$$v_0 = 0$$
$$v_1 - \delta_1 v_0 = 0$$
$$v_2 - \delta_1 v_1 = 0$$
$$v_3 - \delta_1 v_2 = \omega_0$$
$$\vdots$$
$$v_k - \delta_1 v_{k-1} = 0, \qquad k > 3 \qquad (4.3.14)$$

Thus for the sales and leading indicator model, from the first three equations in (4.3.14) we have $\hat{v}_0 = \hat{v}_1 = \hat{v}_2 = 0$. From the third equation, $\hat{v}_3 = \hat{\omega}_0 = 4.82$. All remaining equations $(k > 3)$ show $\hat{v}_k = \hat{\delta}_1 \hat{v}_{k-1} = 0.72 \hat{v}_{k-1} = 0.72 \hat{v}_{k-1}$, so $\hat{v}_4 = (0.72)\, 4.82 = 3.47$; $\hat{v}_5 = (0.72)\, 3.47 = 2.50$; and so forth. For this example notice that these $\hat{v}$ weights are identical to those shown in (4.2.11), obtained from the Koyck pattern in (4.2.7). Using (4.3.11) leads to the same results we have obtained here.

4.3.6 Multiple Inputs

It is easy to extend the rational lag form to multiple inputs. For example, with two inputs ($X_{1,t}$ and $X_{2,t}$), we have

$$Y_t = v_1(B)X_{1,t} + v_2(B)X_{2,t}$$
$$= \frac{\omega_1(B)B^{b_1}}{\delta_1(B)} X_{1,t} + \frac{\omega_2(B)B^{b_2}}{\delta_2(B)} X_{2,t} \quad (4.3.15)$$

where

$$v_i(B) = v_{i,0} + v_{i,1}B + v_{i,2}B^2 + \cdots = \omega_i(B)B^{b_i}/\delta_i(B), \quad i = 1, 2$$
$$b_i = \text{dead time for input } X_{i,t}, \quad i = 1, 2$$
$$\omega_i(B) = \omega_{i,0} + \omega_{i,1}B + \cdots + \omega_{i,h_i}B^{h_i}, \quad \text{for } i = 1, 2$$
$$\delta_i(B) = 1 - \delta_{i,1}B - \delta_{i,2}B^2 - \cdots - \delta_{i,r_i}B^{r_i}, \quad \text{for } i = 1, 2$$
$$h_i = \text{order of } \omega_i(B), \quad i = 1, 2$$
$$r_i = \text{order of } \delta_i(B), \quad i = 1, 2$$

Extending this framework to M inputs, $i = 1, 2, \ldots, M$, is straightforward. The result may be written compactly as

$$Y_t = \sum_{i=1}^{M} v_i(B)X_{i,t}$$
$$= \sum_{i=1}^{M} \frac{\omega_i(B)B^{b_i}}{\delta_i(B)} X_{i,t} \quad (4.3.16)$$

Cases 2, 3, 5, and 6 include examples of multiple input models.

4.4 THE COMPLETE RATIONAL FORM DR MODEL AND SOME SPECIAL CASES

Equation (4.3.16) shows the rational form of a model with M inputs and M transfer functions. A complete dynamic regression may also include a constant term (C), and it will have a disturbance series (N_t):

$$Y_t = C + \sum_{i=1}^{M} \frac{\omega_i(B)B^{b_i}}{\delta_i(B)} X_{i,t} + N_t \quad (4.4.1)$$

As discussed in Chapter 3, the disturbance may have a time structure that can be described by an ARIMA mechanism. Using the general notation set forth in Chapter 2, this is written as

$$\phi(B^s)\phi(B)\nabla_s^D \nabla^d N_t = \theta(B^s)\theta(B)a_t$$

or

$$N_t = \frac{\theta(B^s)\theta(B)}{\phi(B^s)\phi(B)\nabla_s^D \nabla^d} a_t \qquad (4.4.2)$$

Substitute (4.4.2) into (4.4.1); then the *combined* multiple input transfer function plus disturbance model (i.e., the complete dynamic regression model) is

$$Y_t = C + \sum_{i=1}^{M} \frac{\omega_i(B) B^{b_i}}{\delta_i(B)} X_{i,t} + \frac{\theta(B^s)\theta(B)}{\phi(B^s)\phi(B)\nabla_s^D \nabla^d} a_t \qquad (4.4.3)$$

4.4.1 Ordinary Multiple Regression

DR models include several other types of models as special cases. For example, consider the following multiple regression model for time series data with two independent variables:

$$Y_t = C + \omega_{1,0} X_{1,t} + \omega_{2,0} X_{2,t} + a_t \qquad (4.4.4)$$

This ordinary regression model is a special case of a dynamic regression with two inputs ($M = 2$), where $b_1 = b_2 = h_1 = h_2 = r_1 = r_2 = 0$, and where $N_t = a_t$ has no autocorrelation pattern.

4.4.2 Distributed Lag Regression

Now consider a regression model with one independent variable that enters the equation with a free-form distributed lag. This model is commonly considered in econometrics textbooks (see, e.g., Gujarati, 1988, Chapter 16):

$$Y_t = C + \omega_0 X_t + \omega_1 X_{t-1} + \omega_2 X_{t-2} + \cdots + a_t \qquad (4.4.5)$$

This is a single-input ($M = 1$) dynamic regression where $N_t = a_t$ is not autocorrelated; $b = 0$; $r = 0$; and h is the number of *lagged* input terms on the right side of the equation (i.e., disregarding $\omega_0 X_t$).

4.4.3 Regression with Autocorrelated Disturbance

Econometrics texts also consider regression models with autocorrelated disturbances (see, e.g., Gujarati, 1988, Chapter 16). For example, if the distributed lag model (4.4.5) has a disturbance term N_t that follows an AR(1) process, we have

$$Y_t = C + \omega_0 X_t + \omega_1 X_{t-1} + \omega_2 X_{t-2} + \cdots + N_t \qquad (4.4.6)$$

where

$$(1 - \phi_1 B)N_t = a_t \qquad (4.4.7)$$

Solving (4.4.7) for N_t and substituting the result into (4.4.6), we find

$$Y_t = C + \omega_0 X_t + \omega_1 X_{t-1} + \omega_2 X_{t-2} + \cdots + \frac{1}{1 - \phi_1 B} a_t \qquad (4.4.8)$$

This is a dynamic regression with $b = 0$; $r = 0$; $\phi(B) = 1 - \phi_1 B$; $\theta(B) = 1$; $\phi(B^s) = 1$; $\theta(B^s) = 1$; $D = 0$; $d = 0$; and h is the number of *lagged* input terms on the right side of the equation (i.e., disregarding $\omega_0 X_t$).

4.4.4 Sales and Leading Indicator Example

As an empirical example of a complete DR model, the following model seems appropriate for the sales and leading indicator data:

$$\nabla Y_t = 0.035 + \frac{4.82}{1 - 0.72B} \nabla X_{t-3} + (1 - 0.54B)\hat{a}_t \qquad (4.4.9)$$

In this case the estimated constant $\hat{C}$ is 0.035. The estimated rational lag transfer function $\hat{\omega}_0 B^3/(1 - \hat{\delta}_1 B)$, with $b = 3$, is $4.82B^3/(1 - 0.72B)$. The stationary (after nonseasonal first differencing, $d = 1$) disturbance term has an MA(1) structure, with $\hat{\theta}_1 = 0.54$. We will consider in detail in Chapter 5 how we arrive at this model by studying the data.

QUESTIONS AND PROBLEMS

1. Box and Jenkins (1976) report the following estimated DR model, where Y_t = percent concentration of carbon dioxide in the output from a gas furnace, and X_t = methane gas input in cubic feet per minute:

$$Y_t = \frac{(-0.53 + 0.37B + 0.51B^2)B^3}{1 - 0.57B} X_t + \frac{1}{1 - 1.53B + 0.63B^2} \hat{a}_t$$

a. What are the values of b, r, and h for the transfer function?

b. What are the values of d, D, p, P, q, and Q for the disturbance series ARIMA model?

c. State the values of the estimates for the following coefficients: ω_0, ω_1, ω_2, ω_3, ω_4, δ_1, δ_2, θ_1, θ_2, ϕ_1, and ϕ_2.

d. Write the transfer function using Box and Jenkins' convention of expressing the numerator of a rational transfer function as $\omega_0 - \omega_1 B - \cdots - \omega_h B^h$. How would using this convention change your answer to part (c)?

e. Find the first 10 estimated v weights for the estimated transfer function. Sketch the estimated impulse response weight graph.

f. What is the estimated numerical value of the steady-state gain (g) for this model? How do you interpret this number?

g. Suppose δ_1 were estimated to be 1.0. What would be the steady-state gain?

h. Does this transfer function satisfy the relevant stability condition(s)? Explain.

i. Suppose δ_1 were estimated to be 1.0. Would the model then be stable? How are your answers to this question and to part (h) related to your answers to parts (f) and (g)?

j. Why is $\hat{a}_t$ shown with a circumflex (^)?

2. Sketch the impulse response weight graphs for the following rational transfer functions:

a. $Y_t = (10 - 10B)BX_t$

b. $Y_t = (10 + 10B - 20B^2)B^2 X_t$

c. $Y_t = \dfrac{100B^2}{1 - 0.5B} X_t$

d. $Y_t = \dfrac{-100B}{1 - 0.5B} X_t$

e. $Y_t = \dfrac{100B^3}{1 + 0.5B} X_t$

f. $Y_t = \dfrac{-100}{1 + 0.5B} X_t$

CHAPTER 5

Building Dynamic Regression Models: Model Identification

5.1 OVERVIEW

Let's summarize the main points from the previous chapters. A dynamic regression (DR) model with one input consists of a transfer function plus a disturbance. This may be written as

$$Y_t = C + v(B)X_t + N_t \qquad (5.1.1)$$

where X_t is the input; Y_t is the output; C is the constant term; $v(B)X_t$ is the transfer function; and N_t is the stochastic disturbance, which may be autocorrelated. N_t is assumed to be independent of X_t. Our goals are as follows:

1. To find a parsimonious expression for the polynomial $v(B)$ as a ratio of two other polynomials of low order, $\omega(B)B^b/\delta(B)$.
2. To find a parsimonious expression for the time structure of the disturbance series N_t in the form of an autoregressive integrated moving average (ARIMA) model, $N_t = [\theta(B^s)\theta(B)/\phi(B^s)\phi(B)\nabla_s^D \nabla^d]a_t$, where a_t is zero-mean and normally distributed white noise.

The first stage in the Box–Jenkins modeling strategy for achieving these goals is to *identify* one or more *tentative* DR models based on a preliminary analysis of the data patterns. At the second and third stages we *estimate* the parameters of these models and then *check* the models for adequacy. If a model is inadequate, we identify another tentative model. Once a satisfactory model is found, we may forecast using that model.

In this chapter we focus on the identification part of the modeling strategy. Then in Chapter 6 we consider model checking. Later, in Chapter 9, we consider model estimation and forecasting. In two appendixes at the end of this chapter we discuss two further transfer function identification methods, the corner table method and the prewhitening method. Before discussing model identification in detail in this chapter, we first consider some preliminary modeling steps.

5.2 PRELIMINARY MODELING STEPS

5.2.1 Inspect for Outliers and Correct Data Errors

As a first step, we inspect a plot and a listing of all data series for obvious outliers. Outliers are unusual data points that deviate from the pattern of the rest of the data. Outliers may result from recording errors, or from identifiable external events called "interventions." Examples of interventions are redefinitions of the variable, labor strikes, breakage of control devices in engineering processes, policy decisions by management, natural disasters, or changes in laws. Outliers in the output series can also arise due to changes in the input variables whose effects are "passed through" to the output series.

The study of outliers is important in time series analysis. Outliers are clues that can lead us to uncover the causes of changes in the behavior of the data. This may lead us to improve our model and our forecasts. We can look for, think about, and attempt to deal with outliers at any stage of our modeling procedure. At the preliminary stage, at least we want to be alert for unusual observations.

Clearly, any misrecorded observations should be corrected, if possible. On the other hand, outliers in the output series that are caused by changes in input variables will (we hope) be "explained" within the subsequent DR model. That is, if an outlying value in the output series is explained by the behavior of the inputs, then it ceases to be an outlier in the context of the DR model. Any remaining outliers in the output series may call for intervention analysis, as discussed in Chapter 7. Related methods to detect and correct for outliers are discussed in Chapter 8.

Consider again the sales and leading indicator data in Figures 1.4 and 1.5. It does not appear that either series has obvious outliers. However, consider again the saving rate data in Figure 1.3. Recall that the observation at $t = 82$ deviates sharply from the rest of the data. A check of the original source reveals that this observation has been correctly recorded. As discussed in Chapter 1, this outlier may have been caused by an external event (a new tax rebate law). Chapter 7 shows how we can estimate the size of the effect of this external event using a special type of DR model called an intervention model.

Outliers or level shifts in the input series may be dealt with in the same way as those in the output series. However, we cannot rely on the DR model to explain these phenomena since we are assuming a one-way relationship. That is, we are assuming that movements in the input series could explain

PRELIMINARY MODELING STEPS 169

movements in the output series, but not vice versa. Therefore we may have to place greater reliance on the intervention methods of Chapters 7 and 8 to explain outliers in the input series.

5.2.2 Transform to Induce Stationary Variances

We should inspect each series to determine if its variance is approximately constant at all levels of the data. The sales and leading indicator series in Figures 1.4 and 1.5 each appear to have a stationary variance. The government receipts series in Figure 2.2, on the other hand, has a variance that increases with the level of the series.

As discussed in Chapter 2, often we can induce a stationary variance by transforming the data. For example, the natural logs or square roots of the original series may display a stationary variance. These transformations are members of the Box–Cox family of power transformations in Equation (2.2.1). Since the sales and leading indicator data in Figures 1.4 and 1.5 already seem to have constant variances, no transformations are needed. On the other hand, we saw in Chapter 2 that the natural logs of the government receipts data (Figure 2.3) have approximately a constant variance.

5.2.3 ARIMA Models

It is good practice to build an ARIMA model for both the output and the input series before attempting to build a DR model. This serves at least four purposes:

1. It gives a baseline model and set of forecasts for the output series. If the DR model does not fit the output data or forecast its future any better than an ARIMA model, it may not be worth the effort to work with a DR model. Of course, in some cases we are more interested in assessing the role of the input variables in explaining the behavior of the output variable than we are in forecasting. In this case we might continue to work with a DR model even though it does not fit the data or forecast any better than an ARIMA model.
2. It gives models that can be used to forecast the input series, since forecasts of the inputs may be necessary to forecast the output. Examples are shown in Cases 3, 4, and 5.
3. It may reveal things about the behavior of the data that lead us to alter our DR model. For example, as discussed in Chapter 2 and Cases 1 and 2, suppose we find an AR(2) component with a stochastic cycle that repeats about every 3 months in an ARIMA model for cumulated monthly data. This suggests that we might use a DR model with trading day inputs.
4. ARIMA models for the inputs provide information needed at the DR

model checking stage (discussed in Chapter 6) and for the computation of the DR model forecast standard errors and confidence intervals (discussed in Chapter 9).

The modeling strategy discussed in Chapter 2 leads to the following ARIMA models for the sales (Y_t) and leading indicator (X_t) data, using the first $n = 140$ observations:

$$(1 - 0.8857B)\nabla Y_t = (1 - 0.6445B)\hat{a}_t$$
$$(12.56) \qquad\qquad (5.61)$$

$$\nabla X_t = (1 - 0.4483B)\hat{\alpha}_t$$
$$(5.87)$$

with $\hat{\sigma}_a = 1.35707$ and $\hat{\sigma}_\alpha = 0.387$.

5.2.4 Consider Relevant Theory

Presumably we are guided by some theory or experience in deciding which inputs to include in a DR model. Ideally, we include all relevant and important inputs. As pointed out in Chapter 3, if we exclude relevant inputs, and if they are correlated with the included inputs, then our DR model coefficient estimates are biased and inconsistent.

In some cases we will also have prior ideas about the nature of the relationship between the inputs and the output. Such ideas may arise from our understanding of a physical mechanism that connects the output to the input, as might occur in the physical or engineering sciences. Or, there may be some economic (or other) theory that suggests the nature of the relationship. At the very least, we may be fairly sure that the *level* of the output is related to the *level* of the input, or that the *level* of the output is related to *changes* in the input.

5.2.5 Check for Feedback

An important assumption in building a *single-equation* DR model is that there is *no feedback* from earlier values of the output to current values of the inputs. The possibility of feedback arises whenever the inputs are stochastic. We introduced this matter in Chapter 1 with reference to Figure 1.6. For example, it may be that a company's quarterly product sales (Y_t) are related to current and past quarterly advertising outlays (X_t) according to the following mechanism:

$$Y_t = C + v_0 X_t + v_1 X_{t-1} + v_2 X_{t-2} + N_t$$

But it is also possible that higher sales will later on induce management to increase advertising outlays, as follows:

$$X_t = C^* + w_1 Y_{t-1} + M_t$$

If X_t responds this way to earlier values of Y_t, we have feedback from the past of Y_t to the current X_t; estimating the first equation as a single-equation DR model will produce inconsistent coefficient estimates.

Feedback might reflect a logical causal connection, as suggested by the sales and advertising examples. Unfortunately, feedback can also arise artificially when the data are temporally aggregated; see Tiao and Wei (1976), Wei (1982), and Wei (1990). This could happen with quarterly sales and advertising data, for example, since the data are quarterly sums of daily figures. Thus a feedback check is important if the data are temporal aggregates, even when we are quite sure that the DR model output does not logically cause the input(s).

In Chapter 10 we consider a procedure to check for feedback in the context of vector ARMA models, which are multiple-equation models. Essentially the same results may be obtained using single-equation regression methods. To check for feedback, each input series is regressed on its own past, on the past of each other input, and on the past of the output series. [This feedback check is in the spirit of the "Granger causality test"; for a detailed introduction, see Granger and Newbold (1986; Chapters 7 and 8).] For example, for a possible DR model with two inputs ($X_{1,t}$ and $X_{2,t}$), the following equation may be estimated to check for feedback from the past of Y_t to $X_{1,t}$:

$$X_{1,t} = C + b_1 X_{1,t-1} + b_2 X_{1,t-2} + \cdots + b_K X_{1,t-K}$$
$$+ c_1 X_{2,t-1} + c_2 X_{2,t-2} + \cdots + c_K X_{2,t-K}$$
$$+ d_1 Y_{t-1} + d_2 Y_{t-2} + \cdots + d_K Y_{t-K} + a_t \quad (5.2.1)$$

Past values of $X_{2,t}$ are included in (5.2.1) since there may be feedback from $X_{2,t}$ to $X_{1,t}$, and $X_{2,t}$ may be correlated with Y_t. If past values of $X_{2,t}$ are not included, the feedback that should be attributed to $X_{2,t}$ might be spuriously attributed to Y_t; see Pindyck and Rubinfeld (1991, Section 9.2). Also, by including past values of $X_{2,t}$ we obtain estimation results identical to those produced by the vector ARMA feedback check method discussed in Chapter 10.

The estimated coefficients $\hat{d}_k$, $k = 1, 2, 3, \ldots, K$, in (5.2.1) contain information about possible feedback from past values of Y to the current values of X_1. One way to check for feedback is to examine the individual t values of the $\hat{d}$ coefficients. If they are all insignificant, we may conclude that there is no statistical feedback from the past of Y to X_1, and we may include X_1 as an explanatory variable in a single-equation DR model. We should not insist on a complete lack of significant $\hat{d}$ coefficients in models like (5.2.1); we should make a reasonable allowance for sampling variation and recognize that the results from a model like (5.2.1) are only a guideline.

Colinearity can produce unreliable t values. It may also be argued that possible feedback from Y_t to $X_{1,t}$ is represented in (5.2.1) by all $d_k Y_{t-k}$ terms, $k = 1, 2, \ldots, K$, as a set. Thus, it is helpful to perform a joint test using the null hypothesis

$$H_0: d_1 = d_2 = \cdots = d_K = 0$$

with alternate H_a: not all d_k are zero, in addition to the individual t tests. This joint H_0 may be tested with the following F statistic:

$$F = \frac{(\text{ESS}_1 - \text{ESS}_0)/K}{\text{RSS}_1/(n_1 - k_1)} \qquad (5.2.2)$$

where ESS_1 is the explained sum of squares (discussed in Chapter 3) found from estimating (5.2.1); ESS_0 is the explained sum of squares found from estimating (5.2.1) with the lagged Y terms excluded from the equation; K is the maximum lag on the right-side terms in (5.2.1); RSS_1 is the residual sum of squares found from estimating (5.2.1); $n_1 = n - K$, where n is the original number of observations; and k_1 is the total number of parameters estimated in (5.2.1).

Estimating an equation like (5.2.1) does not tell us anything about the *contemporaneous* relationship between $X_{1,t}$ and Y_t. If an input and the output are contemporaneously related, this could be due to effects going from the input to the output, or vice versa, or some combination of these two. To use a single-equation DR model, we must be willing to attribute a contemporaneous relationship between an input and the output to effects passing from the input to the output.

The maximum lag K in models like (5.2.1) may be chosen according to the judgment of the analyst. K should be large enough to cover any likely important relationships between the input series in question and the past of the relevant variables. It may be helpful to try several values of K to see if the results are sensitive to the choice of K. A more formal method for choosing K is discussed in Chapter 10.

Setting $K = 15$ for the sales and leading indicator data, and using the first $n = 140$ observations, we estimate this equation to check for feedback:

$$X_t = C + b_1 X_{t-1} + b_2 X_{t-2} + \cdots + b_{15} X_{t-15}$$
$$+ c_1 Y_{t-1} + c_2 Y_{t-2} + \cdots + c_{15} Y_{t-15} + a_t$$

The estimated values of the c_k coefficients are shown in Table 5.1. All of the corresponding t values are small. The sample F from (5.2.2) is

5.3 THE LINEAR TRANSFER FUNCTION (LTF) IDENTIFICATION METHOD

Table 5.1 Feedback Check Results for Sales and Leading Indicator Data

Parameter	c_1	c_2	c_3	c_4	c_5
Estimate	0.11098	−0.14179	0.13763	0.02713	−0.08399
t Value	0.84	−1.00	0.96	0.19	−0.60
Parameter	c_6	c_7	c_8	c_9	c_{10}
Estimate	−0.06360	0.24399	−0.26068	0.12659	−0.04260
t Value	−0.45	1.74	−1.83	0.87	−0.30
Parameter	c_{11}	c_{12}	c_{13}	c_{14}	c_{15}
Estimate	−0.10503	0.17044	−0.03018	−0.01266	−0.01267
t Value	−0.73	1.19	−0.31	−0.39	−0.61

$$F = \frac{(148.554 - 147.383)/15}{7.972/(125 - 31)} = 0.92 \qquad (5.2.3)$$

which is insignificant at the 5% level for (15,94) degrees of freedom. We have no evidence that there is any important feedback from the past of the output to the input for this example. We will proceed to identify a single-equation DR model for these data.

5.3 THE LINEAR TRANSFER FUNCTION (LTF) IDENTIFICATION METHOD

In Chapter 1 we said that estimating the simple regression model, $Y_t = C + v_0 X_t + N_t$, could give misleading results about the relationship between Y_t and X_t. Possible problems involve (1) feedback from the output series to the inputs, (2) omitted time-lagged input terms, (3) an autocorrelated disturbance series, and (4) common autocorrelation patterns shared by Y and X that can produce spurious correlations.

In this section we set out a practical model identification procedure designed to handle problems (2), (3), and (4). The details of the identification procedure given here differ from those in Box and Jenkins (1976), but the main ideas are the same. The chief advantage of the procedure given here is that it can handle multiple inputs with relative ease, unlike the procedure of Box and Jenkins. We discuss the procedure of Box and Jenkins in Appendix 5B.

The strategy we present here uses a single-equation linear transfer function model to estimate the v weights in (5.1.1), *plus* one or more autoregressive terms as *initial proxies* for the disturbance series autocorrelation pattern. For convenience, we will refer to this method as the *LTF method* (for linear

```
                    ┌─────────────────────────────────────────┐
              ┌───→ │ (a)  Specify free-form distributed      │
              │     │      lag for transfer function v(B)X_t. │
              │     │ (b)  Specify low order AR(p)(P)_s for   │
              │     │      disturbance series N_t.            │
              │     └─────────────────────────────────────────┘
              │                       ↓
              │     ┌─────────────────────────────────────────┐
              │     │ (a)  Estimate coefficients.             │
              │     │ (b)  Check estimated disturbance        │
              │     │      series for stationarity.           │
              │     └─────────────────────────────────────────┘
              │                       ↓
   ┌──────────┴──┐          No   ┌──────────────────┐
   │ Difference  │ ←─────────────│ Is disturbance   │
   │ inputs and  │               │ stationary?      │
   │ output      │               └──────────────────┘
   └─────────────┘                       ↓ Yes
                    ┌─────────────────────────────────────────┐
              ┌───→ │ (a)  Specify tentative rational lag     │
              │     │      form transfer function.            │
              │     │ (b)  Specify tentative ARMA model       │
              │     │      for disturbance series.            │
              │     └─────────────────────────────────────────┘
              │                       ↓
              │     ┌─────────────────────────────────────────┐
              │     │ (a)  Estimate coefficients.             │
              │     │ (b)  Check model for adequacy.          │
              │     └─────────────────────────────────────────┘
              │                       ↓
              │          No  ┌────────────────┐  Yes
              └──────────────│   Model OK?    │──────→ Forecast
                             └────────────────┘
```

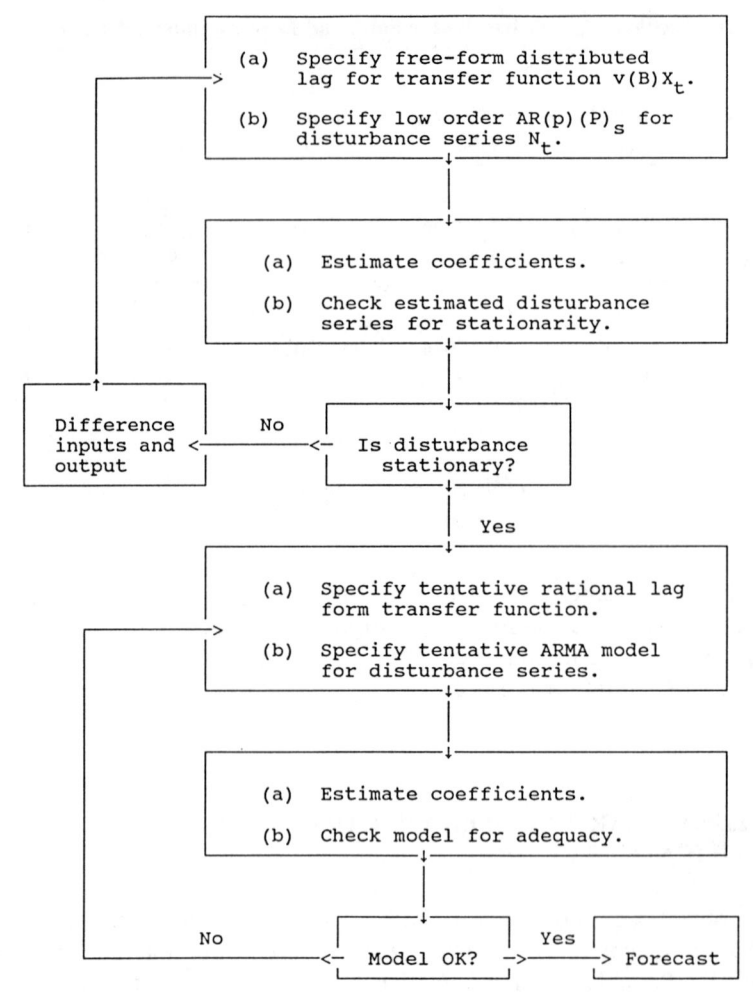

Figure 5.1 A linear transfer function (LTF) modeling strategy for dynamic regression models.

transfer function). The modeling strategy is summarized schematically in Figure 5.1.

The procedure for identifying DR models is similar to the procedure for identifying ARIMA models. We first generate a rather large number of statistics that give clues about the patterns in the data. In the case of a DR model, these statistics are (1) a set of sample impulse response weights (estimated v weights) and (2) a set of sample autocorrelation and partial autocorrelation coefficients for the estimated disturbance series. Then we tentatively choose a parsimonious rational distributed lag model for the transfer function and a parsimonious ARIMA model for the disturbance series. We modify our tentative model as necessary based on the information produced at the estimation and checking stages; these stages are discussed in Chapters 6 and 9.

5.3.1 Free-Form Distributed Lag

At the first step of the LTF identification method, we estimate a *preliminary model* that includes the current and earlier values of X_t through some maximum lag (denoted by K):

$$Y_t = C + v(B) + N_t$$
$$= C + v_0 X_t + v_1 X_{t-1} + v_2 X_{t-2} + \cdots + v_K X_{t-K} + N_t \quad (5.3.1)$$

Model (5.3.1) is called a *linear transfer function* because it has the form of a linear regression. It is also called a *free-form distributed lag* model. A similar free-form distributed lag identification method is discussed by Box and Jenkins (1976, Chapter 11). The order of $v(B)$ (the value of K) in this approach is often chosen arbitrarily. This is a potential weakness with the LTF method since we don't know how many v weights are important. Ideally, K is chosen to include the longest time-lagged response (of Y_t to changes in X_t) that we might reasonably expect to be important. In Chapter 10 we discuss DR models in the context of vector ARMA models; there we consider a more formal criterion for choosing K.

Estimating (5.3.1) may entail the loss of many degrees of freedom compared to estimating a more parsimonious rational form transfer function. But at the start of model identification we often don't know which rational lag model is appropriate, if any. Our hope is that we can get fairly good estimates of the v weights by estimating (5.3.1). Then we tentatively identify a rational form transfer function by *comparing the estimated v-weight pattern with some common theoretical v-weight patterns.* As discussed in Chapter 4, each rational form transfer function is associated with a *theoretical v-weight pattern*. We tentatively choose the rational form model whose theoretical v-weight pattern seems to match our estimated v-weight pattern. This matching procedure is discussed in Section 5.4.

We could estimate (5.3.1) as it stands and proceed as we did in Section 3.9, Case 2, and Case 3. But we will now introduce one more element into the modeling procedure. This added step helps us to get better information from our preliminary LTF model. This added step involves the use of a *proxy ARIMA model* for N_t.

5.3.2 AR Disturbance Proxy

As part of the LTF method we *approximate* the possible autocorrelation pattern of N_t with a *low-order autoregressive (AR) model* such as

$$(1 - \phi_s B^s)(1 - \phi_1 B)N_t = a_t$$

or

$$N_t = [1/(1 - \phi_s B^s)(1 - \phi_1 B)]a_t \qquad (5.3.2)$$

In this case, $p = 1$ and $P = 1$. This approximation procedure is suggested by Liu (1986a) and Liu and Hudak (1986) and is related to the *common-filter* approach of Liu and Hanssens (1982). If we have a large sample, we might use a higher-ordered AR proxy model (say, $p = 2$ and $P = 1$, or $p = 2$ and $P = 2$). Notice in Equation (5.3.2) that we denote the shock term as a_t (an uncorrelated sequence); but because (5.3.2) is *only a proxy* for the possible time structure of the disturbance term, in practice the residual series $\hat{a}_t$ may still be somewhat autocorrelated.

It is acceptable to bypass using an AR proxy to represent the disturbance time structure. However, including an ARIMA model proxy for N_t in the preliminary LTF model has some advantages. As discussed in Chapters 1 and 3, if N_t is autocorrelated, then the estimates of the regression coefficients are *inefficient*, and the standard t tests are *invalid*. But we want to use those t tests to help us decide which lagged X's to include in the model. Including an approximate ARIMA model for N_t can produce more efficient estimates of the v weights, and more informative t values.

Why do we use an $AR(p)(P)_s$ proxy rather than some other ARIMA proxy? If N_t is nonstationary in the mean, then the estimated AR coefficients should capture at least part of this phenomenon. After all, the differencing operators $\nabla = (1 - B)$ and $\nabla_s = (1 - B^s)$ are, in effect, AR operators with $\phi_1 = \phi_s = 1$. In this case using (5.3.2) builds into the preliminary DR model an ARIMA disturbance component that is at least partly correct. If the true disturbance is stationary but AR in form, then we have a good chance of adequately representing that structure using a low-order AR proxy like (5.3.2) since most AR models found in practice are of low order. If the disturbance pattern is moving average (MA), but of low order, the AR terms might act as useful substitutes. (Recall that any invertible MA process may be written as an equivalent AR process of infinitely high order. For an invertible model these equivalent AR-type coefficients tend to become small fairly quickly as we consider values further into the past. Thus an MA model can often be approximated rather well by an AR model of sufficiently high order.)

Substitute (5.3.2) into (5.3.1). We estimate simultaneously the parameters in the following preliminary LTF model:

$$Y_t = C + v_0 X_t + v_1 X_{t-1} + \cdots + v_K X_{t-K} + \frac{1}{(1 - \phi_s B^s)(1 - \phi_1 B)} a_t$$

$$(5.3.3)$$

THE LINEAR TRANSFER FUNCTION (LTF) IDENTIFICATION METHOD

In (5.3.3) we represent the shocks as a_t (an uncorrelated sequence). However, the residual series $\hat{a}_t$ will tend to be somewhat autocorrelated if N_t actually behaves according to something other than an AR(1)(1)$_s$ process.

5.3.3 Differencing

As discussed in Cases 2 and 3, if N_t is nonstationary in the mean, we difference *both the output and the inputs* accordingly. We decide if N_t is nonstationary in the mean by studying the sample autocorrelation function (SACF) of $\hat{N}_t$. That is, we *generate an estimate of the N_t series*, denoted $\hat{N}_t$, and examine its SACF. The estimated disturbance series is computed as

$$\hat{N}_t = Y_t - \hat{C} - \hat{v}(B)X_t$$

For example, consider the simple case where the maximum lag on X_t in (5.3.3) is $K = 2$. Suppose (5.3.3) gives these estimates: $\hat{C} = 10$, $\hat{v}_0 = 4$, $\hat{v}_1 = -3$, $\hat{v}_2 = 2$. Substituting these estimates into $\hat{N}_t = Y_t - \hat{C} - \hat{v}(B)X_t$, and using the observed data $Y_1, Y_2, \ldots, Y_n$, and $X_1, X_2, \ldots, X_n$, the sequence of $\hat{N}_t$ values is

$$\hat{N}_1 = Y_1 - 10 - 4X_1 + 3X_0 - 2X_{-1}$$
$$\hat{N}_2 = Y_2 - 10 - 4X_2 + 3X_1 - 2X_0$$
$$\hat{N}_3 = Y_3 - 10 - 4X_3 + 3X_2 - 2X_1$$
$$\hat{N}_4 = Y_4 - 10 - 4X_4 + 3X_3 - 2X_2$$
$$\vdots$$
$$\hat{N}_n = Y_n - 10 - 4X_n + 3X_{n-1} - 2X_{n-2}$$

Notice that only $n - K$ disturbance values can be computed for this example: $\hat{N}_1$ and $\hat{N}_2$ can't be computed since X_0 and X_{-1} are not observed; and $\hat{N}_2$ can't be computed since X_0 is not observed. In some cases even fewer than $n - K$ disturbance values may be available, depending on the values of d, D, p, and P. In practice the $\hat{N}_t$ series is computed automatically using a software package. For example, the SCA System computes this series as a by-product of estimating model (5.3.3).

After computing the $\hat{N}_t$ series, we then construct its SACF and check how quickly it goes to zero; if it goes to zero only *slowly*, we conclude that the disturbance is not stationary in the mean. If we conclude that N_t is not stationary in the mean, we difference the output series *and* all the input series accordingly. We addressed this issue in Cases 2 and 3. For example, suppose we decide that N_t is not stationary and that regular first differencing ($d = 1$) is needed. Then we consider the differenced disturbance $n_t = \nabla N_t$.

Rewrite this as $N_t = n_t/\nabla$, substitute this expression into (5.3.1), and *multiply through by the differencing operator* to get

$$\nabla Y_t = C + v(B)\nabla X_t + n_t \qquad (5.3.4)$$

Thus model (5.3.1) for the *levels* of Y_t and X_t implies a corresponding model (5.3.4) for the *differenced* series ∇Y_t and ∇X_t. The value of C in (5.3.4) may be quite different from that in (5.3.1) since the series are now differenced. Note that $v(B)$ is the same whether the model is expressed in levels or differences; however the *estimates* of the v weights and their t values may be different when the data are differenced.

After differencing (if needed), we repeat the estimation of a preliminary LTF model, which again includes an AR proxy for the (differenced) disturbance term ARIMA pattern. Now, however, the analysis is applied to the differenced output and input series. For example, suppose we have applied regular first differencing to create the new differenced disturbance $n_t = \nabla N_t$. An approximate ARIMA model for n_t is the AR(1)(1)$_s$

$$(1 - \phi_s B^s)(1 - \phi_1 B)n_t = a_t$$

or

$$n_t = [1/(1 - \phi_s B^s)(1 - \phi_1 B)]a_t \qquad (5.3.5)$$

Substitute the second expression in (5.3.5) into (5.3.4) to get the new preliminary LTF model

$$\nabla Y_t = C + v_0 \nabla X_t + v_1 \nabla X_{t-1} + \cdots + v_K \nabla X_{t-K} + \frac{1}{(1 - \phi_s B^s)(1 - \phi_1 B)} a_t \qquad (5.3.6)$$

Tentatively we represent the shocks in (5.3.6) as a_t (an uncorrelated sequence); however, the residual series $\hat{a}_t$ may still be autocorrelated. The values of C, ϕ_1, and ϕ_s in (5.3.6) may differ from those in (5.3.3) because the data are now differenced.

We estimate the coefficients in (5.3.6). We then construct the estimated disturbance series for the differenced data as

$$\hat{n}_t = \nabla Y_t - \hat{C} - \hat{v}(B)\nabla X_t$$

and examine its SACF to see if it decays rapidly. If the disturbance series n_t does not yet seem stationary in the mean, we then difference both the output and the inputs even further.

Suppose the SACF for $\hat{n}_t$ decays slowly at the seasonal lags ($s, 2s, 3s, \ldots$); we then apply seasonal differencing ($D = 1$) in addition to regular differencing. We create the new differenced disturbance $n_t = \nabla_s \nabla N_t$. Rewrite this as $N_t = n_t/\nabla_s \nabla$, substitute this expression into (5.3.1), and *multiply through by the differencing operators* to get

$$\nabla_s \nabla Y_t = C + v(B)\nabla_s \nabla X_t + n_t \tag{5.3.7}$$

An approximate ARIMA model for the new differenced disturbance n_t is AR(1)(1)$_s$

$$(1 - \phi_s B^s)(1 - \phi_1 B)n_t = a_t$$

or

$$n_t = [1/(1 - \phi_s B^s)(1 - \phi_1 B)]a_t \tag{5.3.8}$$

Substitute the second expression in (5.3.8) into (5.3.7) to get the new preliminary LTF model

$$\nabla_s \nabla Y_t = C + v_0 \nabla_s \nabla X_t + v_1 \nabla_s \nabla X_{t-1} + \cdots + v_K \nabla_s \nabla X_{t-K}$$
$$+ \frac{1}{(1 - \phi_s B^s)(1 - \phi_1 B)} a_t \tag{5.3.9}$$

The values of C, ϕ_1, and ϕ_s in (5.3.9) may differ from those in (5.3.6) because the series have been differenced again. After estimating (5.3.9), we construct the new estimated disturbance $\hat{n}_t = \nabla_s \nabla Y_t - \hat{C} - \hat{v}(B)\nabla_s \nabla X_t$, and study its SACF to see if it decays slowly. Further differencing is applied as needed. This cycle is repeated until we have arrived at a stationary disturbance, induced by differencing.

Once we have achieved a stationary disturbance, we study the most recent set of estimated v weights to tentatively identify a rational form transfer function $\omega(B)B^b/\delta(B)$ to represent $v(B)$. We also study the SACF, SPACF, and ESACF of the most recent estimated (now stationary) disturbance $\hat{n}_t$ to identify a tentative ARMA model for the stationary disturbance n_t.

5.3.4 Application

Let's apply the LTF identification method to the sales and leading indicator data in Figures 1.4 and 1.5. Setting $K = 15$, $p = 1$, and $P = 0$, and combining Equations (5.3.1) and (5.3.2), we estimate the following preliminary LTF model using the first $n = 140$ observations:

$$Y_t = C + v_0 X_t + v_1 X_{t-1} + \cdots + v_{15} X_{t-15} + \frac{1}{1 - \phi_1 B} a_t \tag{5.3.10}$$

Table 5.2 Estimation Results for Preliminary LTF Model (5.3.10) with $d = D = 0$

$$\hat{C} = 30.8768 (|t| = 5.5)$$

$k \rightarrow$	0	1	2	3	4	5	6	7	8
$\hat{v}_k$	−0.116	−0.066	0.042	4.79	3.55	2.40	1.78	1.20	1.09
$|t|$	1.4	0.7	0.4	53.8	39.5	27.3	20.4	13.8	12.6

$k \rightarrow$	9	10	11	12	13	14	15		1
$\hat{v}_k$	0.72	0.48	0.43	0.30	0.23	0.21	0.05	$\hat{\phi}_k$	0.97
$|t|$	8.2	5.5	4.9	3.5	2.6	2.5	0.6		50.0

$$\hat{\sigma}_a = 0.237108$$

(It is known that there is no seasonal pattern in these data so there is no seasonal AR proxy term.)

The results of this estimation are shown in Table 5.2. The SACF of series $\hat{N}_t$ computed from (5.3.10) is shown in Figure 5.2. This SACF decays slowly to zero: The coefficients are more than twice their standard errors as far as lag 10. This suggests that N_t is nonstationary in the mean and that nonseasonal first differencing ($d = 1$) is appropriate. This leads to the new preliminary LTF model for the *differenced* data:

$$\nabla Y_t = C + v_0 \nabla X_t + v_1 \nabla X_{t-1} + \cdots + v_{15} \nabla X_{t-15} + \frac{1}{1 - \phi_1 B} a_t \quad (5.3.11)$$

Since differencing alters the data, the values of C and ϕ_1 in (5.3.11) differ from those in (5.3.10). Likewise, the disturbance in (5.3.11) is now $n_t = \nabla N_t$ rather than N_t; (5.3.11) includes an AR(1) proxy for the ARIMA pattern of n_t, $(1 - \phi_1 B) n_t = a_t$, or $n_t = a_t/(1 - \phi_1 B)$. Estimation results for (5.3.11)

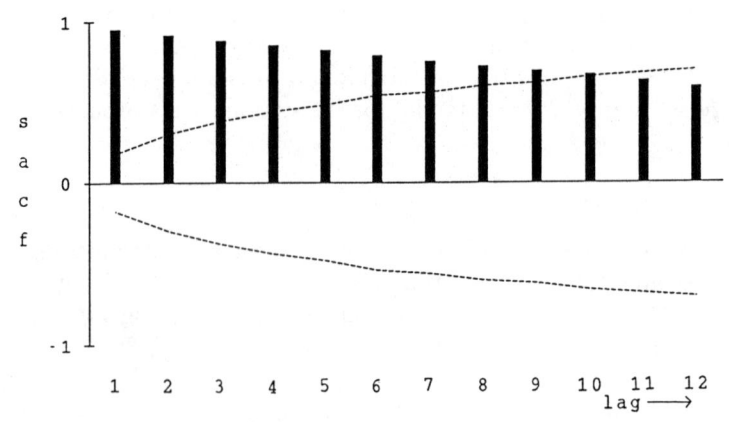

Figure 5.2 SACF for the estimated disturbance $\hat{N}_t$ in model (5.3.10).

THE LINEAR TRANSFER FUNCTION (LTF) IDENTIFICATION METHOD 181

Table 5.3 Estimation Results for Preliminary LTF Model (5.3.11) with $d = 1, D = 0$

$$\hat{C} = 0.0408 (|t| = 2.7)$$

$k \rightarrow$	0	1	2	3	4	5	6	7	8
$\hat{v}_k$	−0.103	−0.079	0.022	4.76	3.52	2.38	1.76	1.20	1.08
$|t|$	1.4	1.1	0.3	65.4	47.8	33.0	24.6	16.8	15.2
$k \rightarrow$	9	10	11	12	13	14	15		1
$\hat{v}_k$	0.70	0.47	0.41	0.28	0.21	0.20	0.05	$\hat{\phi}_k$	−0.43
$|t|$	9.9	6.6	5.7	3.9	3.0	2.8	0.8		5.2

$$\hat{\sigma}_a = 0.216767$$

are shown in Table 5.3. The SACF for $\hat{n}_t$ in Figure 5.3 drops quickly to insignificance; this suggests that n_t is stationary so that no further differencing is needed.

Now that we have a stationary disturbance, we may use the $\hat{v}$ weights to choose the orders (b, r, h) of one (or just a few) *tentative rational form transfer function(s)* to represent $v(B)$. We try to *match* the estimated v-weight pattern with a theoretical v-weight pattern. Each theoretical rational form transfer function is associated with a particular theoretical v-weight pattern. When we find a good match, we choose as a tentative model for our data that rational form transfer function associated with the matching theoretical v-weight pattern.

We may identify the orders (b, r, h) by *visually comparing* the estimated impulse response function with some common theoretical functions; we illustrate this comparison procedure using the sales and leading indicator data in this section. We may also use a set of *practical rules* to identify the orders (b, r, h), as discussed in Section 5.4. Further, we may also use the *corner table*, discussed in Appendix 5A, to identify the orders (b, r, h).

The $\hat{v}$ weights from model (5.3.11), shown in Table 5.3, are graphed in

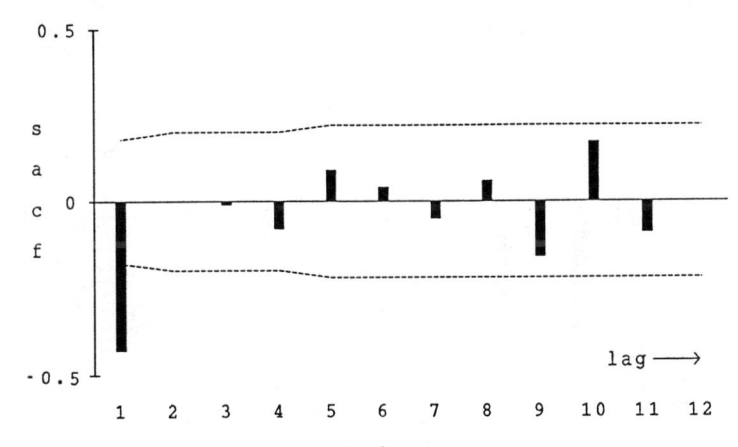

Figure 5.3 SACF for the estimated disturbance $\hat{n}_t$ in model (5.3.11).

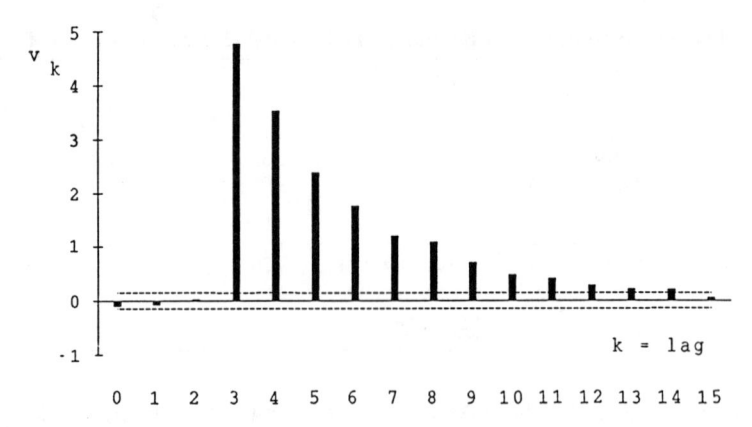

Figure 5.4 Sample impulse response function for leading indicator input in model (5.3.11).

Figure 5.4. In studying sample v weights we must allow for sampling variation. The estimates $\hat{v}_0$, $\hat{v}_1$, and $\hat{v}_2$ are not zero; but since they are small relative to their standard errors, we treat them as zero. As discussed in Chapter 4, dead time (b) is equal to the number of initial zero-valued v weights, starting with v_0. For this example we choose $b = 3$. Inspection shows that Figure 5.4 looks much like the theoretical impulse response function in Figure 4.2(a) (except for the dead time). Since that is a Koyck-type model, with $r = 1$ and $h = 0$, we choose those values for our tentative rational form transfer function. Thus we have identified $(b, r, h) = (3, 1, 0)$.

In addition to identifying a rational form model for the v weights, we also want to identify a tentative ARMA model for the stationary disturbance series n_t. To do this we study the SACF, SPACF, and ESACF of the stationary sample disturbance $\hat{n}_t$ for model (5.3.11); these functions are shown in Figures 5.3 and 5.5 and Table 5.4, respectively. The SACF in

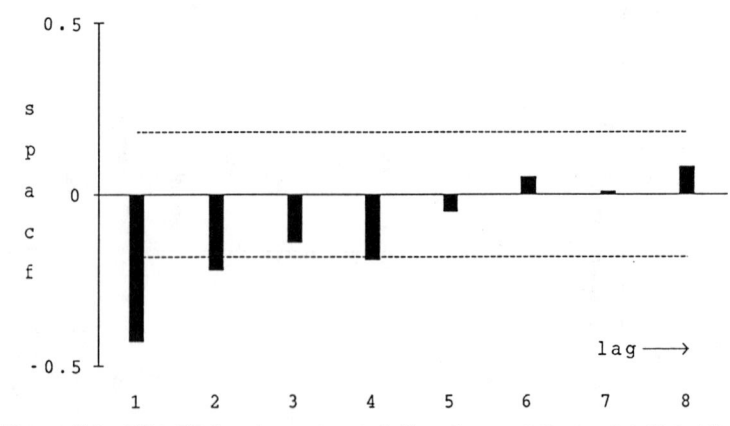

Figure 5.5 SPACF for the estimated disturbance $\hat{n}_t$ in model (5.3.11).

THE LINEAR TRANSFER FUNCTION (LTF) IDENTIFICATION METHOD 183

Table 5.4 Simplified ESACF for the Estimated Disturbance $\hat{n}_t$ in Model (5.3.11)[a]

$(q \rightarrow)$	0	1	2	3	4	5	6	7	8
$(p' = 0)$	X	0	0	0	0	0	0	0	0
$(p' = 1)$	X	0	0	0	0	0	0	0	0
$(p' = 2)$	X	0	0	0	0	0	0	0	0
$(p' = 3)$	X	X	X	X	0	0	0	0	0
$(p' = 4)$	X	X	0	0	0	0	0	0	0
$(p' = 5)$	X	X	0	0	0	0	0	0	0
$(p' = 6)$	X	X	X	0	0	0	0	0	0

[a] X = significant at 5% level.

Figure 5.3 has a significant spike at lag 1 followed by insignificant values; this suggests an MA(1) model for the stationary disturbance n_t. The decay of the SPACF in Figure 5.5 is consistent with this choice. Likewise, the ESACF in Table 5.4 has a triangle of zeros, in the lower right of the table, with its upper left vertex in the $(p' = 0, q = 1)$ cell; this also calls for an MA(1) model. Thus we choose the following tentative model for n_t:

$$n_t = (1 - \theta_1 B)a_t \qquad (5.3.12)$$

We now have a complete tentative DR model for the sales and leading indicator data: $(b, r, h) = (3, 1, 0)$, with an ARIMA $(0, 1, 1)$ model for the disturbance series N_t. The DR model may be written in terms of differenced series as

$$\nabla Y_t = C + \frac{\omega_0 B^3}{1 - \delta_1 B} \nabla X_t + (1 - \theta_1 B)a_t \qquad (5.3.13)$$

It remains for us to estimate this DR model, check it for adequacy, reformulate (reidentify) if necessary, and then forecast with the final model. We discuss these topics further in Chapters 6 and 9.

It is interesting at this point to compare our identified transfer function with the estimated model reported in Table 1.4 (Example 3) in Chapter 1. There we considered the simple regression model

$$Y_t = C + v_0 X_t + N_t \qquad (5.3.14)$$

Estimation results in Table 1.4 show a significant contemporaneous relationship between Y_t and X_t. That is, the absolute t value of $\hat{v}_0$ in Table 1.4 is large (34.13). But model (5.3.14) ignores the possible *time-lagged* response of Y_t to X_t, as well as the possible *autocorrelation structure* of N_t. On the other hand, the LTF identification method takes both of these *dynamic* features into account. In fact, the estimated impulse response function in Table 5.3, graphed in Figure 5.4, shows *no* significant contemporaneous relationship between Y_t and X_t: $\hat{v}_0$ is not significant. The misleading result in Table 1.4 is based on an inadequate identification procedure, one that ignores the possible dynamic relationship between Y_t and X_t, as well as the time structure of the disturbance series. But the LTF identification method takes both of these dynamic matters into account by considering both (1) the possible distributed lag response of Y_t to X_t and (2) the possible autocorrelation pattern of N_t.

5.4 RULES FOR IDENTIFYING RATIONAL DISTRIBUTED LAG TRANSFER FUNCTIONS

In the previous section we used visual comparison of a sample impulse response weight function ($\hat{v}$ weights) with some theoretical impulse response weight functions (those in Figure 4.2) to identify the orders (b, r, h) of a rational form transfer function. Such visual comparison often works quite well in practice. *Experience suggests that rational form transfer functions for economic and business data or other social science data, for example, almost always involve either*

1. $r = 0$ [a few nonzero impulse response weights with no decay pattern, similar to Figure 5.6(*b*)] or
2. $r = 1$ [a simple Koyck-type decay pattern, similar to Figure 5.6(*a*)].

Such patterns are relatively easy to identify visually. However, data from other situations, such as those involving physical mechanisms, may call for more complicated rational form structures, with $r > 1$. And sometimes economic data or other social science data give rise to a sample impulse response function that doesn't seem to match a simple theoretical rational form. In these cases it is useful to have a set of identification rules to help us choose orders (b, r, h) in a systematic fashion.

This section presents five sequential practical rules for identifying the orders (b, r, h) of a rational form transfer function. You may not need these rules to identify relatively simple rational form models. But the rules can be useful if you are having trouble identifying a rational form transfer function.

5.4.1 Identification Rules

For visual clarity we will discuss the identification rules with reference to *theoretical* impulse response functions. Of course, in practice we must study a *sample* function, and compare it with theoretical functions; when studying a sample impulse response function, we must allow for sampling variation. The rules are first stated and explained with reference to the Koyck-model

RULES FOR IDENTIFYING RATIONAL DISTRIBUTED LAG MODELS 185

$v(B)$ function shown in Figure 5.6(a). Then the rules are illustrated and explained further with additional examples in Figure 5.6.

Before discussion of the model identification rules, some preliminary comments are in order:

1. As with other practical rules, these rules are not absolute. They are only meant to help you find a useful starting place in identifying a rational form transfer function.

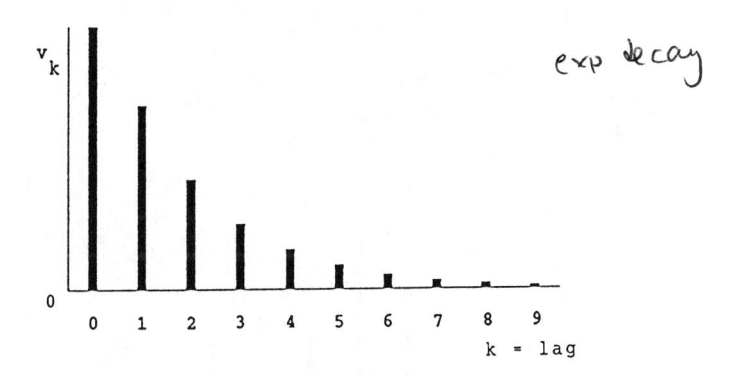

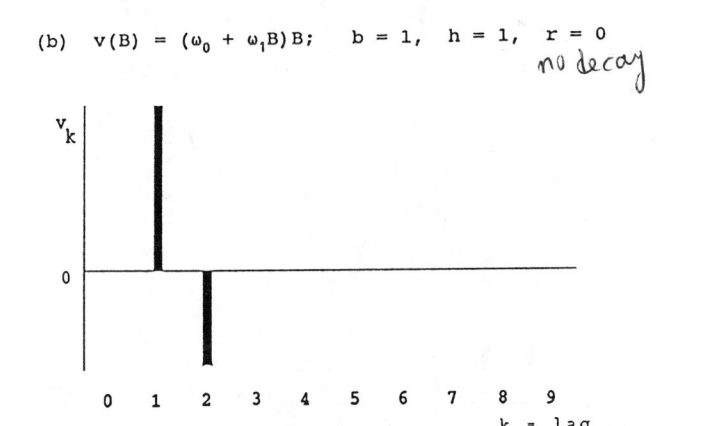

(a) $\quad v(B) = \dfrac{\omega_0}{1 - \delta_1 B}; \quad b = 0, \quad h = 0, \quad r = 1$

exp decay

v_k vs $k = \text{lag}$

(b) $\quad v(B) = (\omega_0 + \omega_1 B) B; \quad b = 1, \quad h = 1, \quad r = 0$

no decay

v_k vs $k = \text{lag}$

Figure 5.6 Examples of rational forms of $v(B)$ and graphs of impulse response functions. (a) $v(B) = \omega_0/(1 - \delta_1 B)$, $b = 0$, $h = 0$, $r = 1$. (b) $v(B) = (\omega_0 + \omega_1 B)B$, $b = 1$, $h = 1$, $r = 0$. (c) $v(B) = (\omega_0 + \omega_1 B)/(1 - \delta_1 B)$, $b = 0$, $h = 1$, $r = 1$. (d) $v(B) = (\omega_0 + \omega_1 B)B^3/(1 - \delta_1 B - \delta_2 B^2)$, $b = 3$, $h = 1$, $r = 2$. (e) $v(B) = (\omega_0 + \omega_1 B)B/(1 - \delta_1 B - \delta_2 B^2)$, $b = 1$, $h = 1$, $r = 2$. (f) $v(B) = (\omega_0 + \omega_1 B + \omega_2 B^2)/(1 - \delta_1 B - \delta_2 B^2)$, $b = 0$, $h = 2$, $r = 2$.

186   BUILDING DYNAMIC REGRESSION MODELS

(c) $v(B) = \dfrac{\omega_0 + \omega_1 B}{1 - \delta_1 B}$; $b = 0,\ h = 1,\ r = 1$ *exp decay*

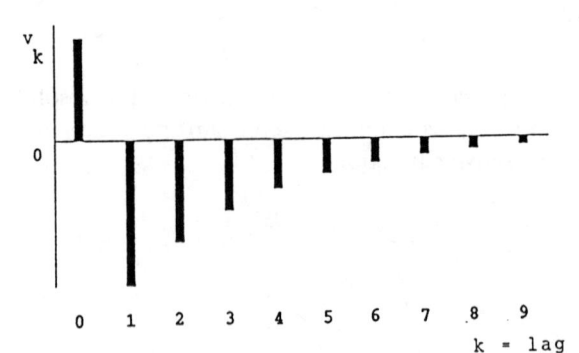

(d) $v(B) = \dfrac{(\omega_0 + \omega_1 B) B^3}{1 - \delta_1 B - \delta_2 B^2}$; $b = 3,\ h = 1,\ r = 2$ *comp exp decay*

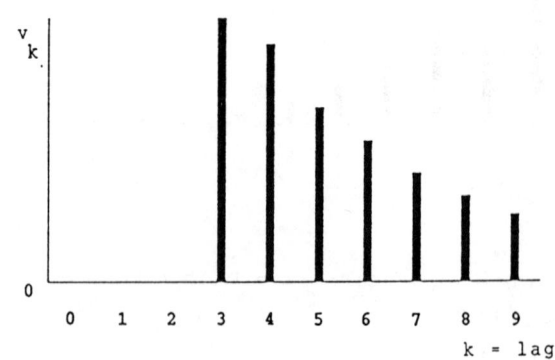

(e) $v(B) = \dfrac{(\omega_0 + \omega_1 B) B}{1 - \delta_1 B - \delta_2 B^2}$; $b = 1,\ h = 1,\ r = 2$ *comp exp decay*

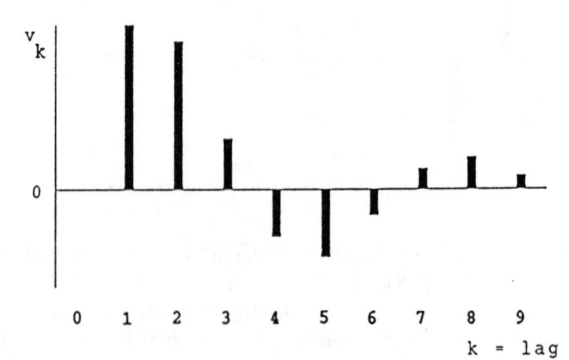

Figure 5.6   (*Continued*)

RULES FOR IDENTIFYING RATIONAL DISTRIBUTED LAG MODELS 187

(f) $\quad v(B) = \dfrac{\omega_0 + \omega_1 B + \omega_2 B^2}{1 - \delta_1 B - \delta_2 B^2}; \quad b = 0, \quad h = 2, \quad r = 2$

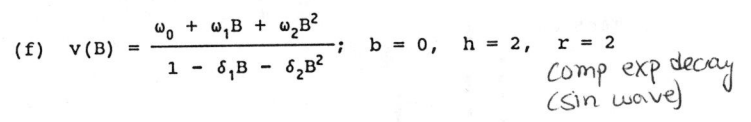

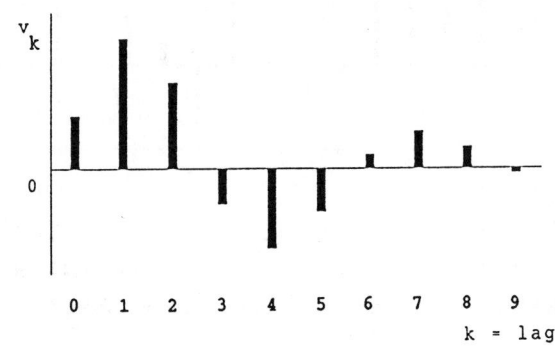

Comp exp decay (sin wave)

Figure 5.6 (*Continued*)

2. As discussed in Chapter 4, the numerator in a rational form transfer function, $\omega(B)B^b$, captures three items: *dead time* (the value of b); *unpatterned v-weight terms* (neither part of the decay start-up nor of the decay pattern); and *decay start-up v-weight terms*.
3. The denominator $\delta(B)$ captures the nature of the *v-weight decay pattern*. See Box and Jenkins (1976, p. 347) for further discussion of the relationships among the v weights, ω weights, and δ weights.
4. These rules apply to *sets* of unpatterned plus decay start-up v-weights that are *contiguous* (all adjoined). See application 3 in Section 5.4.5 for further discussion of this point.

RULE 1. Determine b, the dead time. This is equal to the *number of v-weights that are equal to zero* (insignificant in a sample) *starting from v_0*. That is, we look for a set of zero-valued v weights ($v_0, v_1, \ldots, v_{b-1}$) in the impulse response function. In Figure 5.6(a), reading the graph from left to right, there are *no* initial zero-valued v weights (because $v_0 \neq 0$), so the dead time is $b = 0$.

RULE 2. Determine r, the order of the denominator operator $\delta(B)$. Choose r as follows:

1. If there is *no decay* pattern at all, but rather a group of spikes followed by a *cutoff to zero*, then $r = 0$.
2. If the v weights show a *simple exponential decay*, either immediately or eventually (perhaps after some v weights that don't decay), then $r = 1$.
3. If the v weights show a *compound exponential decay*, or a *damped sine*

188 BUILDING DYNAMIC REGRESSION MODELS

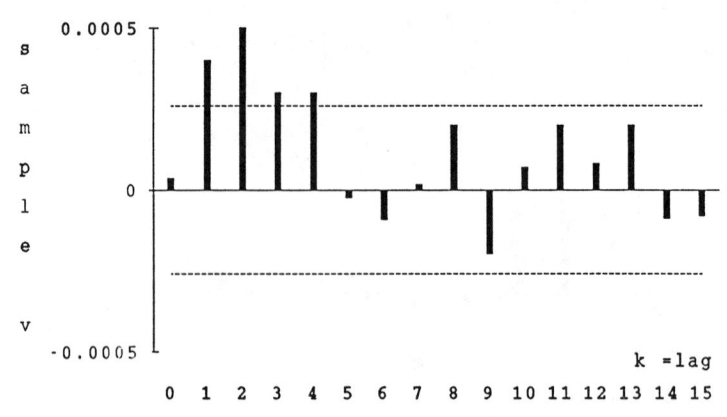

Figure 5.7 Sample impulse response function from Case 5.

wave decay, either immediately or eventually, then $r = 2$. (Experience suggests that models with $r \geq 2$ are rare.

In Figure 5.6(a) the v weights show a simple exponential decay, so the order of $\delta(B)$ is $r = 1$. Therefore there is also one ($r = 1$) decay startup coefficient.

RULE 3. If $r > 0$, determine the decay start-up lags. All decay patterns require one or more start-up values to anchor them. The decay then proceeds from these start-up values. *There are r start-up terms.* Find the largest absolute v weight (denoted max$|v|$) from which the decay proceeds. If $r = 1$, the lag length of this one coefficient is the only start-up lag. If $r = 2$, this lag and the next highest lag are the two start-up lags. (This rule for choosing the decay start-up lags is a practical one that should lead to flexible models. It may sometimes lead to models that are slightly nonparsimonious.) In Figure 5.6(a), $r = 1$ and max$|v| = v_0$. That is, v_0 is the largest absolute v weight from which the decay proceeds. Therefore the decay starts up at lag zero (which is the subscript of v_0).

Finding max$|v|$ in the examples in Figure 5.6 is fairly easy since these are *theoretical* impulse response functions. In practice we must work with *sample* functions, and this can make identification of max$|v|$ more difficult. Consider the estimated impulse response function in Figure 5.7, taken from Case 5. Following rule 3 rigidly would lead us to choose max$|v| = \hat{v}_2$. But if we allow for *sampling variation*, we could easily suppose that parameter v_1 is slightly larger than its estimate $\hat{v}_1$, and that parameter v_2 is slightly smaller than its estimate $\hat{v}_2$. Thus it is reasonable to consider max$|v| = \hat{v}_1$ instead of max$|v| = \hat{v}_2$. This would give a simpler rational form model, with no unpatterned terms. The key point is that rule 3 should not be applied rigidly; its purpose is to help you think clearly so that you can identify just a few reasonable tentative models.

RULE 4. Determine u, the number of unpatterned v weights. (1) If $r > 0$, u is equal to the *number of nonzero (significant in a sample) v weights to the left of the lowest lag decay start-up value*. (2) If $r = 0$, *all nonzero (significant) v weights are unpatterned values*. In Figure 5.6(a), $r = 1 > 0$, and there are *no* nonzero v weights to the left of the decay start-up lag, since v_0 is the leftmost value on the graph. Therefore there are *no* unpatterned v weights, and $u = 0$.

RULE 5. Determine h, the order of the numerator operator $\omega(B)$, as $h = u + r - 1$. [Note that h is the *order* of $\omega(B)$, *not* the number of v weights in $\omega(B)$. The total number of ω weights is $h + 1$, since the first one is always ω_0.] In Figure 5.6(a), $h = u + r - 1 = 0 + 1 - 1 = 0$. Therefore $\omega(B)$ is of zero order with one term, ω_0. For Figure 5.6(a) we have identified a rational transfer function with orders $(b, r, h) = (0, 1, 0)$. This is the Koyck model with $b = 0$ dead time, introduced in Chapter 4.

5.4.2 Examples of Rational Transfer Function Identification Using the Rules

Figure 5.6 shows five additional examples of theoretical impulse response weight graphs, along with their rational forms. We discuss each graph to explain how the five identification rules lead to the rational form functions stated above each graph.

We have already discussed Figure 5.6(a). Consider several questions about this impulse response function: Is ω_0 positive or negative in this example? How do you know? Is δ_1 positive or negative? How do you know? How would the graph be different if each of these coefficients were of opposite sign?

In Figure 5.6(b) there is one initial zero-valued v weight (v_0), so $b = 1$. There is no decay pattern, so $r = 0$ is the order of $\delta(B)$. Therefore *no* ($r = 0$) decay start-up coefficients are present. Thus v_1 and v_2 are both unpatterned terms (neither decay start-up values nor part of a decay pattern), so $u = 2$. Then $h = u + r - 1 = 2 + 0 - 1 = 1$, and $\omega(B) = \omega_0 + \omega_1 B$ has $h + 1 = 1 + 1 = 2$ terms.

In Figure 5.6(c), $b = 0$. (Why?) There is an eventual simple exponential decay pattern, so $r = 1$. There is also *one* decay start-up value (since $r = 1$). The largest absolute v weight from which the decay proceeds is at lag 1, so $\max|v| = v_1$ is the start-up value. (Even if $|v_0|$ were greater than $|v_1|$ in this example, we would still choose $\max|v| = v_1$ because the exponential decay proceeds from v_1, not from v_0.) There is one nonzero v weight to the left of v_1 (at lag 0), so there is one unpatterned spike ($u = 1$). Then $h = u + r - 1 = 1 + 1 - 1 = 1$, and $\omega(B)$ has two terms ($\omega_0 + \omega_1 B$). (How would the rational form transfer function and the impulse response function graph for this example be different for $b = 2$?).

Figure 5.6(d) has three initial zero-valued v weights ($b = 3$). The decay pattern is a compound exponential decay, slower than a simple exponential decay; therefore $r = 2$. Then $\max|v| = v_3$ is the largest absolute v weight from which the decay proceeds, so it is a start-up value. But there are *two* decay start-up values (since $r = 2$), so the next highest-lag v weight (v_4) is also a decay start-up value. There are no nonzero v weights to the left of these two start-up values, so $u = 0$. Then $h = u + r - 1 = 0 + 2 - 1 = 1$, and $\omega(B)$ has two terms ($\omega_0 + \omega_1 B$).

In Figure 5.6(e) the dead time is $b = 1$. (Why?) The damped sine wave decay pattern calls for a second-order denominator ($r = 2$), and there are therefore also two ($r = 2$) decay start-up values. In this case $\max|v| = v_1$, so the decay start-up values are v_1 and v_2. (Why?). There are no unpatterned nonzero v weights to the left of v_1, so $u = 0$. Finally, $h = u + r - 1 = 0 + 2 - 1 = 1$, so $\omega(B)$ is a first-order polynomial with two terms ($\omega_0 + \omega_1 B$).

Rule 3 gives more flexible models in exchange for the estimation of more parameters. (See Box and Jenkins, 1976, Chapter 10). Consider the effect of writing the rational form model for Figure 5.6(e) with $h = 0$: $\omega(B) = \omega_0$. With a decay pattern present ($r > 0$), the last ω weight is necessarily the last decay start-up term. If ω_0 is the only term in $\omega(B)$, it must be the last start-up value (corresponding to v_1, in this case, since $b = 1$). But with $r = 2$, there must be two decay start-up terms. Therefore the first start-up value is forced to be $v_0 = 0$. This constraint gives a less flexible model than setting $h = 1$. Following Rule 3 and setting $h = 1$ in this case will, in practice, allow both start-up values to be estimated from the data.

Figure 5.6(f) is similar to Figure 5.6(e), except the v weight at lag zero is nonzero, so $b = 0$. The damped sine wave decay calls for $r = 2$, and there must also be two ($r = 2$) start-up terms. Since $\max|v| = v_1$, the decay start-up values are v_1 and v_2. There is one nonzero v weight (v_0) to the left of these two decay start-up terms, so $u = 1$. Then $h = u + r - 1 = 1 + 2 - 1 = 2$, and $\omega(B)$ has $h + 1 = 2 + 1 = 3$ terms ($\omega_0 + \omega_1 B + \omega_2 B^2$).

A more parsimonious model for an estimated impulse response function that resembles Figure 5.6(f) could be achieved with $h = 1$, so that $\omega(B) = \omega_0 + \omega_1 B$. Then ω_1 (corresponding to v_1), is the last start-up term, and ω_0 (corresponding to v_0) is the first start-up term. Then ω_0 is not an unpatterned term; instead it is a decay start-up term. Thus, contrary to rule 3, it is acceptable to have the first decay start-up term be something other than ($\max|v|$).

It is difficult in practice to decide if a term like v_0 in Figure 5.6(f) is best treated as an unpatterned term or a decay start-up term. Rule 3 resolves the dilemma in favor of starting the decay from $\max|v|$, with all preceding nonzero v weights treated as unpatterned values. Following rule 3 may at times lead to slightly less parsimonious models, but it tends to give more flexible models. As with other modeling rules of thumb, rule 3 is not unbreakable. In particular, we must be prepared to allow for sampling variation when studying sample v weights. The purpose of identification rules 1 to 5 and these examples is to help you understand the implications of the choices

5.4.3 Application 1

Let's apply the identification rules to the sales and leading indicator data. The sample impulse response function (after differencing) is shown in Table 5.3 and Figure 5.4. There are three initial zero-valued $\hat{v}$ weights (statistically insignificant, with relatively small absolute t values), so $b = 3$. There seems to be an eventual simple exponential decay pattern, so $r = 1$. This also means that there must be one ($r = 1$) decay start-up term. For this example, $\max|v| = \hat{v}_3$ so $\hat{v}_3$ is the decay start-up term. There are no nonzero (statistically significant) $\hat{v}$ weights to the left of $\hat{v}_3$, so there are *no* unpatterned terms ($u = 0$). Then $h = u + r - 1 = 0 + 1 - 1 = 0$, and $\omega(B)$ has one term (ω_0). The identified rational form model then is $\omega_0 B^3/(1 - \delta_1 B)$, which is the Koyck model with $b = 3$. This is similar to Figure 5.6(a), except that the dead time for the present example is $b = 3$ instead of $b = 0$. This is the same model we identified for these data using the visual comparison method at the end of Section 5.3.

5.4.4 Application 2

Consider a similar application of our ideas in this chapter to the valve shipments and valve orders data from Chapter 1, shown in Figures 1.1 and 1.2. Table 1.4 (Example 1) shows the results of estimating the simple regression equation (5.3.14) using these data. There we see an apparently strong contemporaneous relationship between shipments (Y_t) and orders (X_t), reflected in the large absolute t value associated with $\hat{v}_0$. Let's see what happens if we use the LTF method to estimate a free-form distributed lag relationship between Y_t and X_t, with an AR(1) proxy for the disturbance series autocorrelation structure.

The results are shown in Table 5.5. Three features stand out: (1) None of the sample v coefficients is significantly different from zero at the 10% level; (2) the sample ϕ_1 coefficient is significant; and (3) the residual standard deviation ($\hat{\sigma}_N$) is much smaller than the one in Table 1.4. There seems to be much more explanatory power in past values of Y_t (represented by the disturbance AR term) than in current or earlier values of X_t. The results in Table 1.4 are misleading because they ignore the autocorrelation pattern in the disturbance. Once that pattern is taken into account, the explanatory power of the X_t variables is almost nil. This does not mean that there is no relationship between Y_t and X_t at all. It does say that there is little useful forecasting information in current and lagged values of X_t compared to the information in past values of Y_t.

Table 5.5 Estimation Results for Preliminary LTF Model for Shipments and Orders Data, $d = D = 0$

Parameter	Estimate	$\|t \text{ Value}\|$
C	31560.	4.25
v_0	0.1200	1.30
v_1	0.1073	1.17
v_2	0.0375	0.41
v_3	−0.0700	0.78
ϕ_1	0.7943	7.81
$\hat{\sigma}_N = 441.293$		

5.4.5 Application 3

At times we may find a sample impulse response function with several *distinct sets* of unpatterned plus decay start-up v weights; that is, each contiguous set may be separated from each other contiguous set by zero-valued v weights or by decay pattern v weights. An example is found in Koch and Rasche (1988, model C.7).

Functions like this *do not occur often*; when they do, we should consider analyzing each contiguous set of $\hat{v}$ weights *separately*. Consider the impulse response function in Figure 5.8. It may be tempting to identify the rational form for $v(B)$ as

$$v(B) = \frac{\omega_0 + \omega_{10}B^{10} + \omega_{11}B^{11}}{1 - \delta_1 B} \tag{5.4.1}$$

where ω_0 is the decay start-up term for the exponential decay after lag 0; ω_{10} is an unpatterned term; and ω_{11} is the decay start-up term for the exponential decay after lag 11. Doing this, however, forces the two decay

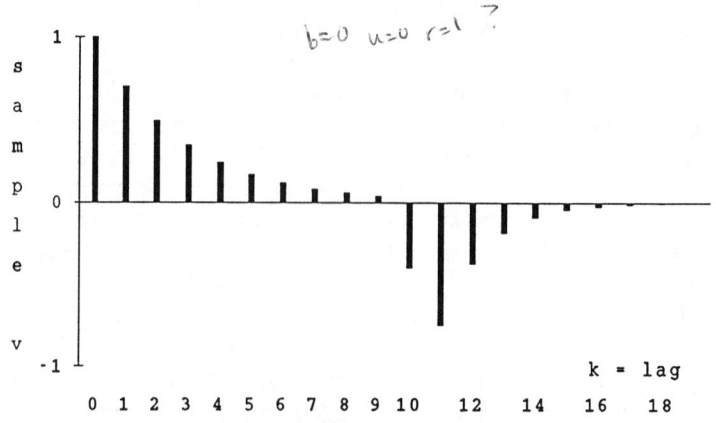

Figure 5.8 Impulse response function with two decay patterns.

patterns to have the same decay rate coefficient (δ_1), which may be undesirable. Instead, we can treat the two decay patterns as separate additive features of the model. Following the identification rules for each decay pattern separately, we identify the transfer function

$$v(B)X_t = \frac{\omega_0}{1 - \delta_1' B} X_t + \frac{\omega_{10} B^{10} + \omega_{11} B^{11}}{1 - \delta_1^* B} X_t \qquad (5.4.2)$$

This form permits different values for δ_1' and δ_1^*; thus we can have two different decay rates, one from a start-up at lag 0 and another from a start-up at lag 11.

QUESTIONS AND PROBLEMS

(You may find it helpful to read Cases 4 and 5 following Chapter 6 before doing the following problems.)

1. In the Data Appendix are two series called U.K. Index of Consumption of Petrochemicals (Y_t, Series 15) and U.K. Index of Industrial Production (X_t, Series 16). The latter series has been seasonally adjusted. Do you find evidence of feedback from Y_t to X_t? What DR model do you identify for these data? Explain.

2. In the Data Appendix are two series called Percent Carbon Dioxide in Output from a Gas Furnace (Y_t, Series 17) and Methane Gas Input in Cubic Feet per Minute (X_t, Series 18). Do you find evidence of feedback from Y_t to X_t? What DR model do you identify for these data? Explain.

3. In the Data Appendix are two series called Capital Expenditures (Y_t, Series 19) and Capital Appropriations (X_t, Series 20). Both series are seasonally adjusted. Since funds must be appropriated before expenditures occur, and since capital construction projects take time to complete, it is expected that current expenditures might respond to past appropriations. Do you find evidence of feedback from Y_t to X_t when the data are in their original metrics? Do you find evidence of feedback from Y_t to X_t when the data are each in the natural log metric? What DR model do you identify for these data? Explain.

4. In the Data Appendix are two series called Lydia Pinkham Sales (Y_t, Series 21) and Lydia Pinkham Advertising (X_t, Series 22). These data have been studied often, with the idea that sales may respond to changes in advertising outlays; we study these data further in Chapter 10. Do you

find evidence of feedback from Y_t to X_t? What DR model do you identify for these data? Explain.

APPENDIX 5A. THE CORNER TABLE

When $r > 0$, it is sometimes difficult to determine the value of r or to decide which v weight lags (if any) should be treated as unpatterned. In this case an identification tool called the *corner table* (C table) can be helpful. The C table also provides information about the order b. The C table suggests how we can approximate polynomials such as $v(B)$, which may be of infinitely high order, by the ratio of two polynomials of finite degree, such as $\omega(B)B^b/\delta(B)$. These are known as Padé approximations. Baker (1975) gives proofs of the properties of the C table. Liu and Hanssens (1982) and Tsay (1985) discuss the use of the C table in identifying the orders of rational form transfer functions.

The C table contains elements denoted $c(f, g)$, where f refers to rows ($f = 0, 1, 2, \ldots$) and g refers to columns ($g = 1, 2, 3, \ldots$). Element $c(f, g)$ in the C table is the *determinant* of a $g \times g$ matrix:

$$c(f, g) = \begin{vmatrix} v_f & v_{f-1} & \cdots & v_{f-g+1} \\ v_{f+1} & v_f & \cdots & v_{f-g+2} \\ \vdots & \vdots & \vdots & \vdots \\ v_{f+g-1} & v_{f+g-2} & \cdots & v_f \end{vmatrix} \quad (5A.1)$$

where the vertical lines ($|\ |$) denote the determinant of the array; $f \geq 0$; $g > 0$; and $v_j = 0$ if $j < 0$. For computational reasons it is helpful to standardize the v weights. To do this, replace v_i in (5A.1) with $v_i/\max|v|$, where $\max|v|$ is the largest absolute v weight in the set $v_1, v_2, \ldots, v_K$.

For *known* b, r, and h the C table has the pattern shown in Table 5A.1. (In practice we use sample v weights to construct the C table; then we tentatively identify the values of b, r, and h from the table.) Table 5A.1 has b *rows of zeros*, followed by $h + 1$ *rows of nonzero values* (denoted by X). Then there is a *rectangle of zeros* with its *upper left corner* in the ($f = b + h + 1$, $g = r + 1$) cell. Taking the values of f and g from this cell, we identify the rational form orders as $r = g - 1$ and $h = f - b - 1$. The remaining C-table elements are either nonzero, *especially at the border of the rectangle of zeros* or unpatterned (denoted by *).

In practice we have only sample v weights, and therefore only sample C-table values from (5A.1). Sample C-table values are subject to sampling variation, and we need their standard errors to test for significance. The finite-sample standard errors of the C-table values are too complicated for practical use. However, we can find asymptotic approximations as shown by

APPENDIX 5A: THE CORNER TABLE

Table 5A.1 C-Table Pattern for Known b, r, and h

	\multicolumn{6}{c}{g}
	1	2	$\cdots$	r	$r+1$	$r+2$	$\cdots$
($f=0$)	0	0	$\cdots$	0	0	0	$\cdots$
$\vdots$	$\vdots$	$\vdots$		$\vdots$	$\vdots$	$\vdots$	
($f=b-1$)	0	0	$\cdots$	0	0	0	$\cdots$
($f=b$)	X	X	$\cdots$	X	X	X	$\cdots$
$\vdots$	$\vdots$	$\vdots$		$\vdots$	$\vdots$	$\vdots$	
($f=b+h$)	X	X	$\cdots$	X	X	X	$\cdots$
($f=b+h+1$)	*	*		X	0	0	$\cdots$
($f=b+h+2$)	*	*		X	0	0	$\cdots$
$\vdots$	$\vdots$	$\vdots$		$\vdots$	$\vdots$	$\vdots$	

Koreisha and Taylor (1985). To find these approximations we need the variance–covariance matrix (**V**) of the $\hat{v}$ weights. This is available as a by-product of estimating the last LTF model (after differencing to achieve a stationary disturbance).

The variance of each $\hat{c}(f, g)$ is a function of the variances and covariances of certain $\hat{v}$ weights. For example, from (5A.1) the variance of $\hat{c}(f, 2)$ is a function of the variances and covariances of $\hat{v}_{f-1}$, $\hat{v}_f$, and $\hat{v}_{f+1}$. A first-order approximate variance of $\hat{c}(f, g)$ may be found using the Taylor series approximation for variances of functions of random variables; see, for example, Kmenta (1986, p. 486). The result is

$$\text{var}[\hat{c}(f, g)] = \mathbf{F}'(f, g)\mathbf{M}(f, g)\mathbf{F}(f, g) \tag{5A.2}$$

where $\mathbf{M}(f, g)$ is the variance–covariance matrix of $\mathbf{S}(f, g) = (\hat{v}_{f-g+1}, \hat{v}_{f-g+2}, \ldots, \hat{v}_{f+g-1})'$, with $\hat{v}_i = 0$ if $i < 0$. The elements of **M** are obtained from **V** (with element $m_{ij} = 0$ if the corresponding $\hat{v}_k = 0$ due to $k < 0$.) $\mathbf{F}(f, g)$ is defined in the same way as $\mathbf{S}(f, g)$, but with $\hat{v}_i$ in $\mathbf{S}(f, g)$ replaced by the sum of its cofactors in the determinantal array used to find $\hat{c}(f, g)$. [The sum of these cofactors is the partial derivative of $\hat{c}(f, g)$ with respect to $\hat{v}_i$.]

For example, consider $\hat{c}(0, 2)$. From (5A.1) this is

$$\hat{c}(0, 2) = \begin{vmatrix} \hat{v}_0 & 0 \\ \hat{v}_1 & \hat{v}_0 \end{vmatrix} = \hat{v}_0^2 - (0)\hat{v}_1 = \hat{v}_0^2 \tag{5A.3}$$

Then $\mathbf{S}(f, g) = \mathbf{S}(0, 2) = (\hat{v}_{0-2+1}, \hat{v}_{0-2+2}, v_{0+2-1})' = (0, \hat{v}_0, \hat{v}_1)'$, and

$$\mathbf{M}(0, 2) = \begin{bmatrix} 0 & 0 & 0 \\ 0 & \text{var}(\hat{v}_0) & \text{cov}(\hat{v}_0, \hat{v}_1) \\ 0 & \text{cov}(\hat{v}_1, \hat{v}_0) & \text{var}(\hat{v}_1) \end{bmatrix}$$

Finally, $\mathbf{F}(f, g) = \mathbf{F}(0, 2) = (-\hat{v}_1, 2\hat{v}_0, 0)'$ is the set of sums of cofactors of the elements in $\mathbf{S}(0, 2)$ computed from the determinantal array used to find $\hat{c}(0, 2)$ in (5A.3). In this case there are three sums of cofactors since there are three elements in $\mathbf{S}(0, 2)$. For example, the first element in $\mathbf{F}(0, 2)$, equal to $-\hat{v}_1$, is the sum of the cofactors of zero [the first element in $\mathbf{S}(0, 2)$], where the cofactors are computed from the determinantal array defining $\hat{c}(0, 2)$ in (5A.3). Other elements in $\mathbf{F}(0, 2)$ are found similarly.

The approximate standard deviations of the C-table elements are the square roots of the variances in (5A.2). We use these standard errors to look for a rectangle of insignificant values (small relative to their standard errors) in the lower right part of the C table. As a rough rule of thumb, $\hat{c}(f, g)$ is significant if it is two or more times its standard error in absolute value.

Let's use the C table to identify a rational transfer function for the sales and leading indicator data. Initial estimation results for these data after differencing show a decay pattern in the $\hat{v}$ weights (Figure 5.4). This suggests that $r > 0$, and the C table may therefore be useful. The C table for the $\hat{v}$ weights in Figure 5.4 is shown in Table 5A.2. The top table shows the computed values; the bottom table simplifies the top table by putting an "X" in each cell where the absolute value of the C-table element is two or more times its standard error. The C-table values (though not their standard errors) are computed automatically by the SCA System using appropriate commands.

In this example there are three rows of insignificant values, so $b = 3$. There is a rectangle of insignificant values in the lower right of the table, with its upper left corner in the ($f = 4$, $g = 2$) position. Therefore $r = g - 1 = 2 - 1 = 1$, and $h = f - b - 1 = 4 - 3 - 1 = 0$. Therefore we identify $(b, r, h) = (3, 1, 0)$ as the tentative orders of a rational form transfer function for these data. This is the same result we found in Chapter 5 using identification rules 1 to 5. This example is unusually clear. Often the C table evidence is more ambiguous.

APPENDIX 5B. TRANSFER FUNCTION IDENTIFICATION USING PREWHITENING AND CROSS CORRELATIONS

In this appendix we present the identification method emphasized by Box and Jenkins (1976); it involves a procedure called *prewhitening* and *cross correlation functions* (CCF). Since this method is mentioned often in the literature on DR models, we give an explanation for the reader who wants

APPENDIX 5B: PREWHITENING AND CROSS CORRELATIONS

Table 5A.2 Corner Table for the Impulses Response Weights in Table 5.3[a]

	\[g\] 1	2	3	4	5	6	7
(f = 0)	−0.02	0.00	0.00	0.00	0.00	0.00	0.00
	(.05)	(.00)	(.00)	(.00)	(.00)	(.00)	(.00)
(f = 1)	−0.02	0.00	0.00	0.00	0.00	0.00	0.00
	(.05)	(.00)	(.00)	(.00)	(.00)	(.00)	(.00)
(f = 2)	0.00	0.02	−0.02	0.00	0.00	0.00	0.00
	(.05)	(.05)	(.05)	(.00)	(.00)	(.00)	(.00)
(f = 3)	1.00	1.00	0.99	0.99	0.99	0.98	0.98
	(.05)	(.10)	(.16)	(.21)	(.27)	(.33)	(.38)
(f = 4)	0.74	0.05	0.03	0.02	0.06	0.05	0.01
	(.05)	(.09)	(.09)	(.09)	(.09)	(.09)	(.09)
(f = 5)	0.50	−0.02	0.00	0.00	0.00	0.00	0.00
	(.05)	(.06)	(.00)	(.00)	(.01)	(.01)	(.00)
(f = 6)	0.37	0.01	0.00	0.00	0.00	0.00	0.00
	(.05)	(.04)	(.00)	(.00)	(.00)	(.00)	(.00)
(f = 7)	0.25	−0.02	0.00	0.00	0.00	0.00	0.00
	(.05)	(.03)	(.00)	(.00)	(.00)	(.00)	(.00)

Simplified Corner Table[b]

	\[g\] 1	2	3	4	5	6	7
(f = 0)	0	0	0	0	0	0	0
(f = 1)	0	0	0	0	0	0	0
(f = 2)	0	0	0	0	0	0	0
(f = 3)	X	X	X	X	X	X	X
(f = 4)	X	0	0	0	0	0	0
(f = 5)	X	0	0	0	0	0	0
(f = 6)	X	0	0	0	0	0	0
(f = 7)	X	0	0	0	0	0	0

[a]Standard errors in parentheses.
[b]X = absolute value more than 2 times standard error.

a more complete background in DR modeling. But we do not emphasize this approach because it is difficult to apply with more than one input. By contrast, the LTF method discussed in Chapter 5 can more easily handle multiple inputs.

Consider the case of a single-input DR model. It is assumed that both X_t and Y_t have been transformed to induce constant variances. If differencing is needed to induce stationary means, we consider the differenced series x_t and y_t, with constant means μ_x and μ_y, respectively. Consider the following DR model for y_t and ARIMA model for x_t:

$$y_t = \frac{\omega(B)B^b}{\delta(B)} x_t + \frac{\theta(B)}{\phi(B)} a_t$$

$$x_t = \frac{\theta_x(B)}{\phi_x(B)} \alpha_t \tag{5B.1}$$

where a_t and α_t are each a sequence of zero-mean and normally distributed white-noise random shocks, uncorrelated with each other. For simplicity we assume that the constant term in the DR model is zero and that no seasonal elements appear in either the ARIMA model for x_t or the ARIMA model for the disturbance series in the DR model; the results we show here can be extended to include a constant term and seasonal series. First, we discuss cross correlations; then we consider the idea of prewhitening.

Cross Correlations

The main tool in the prewhitening identification method is the cross correlation function (CCF), which consists of a set of cross correlation coefficients. The population CCF for variables x_t and y_t is defined as

$$\rho_{xy}(k) = \gamma_{xy}(k)/\sigma_x \sigma_y \qquad k = 0, \pm 1, \pm 2, \ldots \tag{5B.2}$$

where $\gamma_{xy}(k)$ is the population cross covariance between x_t and y_t at lag k, and σ_x and σ_y are population standard deviations of x_t and y_t, respectively. Thus a cross correlation coefficient is a standardized cross covariance. The cross covariance is defined as the following expected value:

$$\gamma_{xy}(k) = E[(x_t - \mu_x)(y_{t+k} - \mu_y)] \qquad k = 0, \pm 1, \pm 2, \ldots \tag{5B.3}$$

Sample Cross Correlations

Suppose we have n observations after differencing the series. An estimate, denoted $c_{xy}(k)$ of the cross covariance in (5B.3), is given by

$$c_{xy}(k) = \begin{cases} n^{-1} \sum_{t=1}^{n-k} (x_t - \bar{x})(y_{t+k} - \bar{y}) & k = 0, 1, 2, \ldots \\ n^{-1} \sum_{t=1}^{n+k} (y_t - \bar{y})(x_{t-k} - \bar{x}) & k = 0, -1, -2, \ldots \end{cases} \tag{5B.4}$$

where $\bar{x}$ and $\bar{y}$ are the sample means of the two series. The statistics $s_x = [c_{xx}(0)]^{1/2}$ and $s_y = [c_{yy}(0)]^{1/2}$ are sample estimates of σ_x and σ_y, respectively. Substitute these estimates, and the values $c_{xy}(k)$ from (5B.4), into (5B.2) to obtain an estimate of $\rho_{xy}(k)$, denoted $r_{xy}(k)$:

APPENDIX 5B: PREWHITENING AND CROSS CORRELATION

$$r_{xy}(k) = c_{xy}(k)/s_x s_y \qquad k = 0, \pm 1, \pm 2, \ldots \tag{5B.5}$$

Since the $r_{xy}(k)$ values are sample statistics, we need their standard errors to decide if any of them differ significantly from zero. From Bartlett (1946, 1966) as discussed by Box and Jenkins (1976, pp. 376–377), if x_t and y_t are *not cross correlated, and if one of the series is white noise, and using the normality assumption*, then approximately

$$\text{var}[r_{xy}(k)] = (n - k)^{-1} \tag{5B.6}$$

Prewhitening

The ARIMA process for the stationary input is shown in (5B.1). Rewriting this process, we may think of the AR and MA operators as a filter that, when applied to x_t, produces an uncorrelated "residual" series:

$$\alpha_t = \theta_x^{-1}(B)\phi_x(B)x_t \tag{5B.7}$$

The series α_t (in practice $\hat{\alpha}_t$) is called the *prewhitened* x_t series. Now suppose we apply the *same filter* to y_t; this will produce another "residual" series:

$$e_t = \theta_x^{-1}(B)\phi_x(B)y_t \tag{5B.8}$$

Now use (5B.1) to substitute into (5B.8) the DR expression for y_t and the ARIMA expression for x_t to obtain

$$e_t = v(B)\alpha_t + u_t \tag{5B.9}$$

where $v(B) = \omega(B)B^b/\delta(B)$ and $u_t = \theta_x^{-1}(B)\phi_x(B)\phi^{-1}(B)\theta(B)a_t$. Now multiply both sides of (5B.9) by α_{t-k} and take expected values to get the cross covariance between α_t and e_t. Divide the result by σ_e and σ_α to get the cross correlation between α_t and e_t at lag $+k$:

$$\rho_{\alpha e}(k) = v_k \sigma_\alpha / \sigma_e \qquad k = 0, 1, 2, \ldots \tag{5B.10}$$

Therefore, from (5B.10) we find

$$v_k = \rho_{\alpha e}(k)\sigma_e/\sigma_\alpha \qquad k = 0, 1, 2, \ldots \tag{5B.11}$$

Equation (5B.11) shows that if we prewhiten the input, and apply the same filter to the output, then the v weights are *proportional to the cross correlations of the residuals from these two filtering procedures*. Therefore we may study the sample CCF, or obtain estimates of the v weights from the CCF using (5B.11), to identify a transfer function for the relationship between Y_t and X_t. In practice we don't know the parameters on the right side of

Equation (5B.11). Instead we substitute estimates of these parameters obtained from the data to arrive at the following estimated v weights:

$$\hat{v}_k = r_{\alpha e}(k)\hat{\sigma}_e/\hat{\sigma}_\alpha \qquad k = 0, 1, 2, \ldots \qquad (5B.12)$$

Application

Let's apply the ideas presented in this appendix to the sales and leading indicator data. In this case no transformations are needed to induce constant variances since the variances of both Y_t (sales) and X_t (leading indicator) appear to be stationary already. Next, we difference each series as needed to achieve a stationary mean. The differenced input and output series are denoted x_t and y_t, respectively. For each series nonseasonal first differencing ($d = 1$) appears to induce a constant mean.

Next, we find an ARIMA model for the stationary input series x_t. The usual identification, estimation, and checking steps give this ARMA(0, 1) for x_t based on the first $n = 140$ observations:

$$x_t = (1 - 0.4483B)\hat{\alpha}_t \qquad (5B.13)$$

with $\hat{\sigma}_\alpha = 0.2835$. We can use (5B.13) to compute the estimated residuals $\hat{\alpha}_t$. Then we filter the y_t series with the ARMA(0, 1) model for x_t in (5B.13) and obtain the values $\hat{e}_t$ from $y_t = (1 - 0.4483B)\hat{e}_t$. The sample CCF for $\hat{\alpha}$ and $\hat{e}$, computed using the SCA System, is shown in Figure 5B.1. The left side of this CCF ($k = -1, -2, \ldots$) provides information about feedback from earlier values of Y_t to X_t; in this case there is no evidence of feedback. The right side of this CCF ($k = 0, 1, 2, \ldots$) provides information about how Y_t is related to current and earlier values of X_t.

The sample CCF coefficients may now be used along with the sample standard deviations $\hat{\sigma}_e = 1.99$ and $\hat{\sigma}_\alpha = 0.2835$ to find the $\hat{v}$ weights using

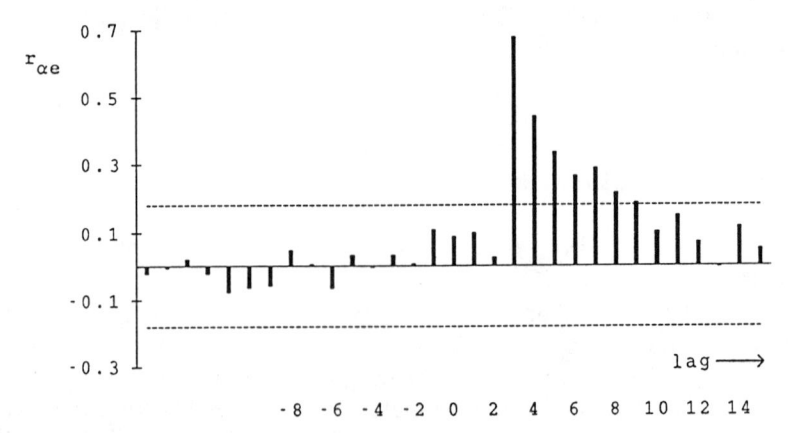

Figure 5B.1 Sample cross correlation function for the sales and leading indicator data.

APPENDIX 5B: PREWHITENING AND CROSS CORRELATIONS

Table 5B.1 Sample v Weights from the Prewhitening Method Applied to the Sales and Leading Indicator Data

k	0	1	2	3	4	5	6	7
$\hat{v}_k$	0.602	0.698	0.171	4.743	3.104	2.352	1.850	2.012
k	8	9	10	11	12	13	14	15
$\hat{v}_k$	1.507	1.306	0.707	1.046	0.489	−0.031	0.811	0.348

(5B.12). The $\hat{v}$ weights (for $k = 0, 1, 2, \ldots$) are shown in Table 5B.1. Their sizes and pattern are similar to those for the $\hat{v}$ weights found using the LTF method in Chapter 5 (Table 5.3).

Using (5B.6) we may compare the sample CCF values with their standard errors to check the significance of the corresponding $\hat{v}$ weights. In this case the CCF standard errors are all about 0.09; the dashed lines in Figure 5B.1 are approximate two standard error limits. Applying the identification rules from Chapter 5 to the sample CCF in Figure 5B.1 (or to the $\hat{v}$ weights in Table 5B.1) we identify the orders of the transfer function as $(b, r, h) = (3, 1, 0)$; this is the same model we identified in Chapter 5 using the LTF method and in Appendix 5A using the C table.

CHAPTER 6

Building Dynamic Regression Models: Model Checking, Reformulation, and Evaluation

6.1 DIAGNOSTIC CHECKING AND MODEL REFORMULATION

At the identification stage we tentatively specify a rational form transfer function model of orders (b, r, h) and a disturbance series autoregressive integrated moving average (ARIMA) model of orders $(p, d, q)(P, D, Q)_s$. For example, for the sales and leading indicator data presented in Chapter 1 and discussed in Chapters 4 and 5, we tentatively identified the following DR model:

$$\nabla Y_t = C + \frac{\omega_0 B^3}{1 - \delta_1 B} \nabla X_t + (1 - \theta_1 B)a_t \qquad (6.1.1)$$

The tentative transfer function model has orders $(b, r, h) = (3, 1, 0)$. The tentative disturbance series ARIMA model is an ARIMA(0, 1, 1): $\nabla N_t = (1 - \theta_1 B)a_t$.

At stage 2 of our modeling strategy we *estimate* the parameters of the identified DR model using the available data. The details of estimation are discussed in Chapter 9. Estimation of model (6.1.1) for the sales and leading indicator data, using the SCA System and the first 140 observations, gives these (rounded) results, with absolute t values in parentheses:

$$\nabla Y_t = 0.033 + \frac{4.716 B^3}{(90.5)} \nabla X_t + (1 - 0.620 B)\hat{a}_t \qquad (6.1.2)$$
$$ (3.9) \quad 1 - 0.725 B (9.0)$$
$$ (200.3)$$

with residual standard error $\hat{\sigma}_a = 0.217622$. In addition, an ARIMA model for X_t is this ARIMA(0, 1, 1):

$$\nabla X_t = (1 - 0.4483B)\hat{\alpha}_t$$

with residual standard deviation $\hat{\sigma}_\alpha = 0.287$.

In this section we discuss several *diagnostic checks* that help us to decide if a DR model is adequate. If it is not, we attempt to reformulate the model; that is, we tentatively identify another DR model, estimate it, and check it. The discussion of model checking in this section complements the material in Section 6.2 on evaluating estimation stage results. We will use model (6.1.2) to illustrate the ideas.

The procedures presented here are useful in checking on the statistical adequacy of a DR model. But keep in mind that these procedures cannot ensure that we have chosen appropriate inputs or that we have included all appropriate inputs. On these matters we must rely on the underlying theory (economic, biological, psychological, and so forth) regarding the behavior of the output.

6.1.1 Residual Cross Correlations

A chief tool used at the checking stage is the *residual cross correlation function* (residual CCF). The residual CCF gives the relationship between the *residuals from the DR model* ($\hat{a}_t$) and the *current and time-lagged residuals from the ARIMA model for the input series* ($\hat{\alpha}_t, \hat{\alpha}_{t-1}, \hat{\alpha}_{t-2}, \ldots$).

For example, suppose that a correct transfer function is $(\omega_0 + \omega_1 B + \omega_2 B^2)X_t$, but instead we estimate $(\omega_0 + \omega_1 B)X_t$. Since $\omega_2 B^2 X_t$ is incorrectly excluded from the model, then part of the variation in Y_t that is related to the input two periods earlier (X_{t-2}) is likely to be *unexplained* by the estimated transfer function. Instead, this unexplained part of Y_t is likely to appear in the DR model residual series. We might expect to see a correlation between the DR model residual series ($\hat{a}_t$) and the residual series from the X_t ARIMA model two periods earlier ($\hat{\alpha}_{t-2}$). Thus the residual CCF might warn us of an inadequate transfer function and provide clues about how the transfer function may be improved.

The residual CCF is computed as follows:

$$r_k^* = \frac{c_{\alpha a}(k)}{\hat{\sigma}_\alpha \hat{\sigma}_a} \qquad k = 1, 2, 3, \ldots \qquad (6.1.3)$$

where $\hat{\sigma}_\alpha$ and $\hat{\sigma}_a$ are the sample residual standard deviations for $\hat{\alpha}_t$ and $\hat{a}_t$, respectively, and $c_{\alpha a}$ is the sample covariance of these two series, computed as

204 BUILDING DYNAMIC REGRESSION MODELS

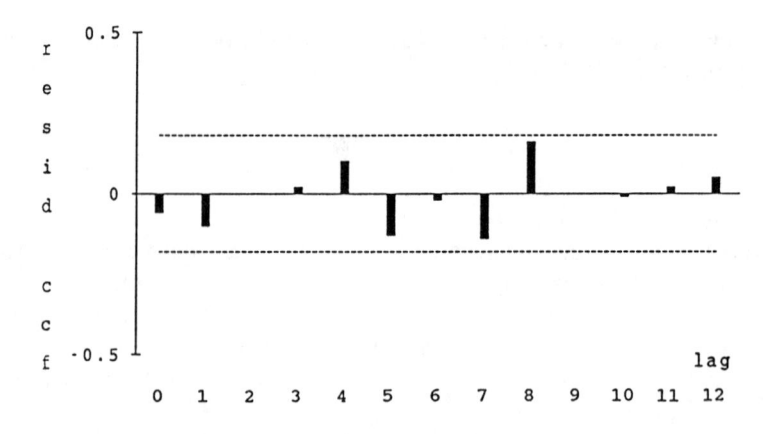

```
r
e
s
i
d      0
 
c
c
f    -0.5                                              lag
        0  1  2  3  4  5  6  7  8  9  10 11 12
```

Chi-squared (S*) = 12.18 for df = 11

k	0	1	2	3	4	5	6	7	8	9	10	11	12
r_k^*	-.06	-.10	.00	.02	.10	-.13	-.02	-.14	.16	-.00	-.01	.02	.05

Figure 6.1 Residual CCF for model (6.1.2).

$$c_{\alpha a}(k) = n^{-1} \sum_{t=1}^{n-k} (\hat{\alpha}_t - \bar{\alpha})(\hat{a}_{t+k} - \bar{a}) \qquad k = 1, 2, 3, \ldots \qquad (6.1.4)$$

where the overbar denotes sample mean.

Each residual CCF coefficient is only an estimate of the corresponding parameter (ρ_k^*) and is subject to sampling variation. To decide if a residual CCF coefficient differs significantly from zero, we may test the null hypothesis

$$H_0: \rho_k^* = 0 \quad \text{for } k = 1, 2, 3, \ldots, K$$

against the alternate $H_a: \rho_k^* \neq 0$. If α_t and a_t are uncorrelated and normally distributed, and if one of these two series is white noise, then r_k^* has the following *approximate* standard error, as discussed by Box and Jenkins (1976, Chapter 11):

$$s(r_k^*) = n^{-1/2} \qquad (6.1.5)$$

where n is the smaller of the number of observations available for $\hat{\alpha}_t$ or $\hat{a}_t$. [The expression in (6.1.5) tends to *overestimate* standard errors at the shorter lags, say $k = 0, 1, 2, 3$.]

Figure 6.1 shows the residual CCF for the sales and leading indicator DR model in (6.1.2). The dashed lines are approximate two standard error limits. None of these estimated coefficients is more than twice its standard error in absolute value. Thus we do not reject H_0 for any $k = 1, 2, 3, \ldots, K$, where

DIAGNOSTIC CHECKING AND MODEL REFORMULATION

$K = 12$. According to this check, model (6.1.2) has adequately captured the distributed lag response of Y_t to X_t.

Overall Test

Another useful statistic involves a test on all K residual CCF coefficients *as a set*. Consider the *joint null hypothesis*

$$H_0: \rho_0^* = \rho_1^* = \cdots = \rho_K^* = 0$$

where H_a is that *not all* of these ρ^* values are zero. Ljung and Box (1978) suggest the test statistic

$$S^* = n^2 \sum_{k=0}^{K} (n-k)^{-1}(r_k^*)^2 \qquad (6.1.6)$$

where n is the smaller of the number of observations available for either $\hat{\alpha}_t$ or $\hat{a}_t$. Under the stated H_0, S^* is approximately χ^2 distributed with $K + 1 - m$ degrees of freedom, where m is the number of parameters estimated in the *transfer function* part of the DR model. Note that m does not include parameters in the disturbance ARIMA model. If all population cross correlations between α_t and a_t are zero, S^* should be small as measured on the appropriate χ^2 distribution.

For the sales and leading indicator model (6.1.2), we have $n = 130$. K in the residual CCF in Figure 6.1 is 12. Using the residual CCF value in Figure 6.1, the test statistic is

$$S^* = (130)^2[(\tfrac{1}{130})(-0.06)^2 + (\tfrac{1}{129})(-0.1)^2 + \cdots + (\tfrac{1}{118})(0.05)^2]$$
$$= 12.18$$

This value is less than the χ^2 critical value (19.68) for $K + 1 - m = 12 + 1 - 2 = 11$ degrees of freedom at the 5% level. Therefore we do not reject the stated H_0; the approximate χ^2 test does not suggest inadequacy in the identified transfer function.

Model Reformulation

It can be difficult to reformulate a transfer function model directly from residual CCF information. Suppose, for example, that the correct transfer function is $(\omega_0 + \omega_1 B)X_t$, but we estimate instead $\omega_0 X_t$. In practice the residual CCF may show a significant value at lag 0, or at lag 1, or both; it may even show a decay pattern. The following guidelines may help you in reformulating an apparently inadequate transfer function:

1. Use the residual CCF primarily to indicate possible model inadequacy, and less as a guide to choosing alternative models.

2. If the residual CCF suggests an inadequate transfer function, return to the sample impulse response function [obtained from the linear transfer function (LTF) identification procedure] to consider alternative models.

3. If the apparently inadequate transfer function still seems reasonable after reconsideration of the LTF sample impulse response function, try adding just one term at a time to the transfer function. Usually, adding terms to the numerator operator $\omega(B)$ will be more productive than adding terms to the denominator operator $\delta(B)$.

6.1.2 Autocorrelation Check

We also check the adequacy of the ARIMA model for the disturbance series in the DR model by examining the residual sample autocorrelation function (SACF) of series $\hat{a}_t$. We may also examine the sample partial autocorrelation function (SPACF) and extended SACF (ESACF) of this series for added information. Using the SACF, we test the null hypothesis

$$H_0: \rho_k(a) = 0 \quad \text{for } k = 1, 2, 3, \ldots, K$$

where $\rho_k(a)$ is the population autocorrelation coefficient of series a_t at lag k. The alternate hypothesis is H_a: $\rho_k(a) \neq 0$. The test statistic is the sample residual autocorrelation coefficient $r_k(\hat{a})$ for $k = 1, 2, 3, \ldots, K$.

Note an important point: A wrong *transfer function* model will also tend to produce significant *residual autocorrelations, even if the disturbance ARIMA model is correct*. Box and Jenkins (1976, pp. 392–393) discuss this phenomenon. Therefore, if *both* the residual CCF and the residual SACF indicate model inadequacy, it is good practice to *start* by *reformulating the transfer function* part of the DR model. This reformulation may also produce an uncorrelated residual series ($\hat{a}_t$) with no change in the disturbance ARIMA model specification.

Figure 6.2 shows the residual SACF for the sales and leading indicator model (6.1.2). The dashed lines are approximate two standard error limits; the standard errors are computed using Equation (2.3.3) in Chapter 2. None of the residual autocorrelations is large compared to its standard error. Therefore the ARIMA model for the disturbance series seems adequate.

Overall Test

There is an approximate χ^2 statistic to test the *joint null hypothesis*

$$H_0: \rho_1(a) = \rho_2(a) = \cdots = \rho_K(a) = 0$$

The alternate hypothesis is that not all of these ρ's are zero. We introduced this statistic in Chapter 2, Equation (2.5.2). This test is useful since there

DIAGNOSTIC CHECKING AND MODEL REFORMULATION

Figure 6.2 Residual SACF for model (6.1.2).

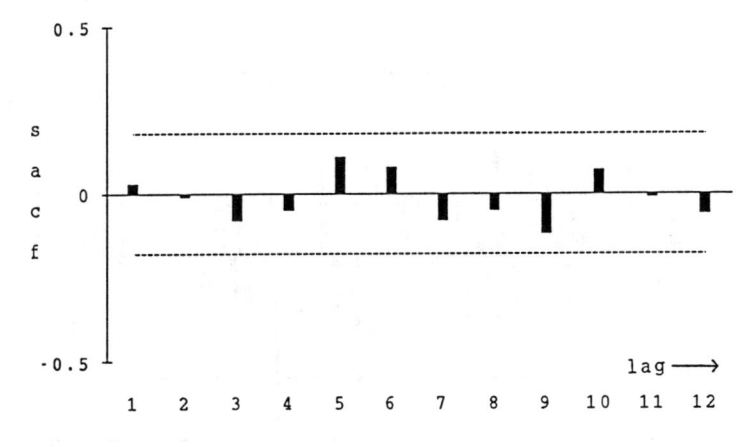

Chi-squared (Q^*) = 8.3 for df = 11

may be a pattern in the residual autocorrelations as a set even though individually each coefficient is fairly small compared to its standard error. Further, the usual formula for the standard errors of the residual autocorrelations [Equation (2.3.3)] tends to overstate the standard errors at shorter lags (e.g., lags 1, 2, and 3). Thus an overall test may be useful in detecting an inadequate disturbance ARIMA model.

The test statistic proposed by Ljung and Box (1978) is

$$Q^* = n(n+2) \sum_{k=1}^{K} (n-k)^{-1} r_k^2(\hat{a}) \qquad (6.1.7)$$

Under the null hypothesis, Q^* is approximately χ^2 distributed with $K - m$ degrees of freedom, where m is the total number of parameters estimated *in the disturbance ARIMA model*. Note that the number of estimated transfer function parameters is not included in m. For the sales and leading indicator model (6.1.2), the residual SACF in Figure 6.2 shows $Q^* = 8.3$ for $K - m = 12 - 1 = 11$ degrees of freedom. Since this is not significant at the 5% level, this test does not reveal inadequacy in the disturbance ARIMA model. Considering both the individual residual autocorrelations relative to their standard errors, and the Q^* statistic, the ARIMA model for the disturbance series in model (6.1.2) seems adequate.

6.1.3 Check for Outliers

It is helpful to check a plot of the standardized DR model residuals for outliers (unusual values). Standardized residuals are found by dividing each residual $\hat{a}_t$ by the residual standard deviation ($\hat{\sigma}_a$). Absolute standardized

std res

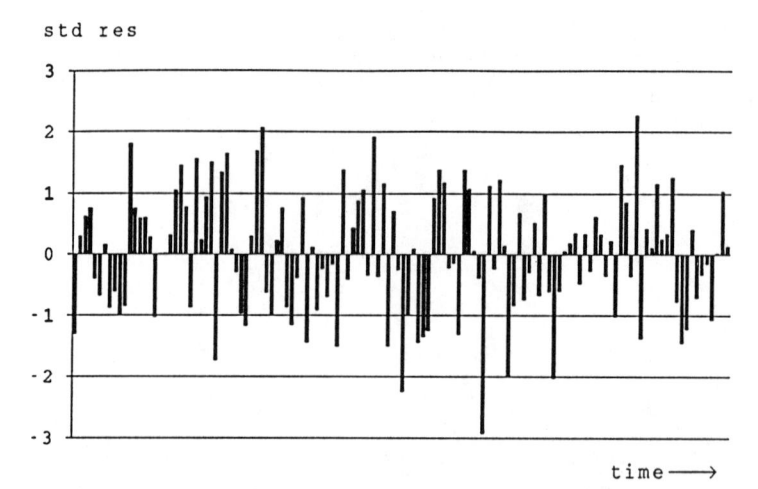

Figure 6.3 Standardized residuals from model (6.1.2).

residuals greater than some number such as 3.0, 3.5, or 4.0 may be considered important. Of course, some residuals may be large just by chance.

If we can associate an outlier with a known external event, then the DR model may be modified to include a binary (0, 1) input variable to represent the event. This is an exercise in intervention analysis and is discussed further in Chapter 7. Chapter 8 also presents a procedure for identifying outliers and the types of intervention models that may be called for.

Figure 6.3 is a plot of the standardized residuals $(\hat{a}_t/\hat{\sigma}_a)$ for DR model (6.1.2). Only four or five of these values are more than two standard deviations from zero, and none falls more than three standard deviations from zero. Thus, informal inspection of this plot suggests that there are no large outliers. The formal outlier detection procedure in Chapter 8 might produce different conclusions.

6.1.4 Normality Check

We may also check the DR model residuals to see if they are approximately normally distributed. The validity of our inferences about the estimated coefficients in a DR model rests on the assumption that the shocks (a_t) are normally distributed. It can be difficult to diagnose nonnormality of the shocks by studying model residuals, as discussed by Weisberg (1980, Section 6.4). Thus the check for normality is often based on informal methods.

An informal check may be performed by examining a *histogram of the residuals* for approximate normality. In addition, we can examine a *normal probability plot*. That is, the ordered standardized residuals are plotted against theoretical normal values from rank-ordered standard normal data. If the residuals are approximately normally distributed, then the normal probability plot is approximately a straight line. See Weisberg (1980) and

```
LOWER      UPPER    FREQ-  0   10   20   30   40   50
BOUND      BOUND    UENCY  +----+----+----+----+----+
-3.000  -  -2.000      4  I>>
-2.000  -  -1.000     20  I>>>>>>>>>>
-1.000  -   0.000     37  I>>>>>>>>>>>>>>>>>>>
 0.000  -   1.000     45  I>>>>>>>>>>>>>>>>>>>>>>>
 1.000  -   2.000     22  I>>>>>>>>>>>
 2.000  -   3.000      2  I>
                          +----+----+----+----+----+
                          0   10   20   30   40   50
```

Figure 6.4 Histogram of standardized residuals from model (6.1.2).

Liu and Hudak (1986) for an introductory discussion of the normal probability plot.

Figure 6.4 shows the histogram of the standardized residuals for DR model (6.1.2); Figure 6.5 shows the normal probability plot of the residuals. The assumption of normally distributed shocks seems reasonable since the residual histogram is at least roughly symmetrical, and the normal probability plot is nearly a straight line.

6.2 EVALUATING ESTIMATION STAGE RESULTS

Estimation stage results provide information that helps us to evaluate a DR model and determine how it might be improved. The discussion in this section complements the discussion of diagnostic checking in Section 6.1, since the purpose of diagnostic checking is also to help us evaluate and possibly improve a model.

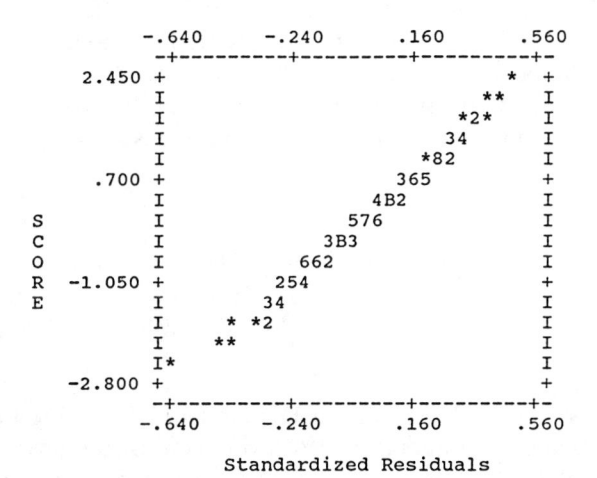

```
                 -.640      -.240       .160       .560
               -+----------+----------+----------+-
        2.450 +                                *   +
              I                               **   I
              I                              *2*   I
              I                              34    I
              I                             *82    I
         .700 +                            365     +
              I                           4B2      I
        S     I                         576        I
        C     I                        3B3         I
        O     I                       662          I
        R   -1.050 +                  254          +
        E     I                         34         I
              I           *  *2                    I
              I          **                        I
              I*                                   I
       -2.800 +                                    +
               -+----------+----------+----------+-
                 -.640      -.240       .160       .560

                        Standardized Residuals
```

Figure 6.5 Normal probability plot of residuals from model (6.1.2).

6.2.1 Parsimony

A good model is parsimonious. At the estimation stage we are especially alert for ways to simplify the model (reduce the number of estimated coefficients) in ways that do not impair the model's ability to adequately describe the data. Parsimony is related to the matter of redundant coefficients, discussed in Section 6.2.5.

The possible relationship in model (6.1.2) between sales and the leading indicator is somewhat complicated. It involves the following:

1. Two nonstationary series.
2. A nonzero dead time ($b = 3$) in the response of Y_t to changes in X_t.
3. After the dead time, a lengthy distributed lag response of Y_t to changes in X_t.
4. A deterministic trend element (the constant term).
5. A time-structured disturbance series.

Model (6.1.2) seems to capture this complicated relationship in a fairly parsimonious manner as it involves only four coefficients.

6.2.2 Coefficient Significance

Ideally, estimated DR model coefficients are statistically significant. A standard procedure is to test the null hypothesis H_0: coefficient parameter $= 0$, with alternate hypothesis H_a: coefficient parameter $\neq 0$. If the random shocks a_t are normally distributed, then the DR model coefficient estimates are treated as approximately t distributed in repeated sampling.

A rule of thumb with a sufficiently large sample (say, $n > 50$) is: Reject the null hypothesis if the approximate t value found by dividing the absolute value of a coefficient estimate by its standard error is 2.0 or larger. This rule *should not be followed rigidly* since:

1. In general the t values are only approximate.
2. A term in the model may provide a useful interpretation for the model even when its coefficient has an absolute t value less than 2.0.

For the sales and leading indicator DR model (6.1.2), all estimated coefficients are significant at about the 5% level. That is, the absolute value of each coefficient is more than twice its standard error. The approximate t values are as follows:

EVALUATING ESTIMATION STAGE RESULTS

Coefficient	t Value
$\hat{C}$	$0.033/0.008 = 3.9$
$\hat{\omega}$	$4.716/0.052 = 90.5$
$\hat{\delta}_1$	$0.725/0.004 = 200.3$
$\hat{\theta}_1$	$0.620/0.069 = 9.0$

You may get slightly different results due to rounding if you compute these t values by hand from the reported estimation results.

6.2.3 Stability, Stationarity, and Invertibility

Ideally the DR model will show that the output responds in a stable manner to changes in the inputs. As discussed in Chapter 4, this means that all roots of each characteristic equation $\hat{\delta}_i(B) = 0$ should lie outside the unit circle; here B is treated as an ordinary algebraic variable.

For $r = 1$, the stability requirement amounts to the condition $|\hat{\delta}_1| < 1.0$. For the sales and leading indicator example in (6.1.2), we have $\hat{\delta}_1 = 0.725$, which satisfies the stability condition. Therefore a finite change in input X_t, which persists indefinitely, produces a finite change in output Y_t. This is confirmed by examination of the steady-state gain, as discussed in Chapter 4. For this model the steady-state gain is $g = \hat{\omega}_0/(1 - \hat{\delta}_1 B)$, where B is treated as an ordinary algebraic variable, with $B = 1$. For the estimated coefficients given earlier, $g = 4.716/(1 - 0.725) = 4.716/0.275 = 17.15$. That is, a permanent one-unit increase in X_t is estimated to produce an ultimate permanent increase in Y_t of 17.15 units.

The autoregressive (AR) coefficients in the DR disturbance ARIMA model may be checked for stationarity, as discussed in Chapter 2. Formally, this requires that all roots of the characteristic equations $\hat{\phi}(B) = 0$ and $\hat{\phi}(B^s) = 0$ lie outside the unit circle. Again, B is treated as an ordinary algebraic variable. After differencing, the sales and leading indicator model disturbance series in (6.1.2) is stationary since it contains no AR terms.

The moving average (MA) coefficients in the DR disturbance ARIMA model may be checked for invertibility, as discussed in Chapter 2. Formally, this requires that all roots of the characteristic equations $\hat{\theta}(B) = 0$ and $\hat{\theta}(B^s) = 0$ lie outside the unit circle. When the disturbance ARIMA model is invertible, this implies that the DR model forecast function gives more weight to recent sample disturbance values ($\hat{N}_t$) than to earlier ones, which is intuitively appealing. Further, if the data are differenced, a noninvertible model implies the presence of a deterministic pattern. This discovery may lead us to modify the DR model. For example, if the data are seasonally differenced and we find $\hat{\theta}_s = 1.0$, we might want to model the seasonal pattern in Y_t using seasonal dummy variables.

For $q = 1$ the invertibility condition is $|\hat{\theta}_1| < 1.0$. The sales and leading indicator model in (6.1.2) has $\hat{\theta}_1 = 0.620$, which clearly satisfies this condition.

6.2.4 Coefficient Correlations

Ideally the estimated DR model coefficients are not highly correlated with each other. We cannot entirely avoid correlated coefficients. But high coefficient correlations suggest that the coefficient estimates may be quite sensitive to the particular sample employed and therefore be unreliable. High correlations can also indicate that the model is not parsimonious enough. A careful study of alternative models may uncover another one that describes the data adequately with fewer estimated coefficients.

As a rule of thumb, coefficient correlations that are larger than about 0.9 in absolute value deserve attention. In some cases, however, we will have to live with correlations in that neighborhood, especially when other evidence suggests that we have identified an appropriate model. This may occur especially when high correlations arise among AR coefficients within the same operator. AR coefficients are often quite highly correlated with each other, yet they may represent an important part of the model. For example, a disturbance series ARIMA model with $p = 2$ or $P = 2$ can be useful for representing stochastic cycles in economic data. (See Chapter 2 for a discussion of stochastic cycles in ARIMA models.) In these situations we may accept a model even with coefficient correlations greater than 0.9 in absolute value.

The sales and leading indicator model (6.1.2) has the following coefficient correlations:

	1	2	3	4
1	1.00			
2	·	1.00		
3	−0.29	−0.75	1.00	
4	·	·	·	1.00

Here the top row and leftmost column numbers correspond to the estimated coefficients in the order in which they appear in the model. A dot (·) entry indicates that the correlation is not significantly different from zero at the 5% level. Only the correlation between coefficients 2 and 3 ($\hat{\omega}_0$ and $\hat{\delta}_1$) is somewhat large (equal to −0.75).

6.2.5 Redundant Coefficients

Sometimes a DR model will have unnecessary or redundant coefficients in the transfer function component or the ARIMA component. Such a model violates the principle of parsimony and can lead to unstable estimates. See Pankratz (1983, pp. 203–205) for a discussion of parameter redundancy in ARIMA models. The following model illustrates the idea of redundant coefficients:

EVALUATING ESTIMATION STAGE RESULTS 213

$$Y_t = 5X_t + \frac{1 - 0.5B}{(1 - 0.5B)\nabla} a_t \qquad (6.2.1)$$

Here the MA operator in the numerator cancels exactly with the AR operator in the denominator, leading to the equivalent, but simpler model

$$Y_t = 5X_t + \frac{1}{\nabla} a_t \qquad (6.2.2)$$

Spotting redundant coefficients is not always this easy. The redundancy may be only approximate, and it may become clear only after one of the operators is stated in factored form. For example, the following model has two redundancies:

$$Y_t = \frac{2 - 0.33B - 0.75B^2}{1 - 0.72B} X_t + \frac{1 - 0.48B}{1 + 0.28B - 0.41B^2} a_t \qquad (6.2.3)$$

The redundancies are seen only when the numerator of the transfer function and the denominator of the ARIMA component are approximately factored:

$$Y_t = \frac{(2 + 1.1B)(1 - 0.7B)}{1 - 0.72B} X_t + \frac{1 - 0.48B}{(1 - 0.5B)(1 + 0.8B)} a_t \qquad (6.2.4)$$

Here there is a near cancellation in each ratio.

One way to reduce the chance of redundancy is to start with the simplest reasonable model, and then *add coefficients one at a time*, if any are needed. In particular, it is wise to avoid adding coefficients to a numerator and denominator simultaneously. Following the identification rules 1 to 5 in Chapter 5 can also help us to avoid models with redundant coefficients. The sales and leading indicator model (6.1.2) has no redundant coefficients. (Why?)

6.2.6 Model Interpretation

Ideally the estimated model has an interpretation that is sensible and appealing. This can be especially important in presenting DR model forecasts to managers or other users of forecasts. Thus we prefer a model in which the signs and sizes of the estimated coefficients (especially the $\hat{v}$ weights implied by the rational form model) are reasonable.

The broadest interpretation of DR models is this: The DR forecasts of Y_t are (1) a weighted sum of the inputs plus (2) a weighted average of the estimated disturbance series; the disturbance is interpreted as the *one-step ahead forecast errors* that would occur *if we would use only the transfer function part of the model to forecast*. Individual DR models may have special interpretations that are more precise than this. For example, if both an input and the output are measured in log units, then the $\hat{v}$ coefficients represent the *percent* response of the output to a one *percent* rise in the input.

The DR model (6.1.2) for the sales and leading indicator data has the following interpretation. When X_t rises by one unit, then Y_t does not respond for three time periods ($b = 3$). Then Y_t rises ($\hat{\omega}_0$ is positive), initially by 4.716 units ($\hat{\omega}_0 = 4.716$). The further time-lagged additions to Y_t get smaller each succeeding period, according to a Koyck-type (first-order exponential) decay pattern, with decay coefficient $\hat{\delta}_1 = 0.725$. The constant term is deterministic trend component in the forecast function, with Y_t rising by 0.033 units each time period in addition to any other movements dictated by the transfer function or the disturbance ARIMA pattern. The forecasts for Y_t are further adjusted to include a first-order exponentially weighted moving average (EWMA) of the sample disturbance $\hat{N}_t$. [See Chapter 2 for a discussion of the ARIMA(0, 1, 1) model as a first-order EWMA.]

6.2.7 Goodness of Fit

Another check involves examining the goodness of fit of the DR model. In particular, we can consider the DR model residual standard error and compare it with the residual standard error of the ARIMA model for the output variable. It may not be worth our trouble to build and maintain a DR forecasting model if the ARIMA model for Y_t is able to explain its behavior nearly as well as the DR model. On the other hand, if we are especially interested in evaluating the role of the input variables rather than in forecasting, we may be less concerned about comparing the goodness of fit of the DR model with that of the ARIMA model for Y_t.

The residual standard error is also useful because it is *related to the size of the forecast error standard deviations*; it is therefore related to the size of the *confidence intervals* that we construct around our forecasts. (DR forecast confidence intervals are discussed in Chapter 9.) For example, if $b > 0$, or if the inputs are deterministic, then the one-step ahead forecast standard deviation is equal to the DR model residual standard error ($\hat{\sigma}_a$). In such a case an approximate 95% confidence interval for the future value Y_{n+1} is given by $\hat{Y}_n(1) \pm 2\hat{\sigma}_a$, where $\hat{Y}_n(1)$ is the one-step ahead forecast of the future value Y_{n+1} from origin $t = n$.

The sales and leading indicator model (6.1.2) residual standard error is $\hat{\sigma}_a = 0.217622$. This is much smaller than the residual standard error for the ARIMA(1, 1, 1) model for Y_t, which is 1.35707. Since $b > 0$, the one-step ahead forecast error standard deviation is the same as the DR model residual standard deviation. Thus an approximate 95% confidence interval for the one-step ahead forecast is $\hat{Y}_n(1) \pm 2(0.217622)$.

6.2.8 Forecast Accuracy

How well a DR model fits the available data offers some information about the model's potential usefulness. However, the ultimate test of a DR forecasting model is its ability to forecast accurately in practice. This requires that we continue to monitor the forecast performance of the model. This monitoring is especially important if we have built the DR model with a relatively small sample. As we acquire additional data, we may find that the DR model needs to be changed. Continued poor forecast performance (worse than we would expect from the confidence intervals) is an important clue that we should consider whether the identified DR model is adequate.

Even a DR model that seems adequate according to the criteria set out in this chapter might not forecast with *enough* accuracy *given the needs of the forecaster*. The forecaster may find it necessary to search for additional input series, or other information, that could improve forecast accuracy to the desired level.

QUESTIONS AND PROBLEMS

1. Twelve residuals are too few for meaningful results in practice. But for the sake of the exercise, consider the following 12 values for the residual series $\hat{\alpha}_t$ and $\hat{a}_t$ based on a single-input DR model with transfer function orders $(b, r, h) = (0, 0, 0)$ and a disturbance ARIMA$(1, 0, 0)$ component:

t	1	2	3	4	5	6	7	8	9	10	11	12
$\hat{\alpha}_t$	-2	5	1	-4	2	0	-2	4	1	-3	-2	2
$\hat{a}_t$	3	2	-1	1	-4	-2	3	-2	0	-2	4	1

   a. Compute the first three residual cross correlations for these two series. What are their approximate standard errors? Are these coefficients significant? How does the small sample size affect your criteria for deciding if the coefficients are significant?

   b. Compute S^* for the residual cross correlations in part (a). Is S^* significant at the 10% level?

   c. Compute the first three residual autocorrelations for series $\hat{a}_t$. What are the approximate standard errors? Are these coefficients significant? How does the small sample size affect your criteria for deciding if the coefficients are significant?

   d. Compute Q^* for the residual autocorrelations in part (c). Is Q^* significant at the 10% level?

2. What interpretation do you give to the following DR models?

   a. $\nabla_{12} Y_t' = 0.017 + 0.005 \nabla_{12} X_t' + (1 - 0.9 B^{12}) \hat{a}_t$

   where Y_t' and X_t' are in the natural log metric.

   b. $\nabla Y_t = \dfrac{10 - 20B}{1 - 0.6B} \nabla X_t + (1 - 0.7B) \hat{a}_t$

CASE 4

Housing Starts and Sales

C4.1 THE DATA

Figures C4.1 and C4.2 show monthly U.S. housing starts and sales, respectively, in thousands of units, from January 1965 to December 1975. The data are listed in the Data Appendix as Series 23 and Series 24, respectively. These series have been studied by Hillmer and Tiao (1979) and Abraham and Ledolter (1983). There are 132 observations available. We will build a DR model with the first $n = 120$ observations and use the last 12 observations to check the forecasts.

C4.2 PRELIMINARY STEPS

Before building a DR model, we go through several preliminary steps as discussed in Chapter 5. Inspection of Figure C4.1 and C4.2 suggests that each series has a constant variance, so we will study them in their original metrics. In addition, we want to check the series for feedback, and build an ARIMA model for each series.

C4.2.1 Feedback Check

We want to build a single-equation DR model, with $Y_t =$ starts as the output, and $X_t =$ sales as the input. It is reasonable to think that a change in sales will lead to a change in starts. Increased sales deplete the stock of unsold homes and suggest increased demand, signaling that more new homes might be sold in the future. Builders thus have reason to think that starting more new homes may be profitable. This implies that housing starts may be positively related to sales, perhaps with a time lag.

It is also possible that a change in starts could lead to a change in sales. Homes are often built on the speculation that a buyer will later appear. Of course, builders can't force anyone to buy a home just because it has been

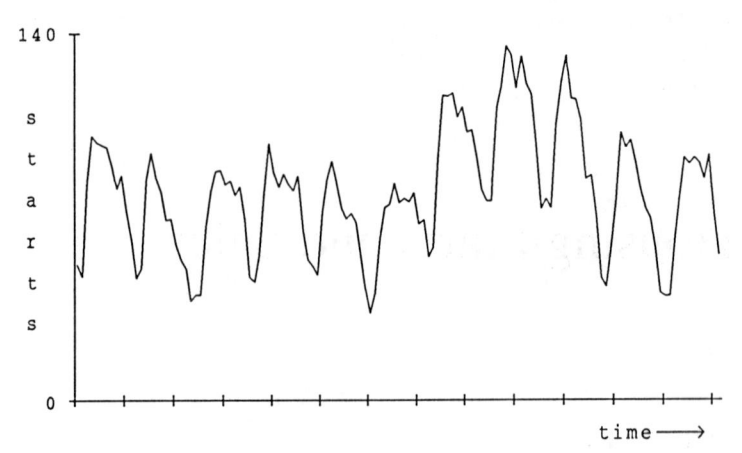

Figure C4.1 Monthly housing starts, 1965–1975.

built. But builders can make the terms of sale more attractive as an inducement to potential buyers. Thus higher starts could lead to higher future sales, again a positive relationship. In other words there might logically be feedback from starts to sales. Further, these data are *temporal aggregates* (monthly totals of daily activity); temporally aggregated data can display statistical feedback even when there is no logical feedback in the underlying mechanism. Feedback from past Y_t values to X_t would produce inconsistent estimates in a single-equation DR model.

As discussed in Chapter 5, we can check for statistical feedback by estimating an appropriate regression model. In this case we estimate how the input X_t is related to its own past and the past of the output Y_t, up to some maximum lag K. Setting $K = 16$, the model we estimate is

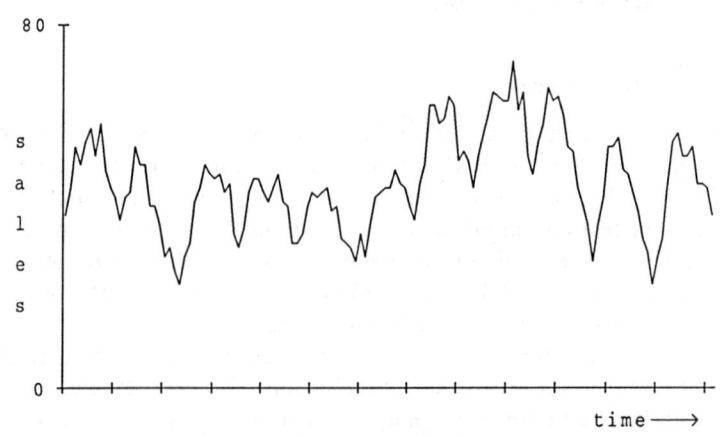

Figure C4.2 Monthly housing sales, 1965–1975.

PRELIMINARY STEPS 219

Table C4.1 Feedback Check Results from Model (C4.2.1)

k	1	2	3	4	5	6	7	8
$\hat{c}_k$	0.0399	−0.0340	0.0775	−0.0142	−0.0465	−0.0867	0.1866	−0.0404
$\|t\|$	0.53	0.44	1.02	0.19	0.62	1.24	2.65	0.56

k	9	10	11	12	13	14	15	16
$\hat{c}_k$	0.0493	0.1925	−0.0296	0.0086	−0.1764	−0.1331	−0.0688	0.0619
$\|t\|$	0.68	2.67	0.39	0.11	2.32	1.69	0.86	0.96

$$X_t = C + b_1 X_{t-1} + \cdots + b_{16} X_{t-16} + c_1 Y_{t-1} + \cdots + c_{16} Y_{t-16} + a_t$$
(C4.2.1)

The estimated coefficients $\hat{c}_k$, $k = 1, 2, \ldots, 16$, provide information about how current sales (X_t) is related to past values of starts (Y_{t-k}) at various lags k. One way to assess the presence of feedback is to look at the t values of these coefficients. The $\hat{c}_k$ values and their t values are shown in Table C4.1.

Large absolute t values suggest possible feedback from starts to sales at lags 7, 10, and 13. The sign of the coefficient at lag 13 is negative, whereas we expect a positive relationship if there is feedback. Perhaps we can dismiss the coefficient at lag 13 as spurious. The coefficients at lags 7 and 10 are more troublesome since they are positive. It is not clear why feedback might occur strongly at lags 7 and 10 but not at other lags. Perhaps contractors who build on speculation wait about six months to one year to start offering strong inducements to buyers when a home remains unsold. Colinearity may have caused this fairly general effect to be assigned somewhat arbitrarily to the coefficients at lags 7 and 10.

We may also test the joint null H_0: $c_1 = c_2 = \cdots = c_{16} = 0$ against H_a: not all c_k are zero, using the F statistic in (5.2.2). The result in this case is $F = 3.12$, which is significant at the 5% level for (16, 71) degrees of freedom. This suggests that we may, indeed, have a feedback problem.

Our feedback check has not provided an especially clean bill of health regarding the use of a single-equation DR model. The best we can say is that the feedback perhaps occurs in a pattern that is not very convincing. We will proceed with a single-equation DR model for the sake of illustration, but we recognize that we may be on shaky ground in using a single-equation model.

C4.2.2 ARIMA Models

The methods discussed in Chapter 2 lead to the following ARIMA(0, 1, 1)(0, 1, 1)$_{12}$ models for each of the two series (t values in parentheses):

$$\nabla_{12}\nabla Y_t = (1 - 0.9558B^{12})(1 - 0.2714B)\hat{a}_t$$
$$(15.90) \qquad (2.91) \qquad \text{(C4.2.2)}$$

$$\hat{\sigma}_a = 6.32424$$

$$\nabla_{12}\nabla X_t = (1 - 1.0055B^{12})(1 - 0.2010B)\hat{\alpha}_t$$
$$(15.80) \qquad (2.12) \qquad \text{(C4.2.3)}$$

$$\hat{\sigma}_\alpha = 3.36462$$

The forecasts from both of these models are interpreted as combinations of a nonseasonal EWMA [the ARIMA(0, 1, 1) part] and a seasonal EWMA [the ARIMA(0, 1, 1)$_{12}$ part]. However, the implied seasonal weights in (C4.2.3) don't decay as we consider observations further into the past: coefficient $\hat{\theta}_{12}$ in (C4.2.3) does not meet its invertibility condition. We discussed this phenomenon in Chapter 3 and saw an example in Case 3. There we explained that EWMA models like this imply that the seasonal pattern is globally constant rather than stochastic. Thus, we could try modeling housing sales (X_t) with seasonal dummies, for example, as discussed in Chapter 3. We will not pursue this matter here; but this result shows how building preliminary ARIMA models can provide insight into the nature of the data and suggest alternative modeling schemes.

In addition, model (C4.2.2) provides a baseline model against which to judge our DR model. We can compare the model fit of our DR model for Y_t, as measured by its residual standard error, as well as its forecasts, with those from (C4.2.2). Further, model (C4.2.3) provides the residual series $\hat{\alpha}_t$ needed at the DR diagnostic checking stage, as discussed in Chapter 6. Model (C4.2.3) can also provide forecasts of the input series that are needed to forecast the output series. And information from (C4.2.3) is needed to produce standard errors for our DR model forecasts, as discussed in Chapter 9.

C4.3 DR MODEL IDENTIFICATION, ESTIMATION, AND REFORMULATION

C4.3.1 Tentative Model 1

The first step in the LTF identification strategy for a DR model is to estimate a preliminary free-form distributed lag model, with regular and seasonal ($s = 12$) AR terms as proxies for the disturbance series ARIMA pattern. The maximum v-weight lag in this case is set at 10. The first preliminary LTF model is

$$Y_t = C + v_0 X_t + v_1 X_{t-1} + \cdots + v_{10} X_{t-10} + N_t \qquad \text{(C4.3.1)}$$

DR MODEL IDENTIFICATION, ESTIMATION, AND REFORMULATION

Table C4.2 Estimation Results for Model (C4.3.1)

$\hat{C} = -7.7452$ ($|t| = 1.18$)

$k \rightarrow$	0	1	2	3	4	5	6
$\hat{v}_k$	0.7234	0.7904	0.4038	0.2835	−0.2797	−0.0446	−0.2764
$|t|$	4.69	4.56	2.21	1.62	1.61	0.26	1.64

$k \rightarrow$	7	8	9	10		1	12
$\hat{v}_k$	0.1166	−0.0442	0.2172	0.0675	$\hat{\phi}_k$	0.3614	0.5524
$|t|$	0.69	0.26	1.34	0.47		3.80	6.07

$\hat{\sigma}_a = 5.90345$

where $N_t = [1/(1 - \phi_{12}B^{12})(1 - \phi_1 B)]a_t$. Estimation results are shown in Table C4.2. Recall that the estimated disturbance series is computed as $\hat{N}_t = Y_t - \hat{C} - \hat{v}(B)$. The SACF for $\hat{N}_t$ is shown in Figure C4.3.

The disturbance SACF in Figure C4.3 falls quickly to insignificant values after lag 1. However, it decays slowly from lag 12 to lag 24. The residual autocorrelations at lags 36 and 48 (equal to 0.37 and 0.34, not shown) continue this slow decay, indicating that seasonal differencing is called for. Seasonal differencing is also consistent with the strong seasonal pattern that is apparent in the plot of the starts data in Figure C4.1. Thus we will analyze the seasonally differenced series. Since $\hat{C}$ in model (C4.3) is insignificant, we exclude the constant from the new preliminary LTF model.

C4.3.2 Tentative Model 2

The new preliminary LTF model is found by starting with the differenced disturbance, $n_t = \nabla_{12} N_t$. Write this as $N_t = n_t/\nabla_{12}$; substitute this expression into (C4.3.1); multiply through the resulting equation by ∇_{12} to get the new preliminary model

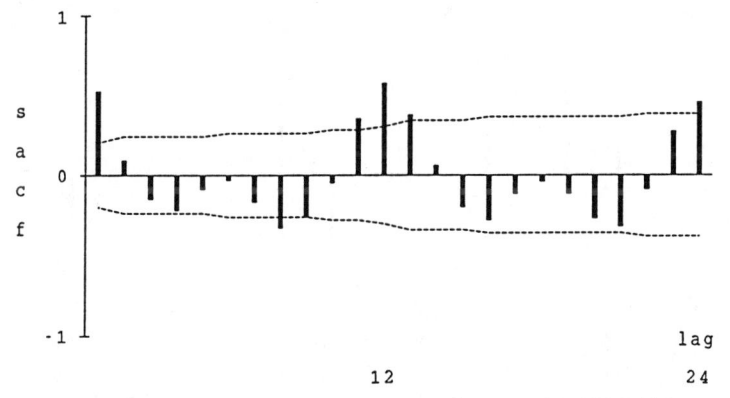

Figure C4.3 SACF for disturbance $\hat{N}_t$ in model (C4.3.1).

Table C4.3 Estimation Results for Model (C4.3.2)

$k \rightarrow$	0	1	2	3	4	5	6
$\hat{v}_k$	0.7683	0.6112	0.1921	0.2377	−0.2309	0.1294	−0.1394
$\|t\|$	4.59	3.33	1.02	1.26	1.19	0.65	0.71

$k \rightarrow$	7	8	9	10		1	12
$\hat{v}_k$	0.2615	−0.0602	0.3428	0.0796	$\hat{\phi}_k$	0.1892	−0.3533
$\|t\|$	1.36	0.31	1.83	0.49		1.72	3.58

$\hat{\sigma}_a = 5.89473$

$$\nabla_{12}Y_t = v_0\nabla_{12}X_t + v_1\nabla_{12}X_{t-1} + \cdots + v_{10}\nabla_{12}X_{t-10} + n_t \quad (C4.3.2)$$

where $n_t = \nabla_{12}N_t = [1/(1 - \phi_{12}B^{12})(1 - \phi_1 B)]a_t$. We have seasonally differenced ($D = 1$) both the input and the output series, and we have inserted a new proxy ARIMA model for the differenced disturbance series.

Estimation results for model (C4.3.2) are shown in Table C4.3. The estimated disturbance is found by solving (C4.3.2) for n_t and inserting estimated values for parameters; that is, we compute $\hat{n}_t = \nabla_{12}Y_t - \hat{v}(B)\nabla_{12}X_t$. The SACF and SPACF of the new estimated disturbance series appear in Figures C4.4 and C4.5; the ESACF of this series is shown in Table C4.4.

No further seasonal differencing is needed since the new disturbance SACF in Figure C4.4 cuts off to insignificance at lag 24. An MA(1)$_{12}$ seems appropriate to represent the remaining seasonal pattern in the differenced disturbance series.

The disturbance SACF in Figure C4.4 decays somewhat slowly; this suggests that further differencing ($d = 1$) might be applied. The combined decays in the SACF and SPACF suggest a mixed model for the seasonally differenced disturbance $n_t = \nabla_{12}N_t$; the ESACF indicates an ARMA(1, 1) since there is a triangle of zeros with its upper left vertex in the $p' = 1$, $q = 1$

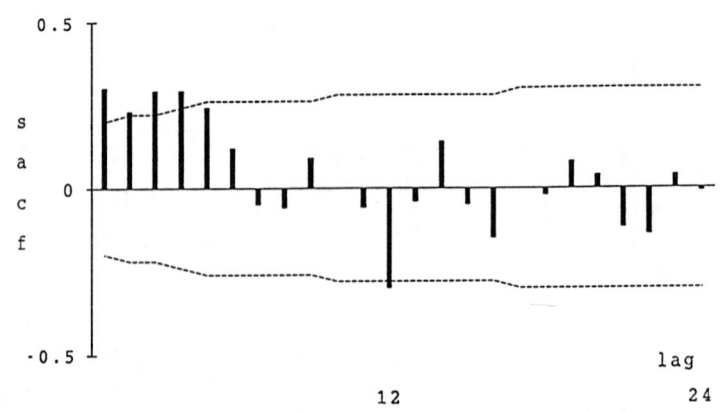

Figure C4.4 SACF for disturbance $\hat{n}_t$ in model (C4.3.2).

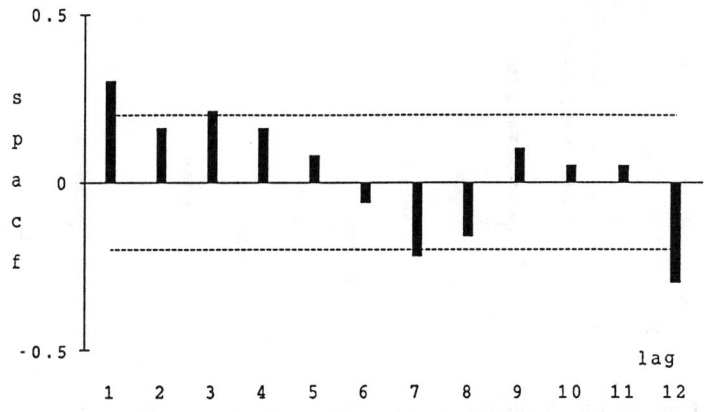

DR MODEL IDENTIFICATION, ESTIMATION, AND REFORMULATION   223

Figure C4.5 SPACF for disturbance $\hat{n}_t$ in model (C4.3.2).

cell. Since the need for further differencing is unclear, we will entertain an ARIMA(1, 0, 1)(0, 1, 1)$_{12}$ model for the original disturbance N_t in (C4.3.1); this is an ARMA(1, 1)(0, 1)$_{12}$ model for the differenced disturbance $n_t = \nabla_{12} N_t$ in (C4.3.2).

We also want to specify a rational form transfer function for the v weights in Table C4.3 (graphed in Figure C4.6). The pattern of the $\hat{v}$ weights suggests two possibilities. Since there are only two $\hat{v}$ weights (at lags 0 and 1) more than twice their standard errors, we might try $(b, r, h) = (0, 0, 1)$. [Compare the estimated impulse response function in Figure C4.6 with the theoretical one in Figure 5.6(b).] Then the identified transfer function is $(\omega_0 + \omega_1 B)\nabla_{12} X_t$. Or, we might interpret the $\hat{v}$ weights as decaying geometrically from lag 0; then we would specify $(b, r, h) = (0, 1, 0)$, which is the Koyck transfer function $[\omega_0/(1 - \delta_1 B)]\nabla_{12} X_t$. To justify this second pattern, we must include in the pattern $\hat{v}$ weights that are less than twice their standard errors, and we must allow for sampling variation. (What rational form transfer function do you identify by following the five identification rules in Chapter 5?)

Table C4.4 Simplified ESACF for Disturbance $\hat{n}_t$ in Model (C4.3.2)[a]

$(q \rightarrow)$	0	1	2	3	4	5	6	7	8
$(p' = 0)$	X	X	X	X	0	0	0	0	0
$(p' = 1)$	X	0	0	0	0	0	0	0	0
$(p' = 2)$	X	X	0	0	0	0	0	0	0
$(p' = 3)$	X	0	0	0	0	0	0	0	0
$(p' = 4)$	X	0	0	X	0	0	0	0	0

[a]X = significant at 5% level.

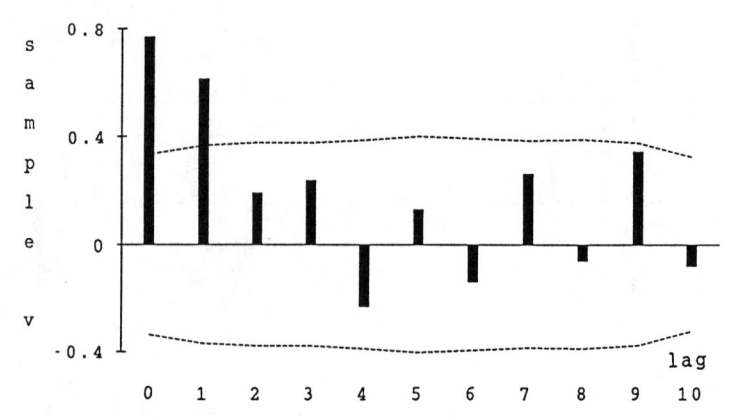

Figure C4.6 Sample impulse response function for model (C4.3.2).

C4.3.3 A Final Model 1

The choice between the two suggested alternative transfer functions is somewhat arbitrary. They are equally parsimonious, each with only two parameters. We will proceed with the Koyck model; along with the identified ARIMA$(1, 0, 1)(0, 1, 1)_{12}$ model for N_t, the identified DR model is

$$\nabla_{12}Y_t = \frac{\omega_0}{1 - \delta_1 B} \nabla_{12}X_t + \frac{(1 - \theta_{12}B^{12})(1 - \theta_1 B)}{1 - \phi_1 B} a_t \qquad \text{(C4.3.3)}$$

Estimation of model (C4.3.3) gives the results shown in Table C4.5. All estimated coefficients are significant at better than the 5% level (all absolute

Table C4.5 Estimation Results and Coefficient Correlations for Model (C4.3.3)

Parameter	Estimate	$\|t\text{ Value}\|$
ω_0	0.7923	6.90
δ_1	0.5635	8.56
θ_1	0.6067	4.25
θ_{12}	0.9969	15.40
ϕ_1	0.8441	8.75

$\hat{\sigma}_a = 4.9277$

Coefficient Correlation Matrix

	1	2	3	4	5
1	1.00				
2	−0.89	1.00			
3	.	.	1.00		
4	.	.	.	1.00	
5	.	.	0.83	.	1.00

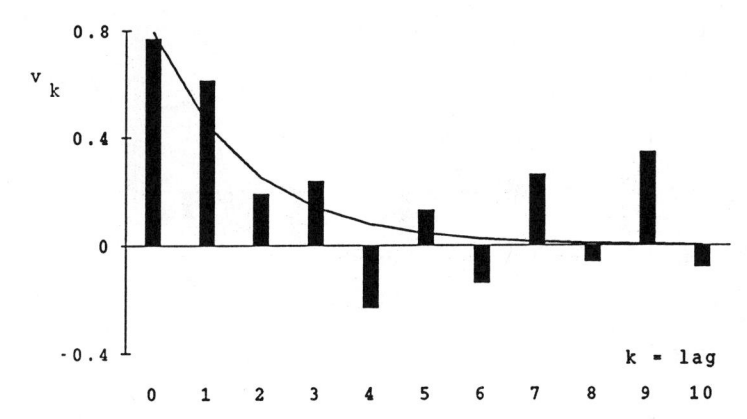

DR MODEL IDENTIFICATION, ESTIMATION, AND REFORMULATION 225

Figure C4.7 Impulse response function implied by the rational form transfer function in model (C4.3.3), superimposed on the sample v weights in Figure C4.6.

t values are larger than 2.0). Coefficient $\hat{\theta}_1$ satisfies its invertibility condition since its absolute value is less than 1.0. Coefficient $\hat{\theta}_{12}$, however, is nearly on the invertibility boundary 1.0. This implies that the data contain a fixed seasonal pattern, which may be represented by a set of seasonal dummy variables, for example.

The estimate $\hat{\phi}_1$ is less than 1.0, but it is not very far from 1.0 in terms of standard errors. This suggests that the regular AR operator is nearly $\nabla = (1 - B)$ so that we could apply regular differencing to both $\nabla_{12}Y_t$ and $\nabla_{12}X_t$. The resulting model would not be very different from (C4.3.3), however, and we may accept (C4.3.3) without additional differencing. The main advantage of differencing is that it permits us to see patterns in an SACF that might be obscured by the nonstationary nature of the data. However, we used the ESACF to help us choose the disturbance model, and the ESACF does not require that the data be differenced.

The estimated transfer function in (C4.3.3) is stable since $|\hat{\delta}_1| < 1.0$. The gain of the transfer function is $g = \hat{\omega}_0/(1 - \hat{\delta}_1 B)$, with $B = 1$; thus $g = 0.7923/(1 - 0.5635) = 1.8151$. Thus, a permanent one-unit increase in X_t is associated with an estimated eventual permanent increase in Y_t of 1.8151 units. The solid line in Figure C4.7 shows the $\hat{v}$ weights that correspond to the rational transfer function in (C4.3.3). These are found from $\hat{\omega}_0$ and $\hat{\delta}_1$ using Equation (4.3.11) in Chapter 4. In Figure C4.7, these $\hat{v}$ weights are superimposed on the LTF $\hat{v}$ weights (reproduced from Figure C4.6). In Figure C4.7 we see that the $\hat{v}$-weight pattern implied by the rational transfer function matches that of the LTF $\hat{v}$ weights fairly well.

The correlation matrix of the estimated coefficients is shown in Table C4.5. The row and column numbers correspond to the order in which the parameters are listed in Table C4.5. Dot ($\cdot$) entries are correlations that are not significantly different from zero at the 5% level. The estimated coefficients $\hat{\omega}_0$ and $\hat{\delta}_1$ are rather highly correlated ($r = -0.89$), but this is

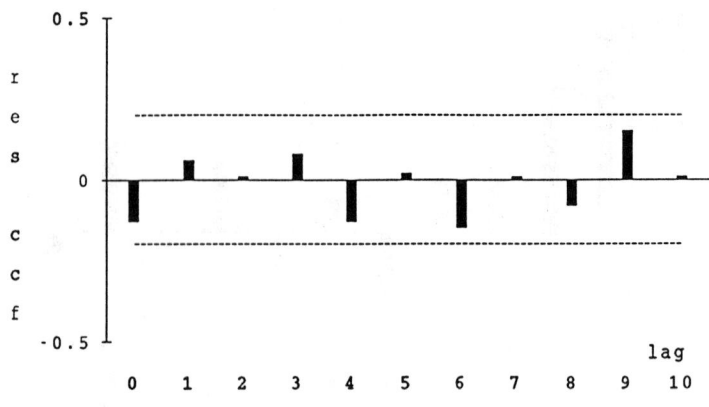

Figure C4.8 Residual CCF for model (C4.3.3).

not unusual for the numerator and denominator terms in a rational transfer function. Likewise, $\hat{\theta}_1$ and $\hat{\phi}_1$ are fairly highly correlated ($r = 0.83$), but this too is not unusual for AR and MA terms at the same lag in a mixed model.

Notice that $\hat{\sigma}_a = 4.9277$ for DR model (C4.3.3) is about 20% smaller than $\hat{\sigma}_a = 6.32424$ for ARIMA model (C4.2.2). This implies that we can expect somewhat smaller forecast errors, on average, using the DR model.

Residual Correlation Check
In Chapter 6 we introduced the residual CCF as a diagnostic checking tool. The residual CCF for model (C4.3.3) is shown in Figure C4.8. The residual CCF offers a check on the adequacy of the specified transfer function. Since none of the residual cross correlations is particularly large compared to its standard error, we conclude that the transfer function in (C4.3.3) adequately captures the response of housing starts to housing sales. (Computation of the approximate χ^2 statistic for the residual CCF in Figure C4.8 is left as an exercise.)

The residual SACF in Figure C4.9 suggests that the ARIMA model for the disturbance series may be acceptable, but it is somewhat troublesome. In particular, the χ^2 statistic (35.0) is significant at the 5% level. However, the residual autocorrelations at the shortest lags (1 through 6) are quite small. The moderately large coefficient at lag 24 indicates that adding a θ_{24} coefficient to (C4.3.3) might improve it, but the improvement is not likely to be large. The other moderately large coefficients (lags 7, 14, 16, and 23) are difficult to explain and do not seem to fit any common pattern. As discussed in Chapter 6, sometimes an inadequate residual SACF reflects an inadequate transfer function; but we have already concluded from the residual CCF in Figure C4.8 that the transfer function in (C4.3.3) is acceptable. Perhaps the alternative transfer function suggested earlier, with $(b, r, h) = (0, 0, 1)$, will fit the data even better; if so, it may improve the residual SACF.

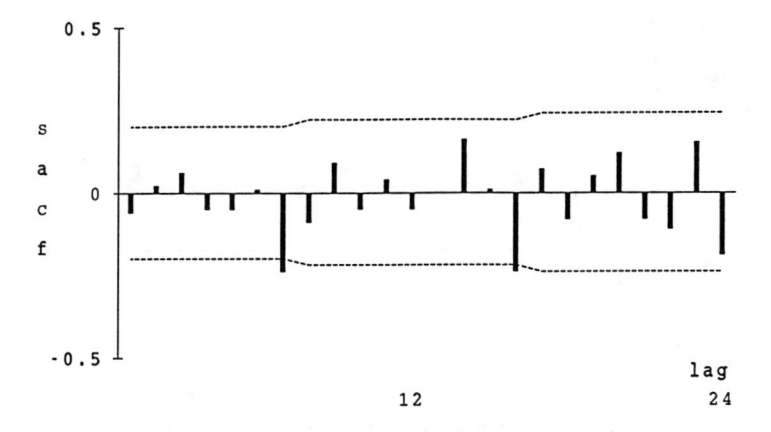

DR MODEL IDENTIFICATION, ESTIMATION, AND REFORMULATION   227

```
  0.5
s
a
c    0
f

 -0.5                                              lag
                     12                             24
```

Chi-squared (Q*) = 35.0   for df = 21

Figure C4.9 Residual SACF for model (C4.3.3).

C4.3.4 A Final Model 2

Consider the alternative transfer function with $(b, r, h) = (0, 0, 1)$. Keeping the same ARIMA model for the disturbance series, we entertain this model:

$$\nabla_{12} Y_t = (\omega_0 + \omega_1 B)\nabla_{12} X_t + \frac{(1 - \theta_{12} B^{12})(1 - \theta_1 B)}{1 - \phi_1 B} a_t \quad (C4.3.4)$$

Estimation results are shown in Table C4.6, the residual CCF is in Figure

Table C4.6 Estimation Results and Coefficient Correlations for Model (C4.3.4)

Parameter	Estimate	$\|t\text{ Value}\|$
ω_0	0.6639	4.59
ω_1	0.8720	6.06
θ_1	0.6208	5.32
θ_{12}	0.9540	15.94
ϕ_1	0.9046	14.26
$\hat{\sigma}_a = 5.0960$		

Coefficient Correlation Matrix

	1	2	3	4	5
1	1.00				
2	−0.59	1.00			
3	·	·	1.00		
4	·	·	·	1.00	
5	·	·	0.74	·	1.00

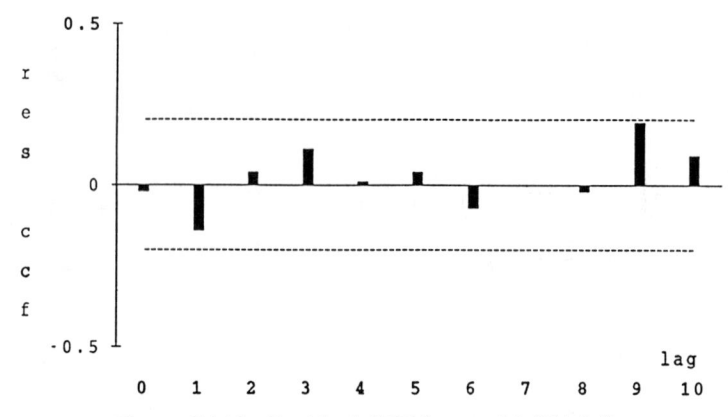

Figure C4.10 Residual CCF for model (C4.3.4).

C4.10, and the residual SACF is in Figure C4.11. Model (C4.3.4) is better than model (C4.3.3) in some ways. Although (C4.3.4) has a slightly larger residual standard deviation than (C4.3.3), the residual SACF for (C4.3.4) is better, and the coefficient correlations are somewhat smaller.

The normal probability plot of the residuals from model (C4.3.4) is shown in Figure C4.12. [This looks much like the corresponding plot for model (C4.3.3).] This plot deviates somewhat from a straight line, though the deviation seems mainly due to a few residuals in the left tail of the distribution. It suggests that the shocks may deviate somewhat from normality, with a slight negative skew. The plot of the standardized residuals ($\hat{a}_t/\hat{\sigma}_a$) in Figure C4.13 shows only three values more than two standard errors from zero, and all are less than three standard errors from zero. From this evidence

Chi-squared (Q^*) = 28.6   for df = 21

Figure C4.11 Residual SACF for model (C4.3.4).

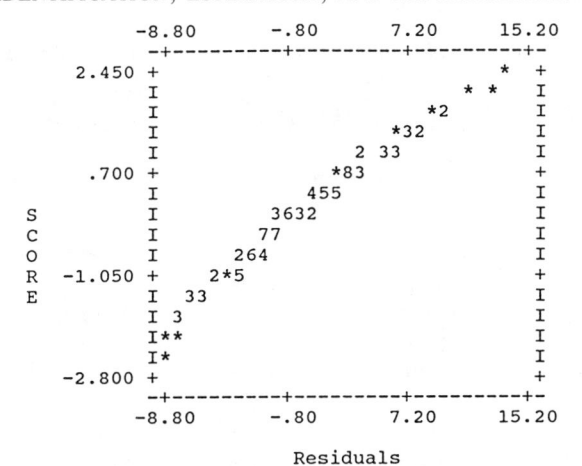

DR MODEL IDENTIFICATION, ESTIMATION, AND REFORMULATION

```
                     -8.80         -.80        7.20       15.20
                    -+----------+----------+----------+-
            2.450 +                                  *   +
                  I                              *  *    I
                  I                           *2         I
                  I                         *32          I
                  I                    2 33              I
             .700 +                  *83                 +
                  I                 455                  I
        S         I                3632                  I
        C         I                77                    I
        O         I             264                      I
        R  -1.050 +         2*5                          +
        E         I    33                                I
                  I   3                                  I
                  I**                                    I
                  I*                                     I
           -2.800 +                                      +
                    -+----------+----------+----------+-
                     -8.80         -.80        7.20       15.20
                                 Residuals
```
Figure C4.12 Normal probability plot of residuals from model (C4.3.4).

we conclude that there are no important outliers. The formal outlier detection procedure discussed in Chapter 8 might lead to different conclusions.

There is little difference between models (C4.3.3) and (C4.3.4). Model (C4.3.4) is somewhat preferable since the χ^2 for its residual SACF is better. These two DR models have identical disturbance ARIMA components, with slight differences in the estimated ARIMA coefficients. It seems that the difference between the two residual SACFs is due mainly to the different transfer functions. This reflects the point raised in Chapter 6: Modifying the transfer function can alter the residual SACF without any change in the disturbance ARIMA model specification.

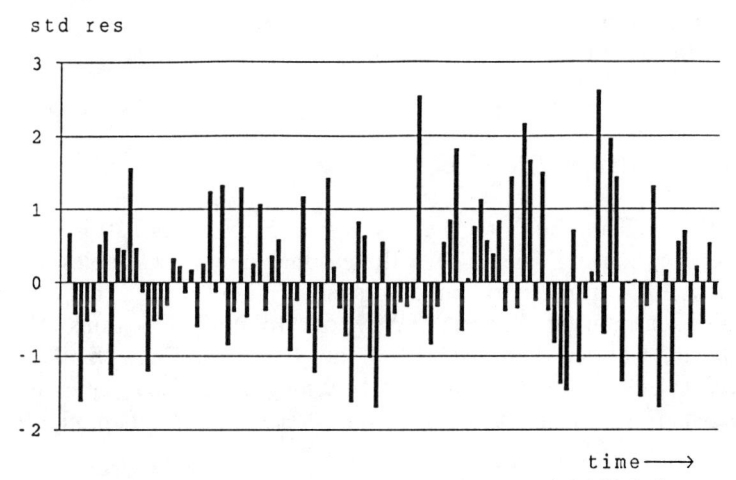

Figure C4.13 Standardized residuals from model (C4.3.4).

Table C4.7 Twelve Forecasts from DR Model (C4.3.4) and ARIMA Model (C4.2.2)

Time	DR Forecast	DR Forecast Standard Error	ARIMA Forecast	ARIMA Forecast Standard Error	Observed
121	34.565	5.5641	37.013	6.3242	39.791
122	38.661	7.4374	39.878	7.8249	39.959
123	66.813	8.6067	68.031	9.0808	62.498
124	82.896	9.6189	84.116	10.1830	77.777
125	84.172	10.5223	85.371	11.1770	92.782
126	81.476	11.3448	82.621	12.0896	90.284
127	75.556	12.1045	76.655	12.9379	92.782
128	74.085	12.8139	75.145	13.7340	90.655
129	66.624	13.4818	67.648	14.4864	84.517
130	66.889	14.1148	67.879	15.2016	93.826
131	53.737	14.7180	54.690	15.8846	71.646
132	39.243	15.2955	40.165	16.5394	55.650

C4.4 FORECASTING

Table C4.7 shows 12 forecasts and standard errors from DR model (C4.3.4), along with 12 forecasts and standard errors from ARIMA model (C4.2.2). Also listed are the 12 future observed values that have not been included in our model estimation. All forecasts are based on information available through $t = 120$. Notice that the forecast standard errors increase at longer forecast lead times. In all cases the DR model forecast standard errors are slightly smaller than the ARIMA model forecast standard errors. The ARIMA model forecasts are more accurate for 10 of the 12 forecast lead times. This emphasizes that the smaller forecast standard errors for the DR model do not guarantee better forecasts all the time; it only suggests that the DR forecasts will be better on average, in repeated use.

QUESTIONS AND PROBLEMS

1. Use the LTF method and the first $n = 120$ observations to build a DR model for housing sales (X_t) with seasonal dummy variables as inputs. What is the rationale for building such a model? Does this DR model require seasonal differencing? (Why, or why not?) Does this DR model fit the data better than ARIMA model (C4.2.3)? Does this DR model forecast series X_t better than ARIMA model (C4.2.3)? Does using the forecasts from this model improve the forecasts from DR model (C4.3.4) for Y_t?

2. Modify model (C4.3.3) by removing the seasonal ARIMA components;

instead, introduce seasonal dummy variables, as discussed in Chapter 3. What is the rationale for this modification? Does the alternative model fit the data better than model (C4.3.3)? Does it forecast better?

3. Consider the residual CCFs in Figures C4.8 and C4.10. The first is based on $n = 102$ effective observations; the second is based on $n = 106$ effective observations. The coefficients are

Figure C4.8:

k	0	1	2	3	4	5	6	7	8	9	10
r_k^*	−0.13	0.06	0.01	0.08	−0.13	0.02	−0.15	0.01	−0.08	0.15	0.01

Figure C4.10:

k	0	1	2	3	4	5	6	7	8	9	10
r_k^*	−0.02	−0.14	0.04	0.11	0.01	0.04	−0.07	0.00	−0.02	0.19	0.09

Is the approximate χ^2 statistic S^* significant at the 10% level for these two functions? What is your null hypothesis? What are the degrees of freedom in each case? What is the critical χ^2 value?

4. Compare the root mean squared forecast errors for the DR forecasts and the ARIMA forecasts in Table C4.7.

CASE 5

Industrial Production, Stock Prices, and Vendor Performance

C5.1 THE DATA

The U.S. Commerce Department publishes a monthly series, the index of leading indicators, that is a composite of 12 economic series. These 12 series are chosen for various reasons, but especially because they appear to *change direction before* the general level of economic activity changes direction. Thus these series may be useful as *leading* economic indicators.

In this case study we build a multiple-input DR model. The output (Y_t) is the Commerce Department's index of industrial production. This index is a measure of current real economic activity. Input $X_{1,t}$ is an index of stock market prices (Standard and Poor's 500 stocks), a component of the index of leading indicators. Input $X_{2,t}$ is vendor performance, also a component of the index of leading indicators. Vendor performance is the percent of companies reporting slower deliveries; the data are obtained from a survey of business purchasing agents. These three series are listed in the Data Appendix as Series 25, 26, and 27, respectively.

We expect to see a positive relationship between the output and each input. Stock prices tend to *rise* as investors anticipate higher profits which, in turn, reflect *higher* expected levels of business sales and production. As the vendor performance series *rises*, more businesses experience slower deliveries; this tends to reflect a rising level of orders for goods and services and therefore *higher* future levels of production and sales.

The monthly data, from January 1947 to September 1980, are plotted in Figures C5.1, C5.2, and C5.3. These data are *not* seasonally adjusted. Seasonal adjustment can obscure the underlying time structures of the individual series as well as the relationships between variables (Neftçi, 1979). There are a total of 405 observations available for this study. We will use the first $n = 400$ to build a DR model and the last 5 to check the forecasts.

THE DATA 233

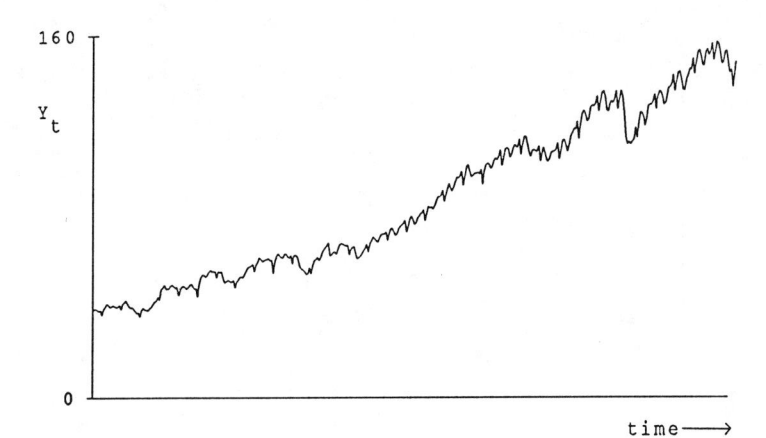

Figure C5.1 Monthly industrial production, January 1947–September 1980.

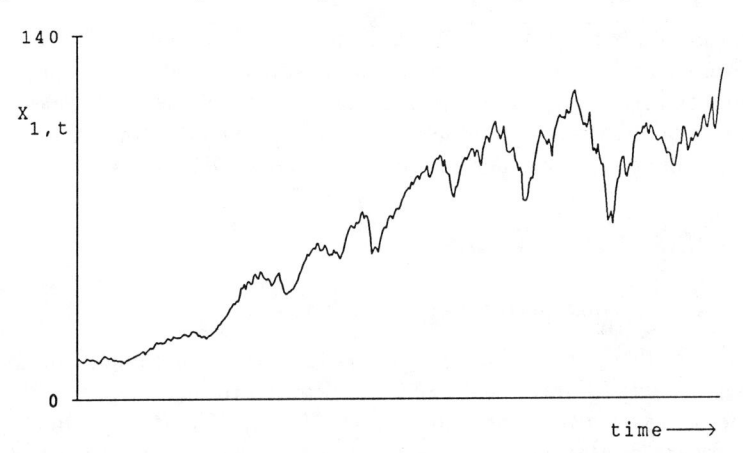

Figure C5.2 Monthly stock prices, January 1947–September 1980.

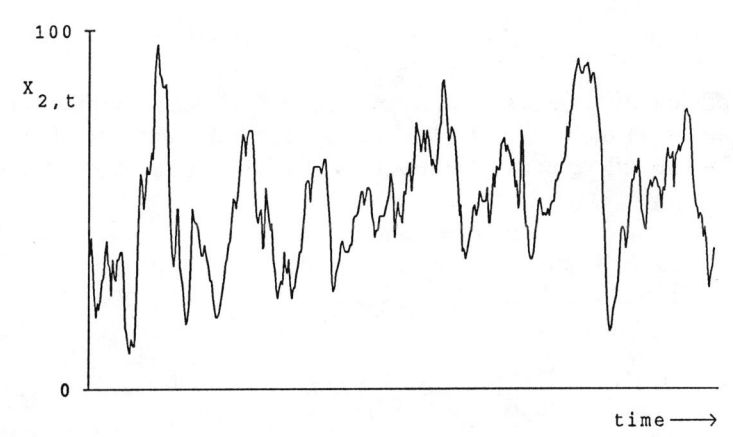

Figure C5.3 Monthly vendor performance, January 1947–September 1980.

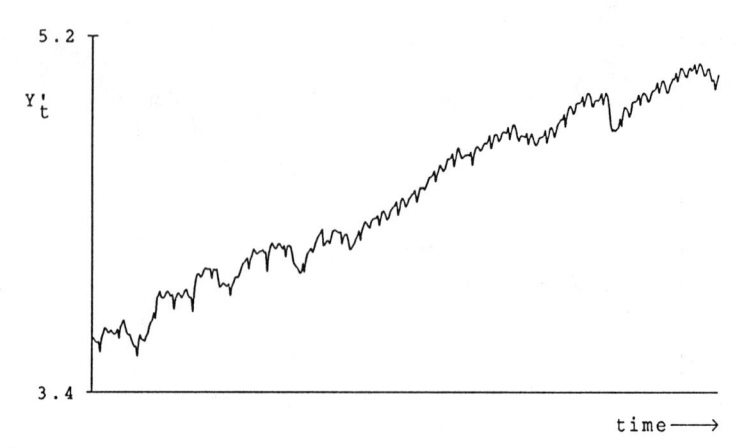

Figure C5.4 Log industrial production.

This case study illustrates two important and related points: (1) Real data may give rise to statistics at the identification stage that do not match theoretical patterns very neatly; and (2) nevertheless, careful use of common sense, practical modeling rules, and the information provided at the various stages of the modeling strategy can lead to a reasonable DR model.

C5.2 PRELIMINARY ANALYSIS

C5.2.1 Transformation to Constant Variance

Inspection of the data indicates that a log transformation induces an approximately constant variance for Y_t and $X_{1,t}$. Therefore our model building will be done using the two transformed series $Y'_t = \ln Y_t$ and $X'_{1,t} = \ln X_{1,t}$. The log series are plotted in Figures C5.4 and C5.5. Series $X_{2,t}$ needs no such transformation.

C5.2.2 ARIMA Models

The modeling procedures outlined in Chapter 2 lead to the following ARIMA(0, 1, 1) model for the log stock price series and ARIMA(2, 0, 0) model for the vendor performance series (absolute t value in parentheses):

$$\nabla X'_{1,t} = (1 + 0.2468B)\hat{a}_{1,t}$$
$$(5.10) \tag{C5.2.1}$$

$$\hat{\sigma}_{a_1} = 0.0325394$$

$$(1 - 1.2257B + 0.2852B^2)X_{2,t} = 0.3048 + \hat{a}_{2,t}$$
$$(25.51) \quad (5.94) \quad\quad (3.83) \tag{C5.2.2}$$

$$\hat{\sigma}_{a_2} = 0.506387$$

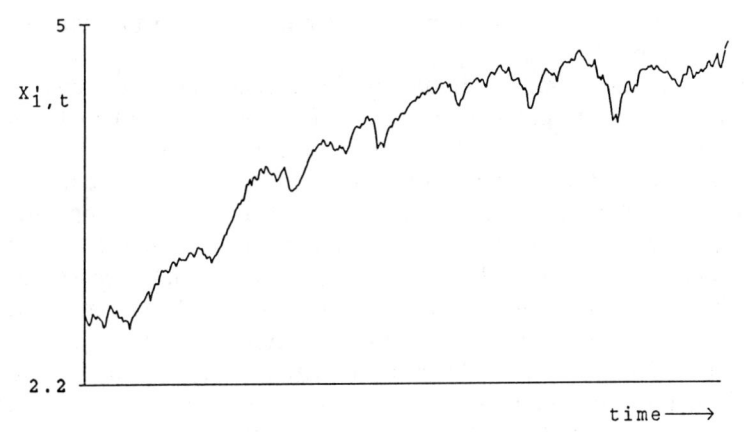

Figure C5.5 Log stock prices.

A constant term could be included in (C5.2.1) for a slightly better fit. (How would you interpret this constant term?)

Models (C5.2.1) and (C5.2.2) are useful in at least three ways:

1. They allow us to generate the series $\hat{a}_{1,t}$ and $\hat{a}_{2,t}$, both of which are needed at the diagnostic checking stage in the DR modeling strategy.
2. They can be used to produce forecasts of the input series; these are needed to forecast the output series.
3. They are needed to compute the standard errors and confidence intervals for forecasts of the output series from the DR model.

An adequate univariate model for the log industrial production series is the following ARIMA$(2, 1, 0)(2, 1, 1)_{12}$ (absolute t values in parentheses):

$$(1 + 0.269B^{12} + 0.2173B^{24})(1 - 0.310B - 0.1318B^2)\nabla_{12}\nabla Y_t'$$
$$\quad (3.94) \quad\;\; (3.60) \qquad\quad\; (5.94) \quad\;\; (2.52)$$

$$= (1 - 0.6304B^{12})\hat{a}_t \qquad\qquad \text{(C5.2.3)}$$
$$\quad\;\; (10.86)$$

$$\hat{\sigma}_a = 0.0132052$$

This model is useful as a baseline model for comparison with the DR model results. If the DR model does not fit or forecast the data any better than ARIMA model (C5.2.3), we may not want to bother with the DR model. Since the data in (C5.2.3) are in logs, we may interpret the residual standard deviation conveniently in the original metric as 1.32052 *percent*. We would expect about two-thirds (± 1 standard deviation) of the one-step ahead forecasts from (C5.2.3), stated in the original metric, to fall within about 1.32% of the observed values.

Notice that the ARIMA model for industrial production contains an AR(2)$_s$ component. Applying the test in Equation (2.11.8) in Chapter 2 (with $\hat{\phi}_{12}$ in place of $\hat{\phi}_1$, and $\hat{\phi}_{24}$ in place of $\hat{\phi}_2$) we find that $\hat{\phi}_{12}^2 + 4\hat{\phi}_{24}$ is negative (-0.797); therefore the forecasts from (C5.2.3) include a *stochastic cycle* component. Using Equation (2.11.9) in Chapter 2, we find that the average period of this cycle is 3.4 time units; since the AR(2)$_s$ appears in the seasonal part of the model, the average period is 3.4 *years*. It appears that we have captured a business cycle pattern in the ARIMA model for Y_t.

The vendor performance series ($X_{2,t}$ in Figure C5.3) lacks the upward trend that characterizes the industrial production series. It may be tempting to think that we should therefore relate the *first differences* of industrial production (which would not trend upward) to the *level* of vendor performance. Perhaps these two variables are, indeed, related in this fashion. However, vendor performance has been chosen as a leading indicator because changes in its *level* appear to precede changes in the *level* of industrial production. In accordance with this perception, we will relate the level of Y_t' to the level of $X_{2,t}$, allowing the disturbance term in the DR model, or the input $X_{1,t}'$, to capture the nonstationary part of Y_t'.

C5.2.3 Feedback Check

In this section we check for feedback from Y_t' to $X_{1,t}'$ or $X_{2,t}'$ using the single-equation regression method discussed in Section 5.2.5 in Chapter 5. Recall that the estimates of the coefficients in a single-equation DR model are inconsistent if there is feedback. Our feedback check will help us to avoid the inappropriate use of a single-equation model if there is feedback. We check for feedback by estimating the following two equations:

$$X_{1,t}' = C + b_1 X_{1,t-1}' + \cdots + b_{15} X_{1,t-15}' + c_1 X_{2,t-1} + \cdots + c_{15} X_{2,t-15} + d_1 Y_{t-1}' + \cdots + d_{15} Y_{t-15}' + a_t$$

$$X_{2,t} = C + e_1 X_{2,t-1} + \cdots + e_{15} X_{2,t-15} + f_1 X_{1,t-1}' + \cdots + f_{15} X_{1,t-15}' + g_1 Y_{t-1}' + \cdots + g_{15} Y_{t-15}' + a_t$$

We are especially interested in the $\hat{d}_k$ and $\hat{g}_k$ coefficients since these measure possible feedback from Y_t' to $X_{1,t}'$ and $X_{2,t}$, respectively. These coefficients and their t values are shown in Table C5.1. Only $\hat{g}_{11}$ is signficant. This one significant coefficient can easily be attributed to sampling variation, especially since it occurs at a long lag (11). The F statistics based on (5.2.2) in Chapter 5 are $F = 0.58$ for the first equation, and $F = 1.42$ for the second; both are insignificant at the 5% level for (15,339) degrees of freedom. It is

DR MODEL IDENTIFICATION, ESTIMATION, AND CHECKING

Table C5.1 Regression Coefficients to Estimate Feedback from Y'_t to $X'_{1,t}$ and $X_{2,t}$

k	1	2	3	4	5	6	7	8
$\hat{d}_k$	−0.047	−0.046	0.024	0.059	0.055	−0.098	0.120	−0.064
$\|t\|$	0.49	0.36	0.24	0.81	0.76	1.35	1.65	0.88

k	9	10	11	12	13	14	15
$\hat{d}_k$	−0.074	−0.005	0.034	−0.124	0.116	0.084	−0.035
$\|t\|$	1.02	0.07	0.47	1.71	1.11	0.67	0.40

k	1	2	3	4	5	6	7	8
$\hat{g}_k$	2.579	−3.755	−10.810	9.250	8.585	3.460	−15.720	4.794
$\|t\|$	0.18	0.20	0.71	0.85	0.79	0.32	1.45	0.44

k	9	10	11	12	13	14	15
$\hat{g}_k$	−18.895	5.771	30.789	−14.029	−0.072	−15.784	13.524
$\|t\|$	1.75	0.53	2.86	1.29	0.00	0.84	1.03

reasonable to conclude that there is no feedback from the output to the inputs; we may proceed to build a single-equation model.

C5.3 DR MODEL IDENTIFICATION, ESTIMATION, AND CHECKING

C5.3.1 Initial Identification

We now apply the LTF identification method as discussed in Chapter 5. We estimate two free-form distributed lag transfer functions, with a maximum lag of fifteen months for each input series, and an AR(1)(1)$_{12}$ proxy for the ARIMA structure of the disturbance. Our first preliminary LTF model is

$$Y'_t = C + v_1(B)X'_{1,t} + v_2(B)X_{2,t} + N_t \tag{C5.3.1}$$

where $v_i(B) = v_{i,0} + v_{i,1}B + \cdots + v_{i,15}B^{15}$, for $i = 1$ and $i = 2$, and $N_t = [1/(1 - \phi_{12}B^{12})(1 - \phi_1 B)]a_t$.

The results of estimating (C5.3.1) are shown in Table C5.2. The SACF of the sample disturbance $\hat{N}_t$ is in Figure C5.6. At this point we are most interested in whether differencing is needed to produce a stationary disturbance. The SACF of $\hat{N}_t$ decays quite slowly, suggesting that nonseasonal differencing ($d = 1$) is called for. Therefore we estimate the new preliminary LTF model

$$\nabla Y'_t = C + v_1(B)\nabla X'_{1,t} + v_2(B)\nabla X_{2,t} + n_t \tag{C5.3.2}$$

Table C5.2 Estimation Results for LTF Model (C5.3.1)

$$\hat{C} = 3.4091\,(|t| = 9.66)$$

$k \to$	0	1	2	3	4	5	6	7
$\hat{v}_{1,k}$	−0.0006	0.0284	0.0724	0.0048	0.0417	0.0655	0.0209	−0.0207
$|t|$	0.03	1.22	3.10	0.20	2.00	3.15	1.00	0.99

$k \to$	8	9	10	11	12	13	14	15
$\hat{v}_{1,k}$	0.0129	0.0042	−0.0148	−0.0013	−0.0082	0.0138	0.0172	0.0361
$|t|$	0.61	0.20	0.71	0.06	0.34	0.58	0.72	1.57

$k \to$	0	1	2	3	4	5	6	7
$\hat{v}_{2,k}$	0.0008	0.0035	0.0051	0.0037	0.0029	0.0010	0.0010	0.0009
$|t|$	0.53	2.21	3.23	2.31	2.13	0.78	0.77	0.67

$k \to$	8	9	10	11	12	13	14	15
$\hat{v}_{2,k}$	0.0023	−0.0028	0.0013	0.0019	0.0007	0.0012	−0.0016	−0.0013
$|t|$	1.73	2.15	1.00	1.42	0.47	0.80	1.05	0.87

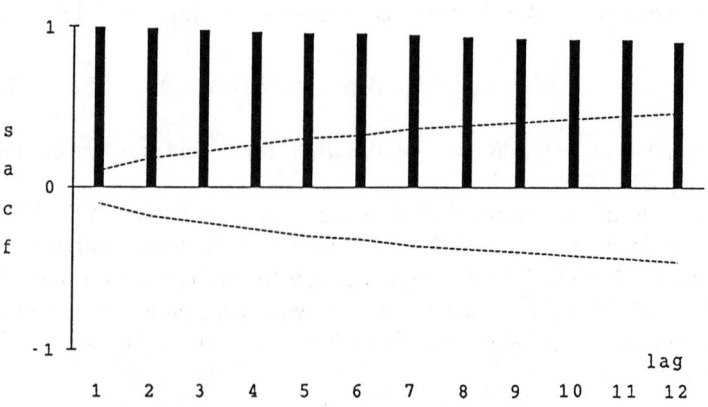

$k \to$	1	12	
$\hat{\phi}_k$	0.9596	0.8655	$\hat{\sigma}_a = 0.0150352$
$|t|$	63.79	32.10	

Figure C5.6 SACF for disturbance $\hat{N}_t$ in model (C5.3.1).

DR MODEL IDENTIFICATION, ESTIMATION, AND CHECKING

Table C5.3   Estimation Results for LTF Model (C5.3.2)

$$\hat{C} = 0.0029\,(|t| = 0.44)$$

$k \to$	0	1	2	3	4	5	6	7
$\hat{v}_{1,k}$	−0.002	0.0265	0.0738	−0.0014	0.0435	0.0660	0.0185	−0.0211
$|t|$	0.01	1.20	3.31	0.06	2.17	3.30	0.92	1.04

$k \to$	8	9	10	11	12	13	14	15
$\hat{v}_{1,k}$	0.0139	0.0021	−0.0185	−0.0010	−0.0010	0.0174	0.0218	0.0297
$|t|$	0.69	0.10	0.92	0.05	0.26	0.77	0.96	1.32

$k \to$	0	1	2	3	4	5	6	7
$\hat{v}_{2,k}$	0.0014	0.0036	0.0054	0.0038	0.0031	0.0008	0.0007	0.0010
$|t|$	0.93	2.41	3.56	2.51	2.35	0.63	0.56	0.76

$k \to$	8	9	10	11	12	13	14	15
$\hat{v}_{2,k}$	0.0023	−0.0030	0.0015	0.0017	0.0008	0.0016	−0.0014	−0.0012
$|t|$	1.80	2.30	1.19	1.34	0.56	1.12	0.98	0.86

$k \to$	1	12
$\hat{\phi}_k$	0.1567	0.8656
$|t|$	3.05	33.29

$\hat{\sigma}_a = 0.0149492$

where $v_1(B)$ and $v_2(B)$ are defined as in (C5.3.2), and $n_t = \nabla N_t = [1/(1 - \phi_{12}B^{12})(1 - \phi_1 B)]a_t$. Of course, the constant and the AR coefficients in the proxy ARIMA model for the disturbance in (C5.3.2) differ from those in (C5.3.1) since the two models have different degrees of differencing.

The results of estimating model (C5.3.2) and the SACF of its estimated disturbance series are shown in Table C5.3 and Figure C5.7, respectively. The main conclusion is that seasonal differencing ($D = 1$) of the disturbance is needed. The SACF of the estimated disturbance in Figure C5.7 has coefficients at the seasonal lags (12,24,36) that decay very slowly. Therefore we form a new preliminary LTF model, with $d = D = 1$. The model is

$$\nabla_{12}\nabla Y_t' = v_1(B)\nabla_{12}\nabla X_{1,t}' + v_2(B)\nabla_{12}\nabla X_{2,t} + n_t \qquad (C5.3.3)$$

where $v_1(B)$ and $v_2(B)$ are defined as in (C5.3.1) and (C5.3.2), and $n_t = \nabla_{12}\nabla N_t = [1/(1 - \phi_{12}B^{12})(1 - \phi_1 B)]a_t$. Again, the new proxy AR coefficients differ from those in the two earlier preliminary LTF models because the models use different degrees of differencing. Notice that (C5.3.3) has no

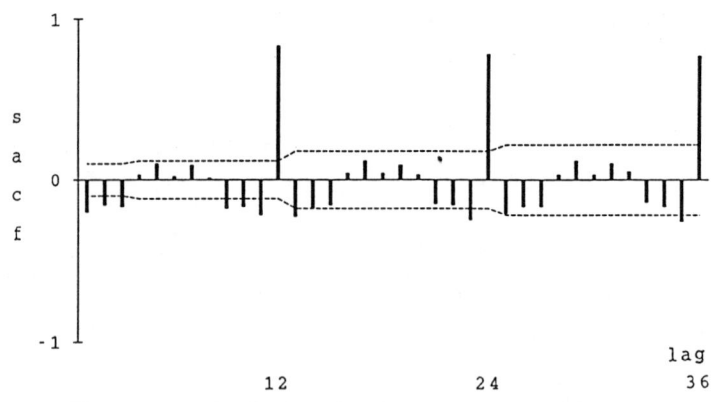

Figure C5.7 SACF for disturbance $\hat{n}_t$ in model (C5.3.2).

constant term since the constant term in (C5.3.2) was insignificant; further differencing will only serve to make the constant term even less important.

Estimation results for model (C5.3.3) are shown in Table C5.4. The SACF of the estimated disturbance series is shown in Figure C5.8; the SPACF is shown in Figure C5.9; the ESACF is shown in Table C5.5. The quick decay (or cutoff) of the SACF at both the nonseasonal and seasonal lags suggests that the disturbance series is now stationary.

Disturbance ARIMA model

The negative spike at lag 12 and the insignificant values at lags 24 and 36 in the SACF (Figure C5.8) suggest an $MA(1)_{12}$ model for the differenced disturbance. The SPACF (Figure C5.9) has decaying values at lags 12, 24, and 36; this tends to confirm the choice of an $MA(1)_{12}$ for the seasonal part of the differenced disturbance.

The nonseasonal part of the disturbance series pattern is more difficult to identify. Together, the SACF and SPACF patterns could call for an AR(1) or an MA(2) or perhaps a mixed model. The ESACF suggests an MA(2) since its first row cuts off to zeros starting at $q = 2$ (disregarding the seasonal effects). We will try both an MA(2) since it seems called for by the ESACF, and an AR(1) since it is a simpler alternative that could be justified from the SACF and SPACF. We have tentatively chosen either an ARIMA $(0, 1, 2)(0, 1, 1)_{12}$ or an ARIMA $(1, 1, 0)(0, 1, 1)_{12}$ model for the disturbance series in our DR model.

Rational Form Transfer Function 1

We must also specify a rational form transfer function for each input. Consider first the stock price input series, $X'_{1,t}$. Table C5.4 shows the estimated impulse response weights ($\hat{v}_{1,k}$) obtained from model (C5.3.3); they are graphed in Figure C5.10 along with the usual dashed lines that are two standard errors from zero. Notice that these $\hat{v}$ weights are mostly positive;

DR MODEL IDENTIFICATION, ESTIMATION, AND CHECKING

Table C5.4 Estimation Results for LTF Model (C5.3.3)

$k \to$	0	1	2	3	4	5	6	7
$\hat{v}_{1,k}$	0.0045	0.0085	0.0643	0.0123	0.0388	0.0555	0.0214	−0.0137
$\|t\|$	0.23	0.44	3.29	0.62	1.97	2.80	1.08	0.69

$k \to$	8	9	10	11	12	13	14	15
$\hat{v}_{1,k}$	−0.0044	0.0188	−0.0002	0.0091	0.0230	0.0140	0.0104	0.0167
$\|t\|$	0.22	0.94	0.01	0.46	1.16	0.70	0.52	0.84

$k \to$	0	1	2	3	4	5	6	7
$\hat{v}_{2,k}$	0.0004	0.0041	0.0048	0.0033	0.0030	−0.0002	−0.0009	0.0002
$\|t\|$	0.26	3.02	3.58	2.49	2.32	0.18	0.72	0.14

$k \to$	8	9	10	11	12	13	14	15
$\hat{v}_{2,k}$	0.0018	−0.0021	0.0007	0.0024	0.0008	0.0016	−0.0009	−0.0008
$\|t\|$	1.43	1.66	0.55	1.91	0.63	1.27	0.69	0.62

$k \to$	1	12						
$\hat{\phi}_k$	0.2232	−0.4637		$\hat{\sigma}_a = 0.0135475$				
$\|t\|$	4.28	10.05						

this is consistent with our expectation that industrial production is positively related to stock prices. To identify a tentative rational form transfer function for these $\hat{v}$ weights, let's apply the identification rules 1 to 5 in Chapter 5 to Figure C5.10:

Table C5.5 ESACF for Disturbance $\hat{n}_t$ in Model (C5.3.3)[a]

$(q \to)$	0	1	2	3	4	5	6	7	8	9	10	11	12
$(p' = 0)$	X	X	0	0	0	0	0	0	0	0	0	X	0
$(p' = 1)$	X	X	0	0	0	0	0	0	0	0	0	X	X
$(p' = 2)$	X	0	0	0	0	0	0	0	0	0	0	X	X
$(p' = 3)$	X	0	X	0	0	0	0	0	0	0	0	X	X
$(p' = 4)$	0	X	X	X	0	0	0	0	0	0	0	X	0
$(p' = 5)$	X	X	X	0	X	0	0	0	0	0	0	X	0
$(p' = 6)$	X	0	X	0	X	0	0	0	0	0	0	X	0

[a] X = significant at 5% level.

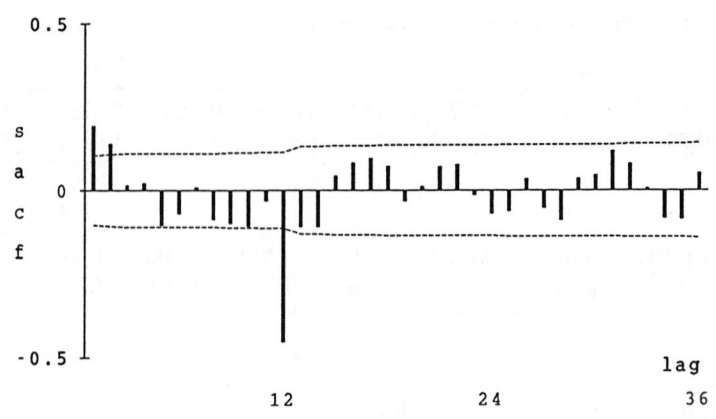

Figure C5.8 SACF for disturbance $\hat{n}_t$ in model (C5.3.3).

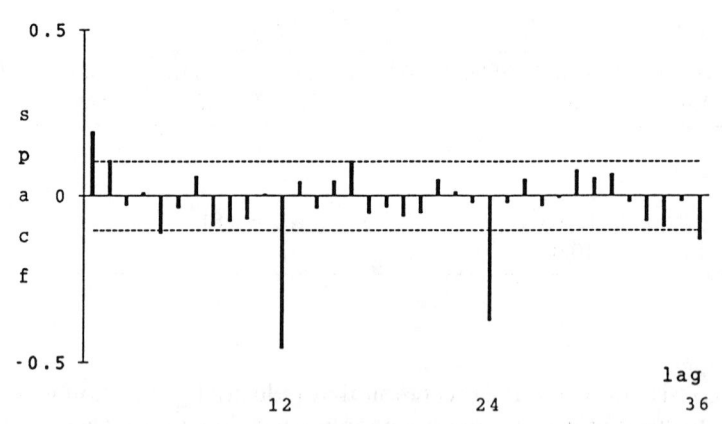

Figure C5.9 SPACF for disturbance $\hat{n}_t$ in model (C5.3.3).

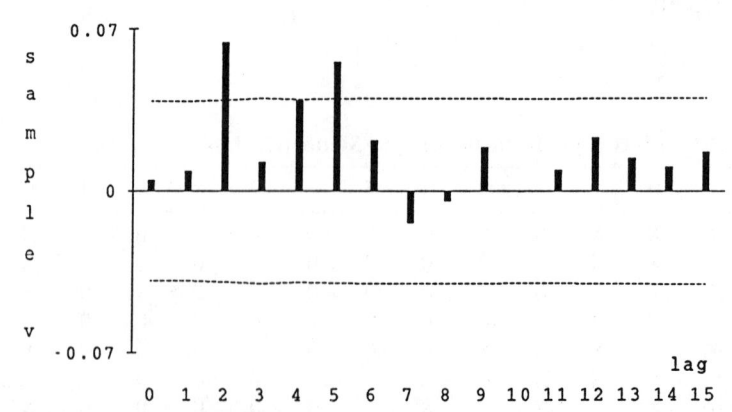

Figure C5.10 Sample impulse response function for input $X'_{1,t}$ in preliminary LTF model (C5.3.3).

DR MODEL IDENTIFICATION, ESTIMATION, AND CHECKING 243

1. There are two insignificant $\hat{v}$ weights, $\hat{v}_{1,0}$ and $\hat{v}_{1,1}$. Their absolute t values are both much less than 2.0. This suggests that the dead time is $b_1 = 2$.

2. The remaining $\hat{v}$ weights could be described as following a simple exponential decay, although they show a lot of irregular variation. (Try drawing a decay line through the peaks at lags 2, 4, 6, 9, 11, and 14.) Alternatively, we might treat all significant $\hat{v}$ weights (at lags 2, 4, and 5 in Figure C5.10) as unpatterned terms and choose $r_1 = 0$ (no decay). It may be difficult to see a decay pattern in Figure C5.10; but it is also difficult to understand why lags 2, 4, and 5 should be especially important, with no patterned relationship to other lags.

For the time being we will interpret the $\hat{v}$-weight pattern in Figure C5.10 as decaying. The decay pattern (if there really is one) is certainly not as clear as the theoretical ones in Figure 5.6, nor as clear as the estimated one for the sales and leading indicator data analyzed in Chapter 5 (Figure 5.4). Of course, in evaluating a sample impulse response function, we must allow for sampling variation and try to focus on the main underlying pattern. In considering tentative decay patterns, we should not ignore the possible role of insignificant $\hat{v}$ weights. All the $\hat{v}$ weights after lag 5 in Figure C5.10 are insignificant, but their behavior is not grossly inconsistent with an exponential decay. While the decay (if there really is one) in Figure C5.10 may be more complicated than a simple exponential pattern, it is best to start with a simpler alternative. For the time being we choose $r_1 = 1$, so $\delta_1(B) = 1 - \delta_{1,1}B$.

3. r_1 is both the order of the denominator in a rational form transfer function *and* equal to the number of decay start-up terms in the numerator of that function. Since we have chosen $r_1 = 1$, we also look for *one* decay start-up term. Rule 3 says: Choose $\max|v|$ as the largest absolute $\hat{v}$ weight from which the decay proceeds. In Figure C5.10, $\max|v|$ occurs at lag 2. (The choice of $\max|v|$ in this example is rather clear. We will see a more controversial example when we discuss the choice of $\max|v|$ for input $X_{2,t}$.)

4. Next we choose u_1, the number of unpatterned $\hat{v}$ weights. With $r_1 > 0$, u_1 is equal to the number of significant $\hat{v}$ weights to the left of the lowest-lag decay start-up term. Our only decay start-up term is $\hat{v}_{1,2}$, and there are *no* significant $\hat{v}$ weights to the left of $\hat{v}_{1,2}$, so $u_1 = 0$.

5. Finally, we find $h_1 = u_1 + r_1 - 1 = 0 + 1 - 1 = 0$. This gives a zero-order operator with one term $(\omega_{1,0})$ in the numerator of the rational form transfer function.

We tentatively choose the orders $(b_1, r_1, h_1) = (2, 1, 0)$ for this transfer function. We may have to restudy our identification information if estimation and checking results show our tentative model to be inadequate.

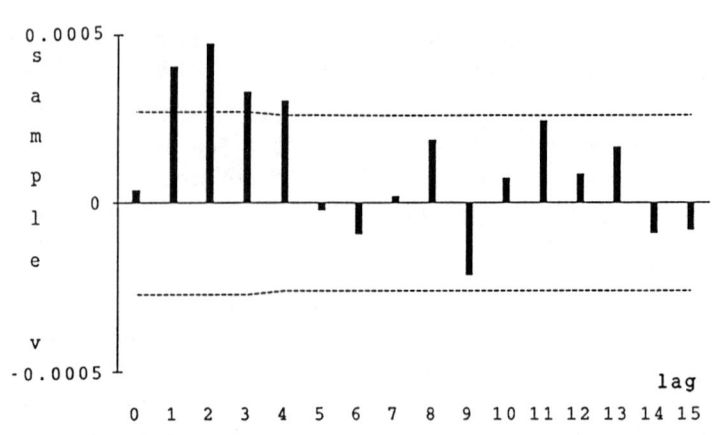

[Figure: bar chart showing sample impulse response, y-axis labeled "sample v" from -0.0005 to 0.0005, x-axis "lag" 0 to 15]

Figure C5.11 Sample impulse response function for input $X_{2,t}$ in preliminary LTF model (C5.3.3).

Rational Form Transfer Function 2

The estimated impulse response weights ($\hat{v}_{2,k}$) obtained from model (C5.3.3) for the vendor performance input ($X_{2,t}$) are also shown in Table C5.4; they are graphed in Figure C5.11. These coefficients, especially the significant ones, are mostly positive; this is consistent with a positive relationship between Y_t and $X_{2,t}$ as we expected. There is one insignificant coefficient ($\hat{v}_{2,0}$), starting from lag 0; this suggests a dead time of $b_2 = 1$ month. Next we must decide what sort of decay pattern, if any, characterizes these $\hat{v}$ weights. Consider these possibilities:

1. Perhaps there is no decay pattern, so that $r_2 = 0$ and $\delta_2(B) = 1$. Then we would focus on the significant terms at lags one through four and specify $\omega_2(B)B = (\omega_{2,0} + \omega_{2,1}B + \omega_{2,2}B^2 + \omega_{2,3}B^3)B$. The remaining $\hat{v}$ weights are insignificant (less than two standard errors from zero) and fairly irregular.

2. Perhaps there is a simple exponential decay, so that $r_2 = 1$ and $\delta_2(B) = 1 - \delta_{2,1}B$. With $r_2 = 1$, we have *one* decay start-up term. If we strictly follow rule 3 in Chapter 5, we choose $\max|v| = \hat{v}_{2,2}$ as the decay start-up term: It is larger in absolute value than any other $\hat{v}$ weight, and there seems to be a decay in the $\hat{v}$ weights after lag 2. This would leave one significant coefficient to the left of lag 2 (at lag 1), so $u_2 = 1$. Then $h_2 = u_2 + r_2 - 1 = 1 + 1 - 1 = 1$, and the tentatively identified transfer function is $[(\omega_{2,0} + \omega_{2,1}B)B/(1 - \delta_{2,1}B)]\nabla_{12}\nabla X_{2,t}$.

3. With $r_2 = 1$, we might choose $\max|v| = \hat{v}_{2,1}$ as the decay start-up term instead of $\hat{v}_{2,2}$. This choice would break rule 3 in Chapter 5; but that rule is proposed only as a guide to thinking. $\hat{v}_{2,1}$ is almost as large as $\hat{v}_{2,2}$, and we can imagine an approximate exponential decay line through the $\hat{v}$ weights after lag 1. (Perhaps we could attribute the larger size of $\hat{v}_{2,2}$ to sampling variation.) This gives a fairly simple model with $(b_1, r_1, h_1) = (1, 1, 0)$. This

DR MODEL IDENTIFICATION, ESTIMATION, AND CHECKING

model has no unpatterned terms, and the decay start-up term is at lag 1. The tentative transfer function for this option is $[\omega_{2,0}B/(1 - \delta_{2,1}B)]\nabla_{12}\nabla X_{2,t}$.

4. Finally, we could suppose that the decay pattern is more complicated, with $r_2 > 0$. However, experience suggests that transfer functions of this type rarely occur, especially with economic data, and we will not pursue it further here.

How shall we choose among these options? Perhaps the best way is to start with the simplest alternative. This appears to be option (3) since it involves the estimation of only two parameters ($\omega_{2,0}$ and $\delta_{2,1}$). This is how we will proceed. Therefore our complete tentative DR model is

$$\nabla_{12}\nabla Y'_t = \frac{\omega_{1,0}B^2}{1 - \delta_{1,1}B}\nabla_{12}\nabla X'_{1,t} + \frac{\omega_{2,0}B}{1 - \delta_{2,1}B}\nabla_{12}\nabla X_{2,t} + n_t \quad (C5.3.4)$$

where we have two alternative ARMA models for the differenced disturbance series $n_t = \nabla_{12}\nabla N_t$. The first is an ARMA(0, 2)(0, 1)$_{12}$; in this instance substitute $n_t = (1 - \theta_{12}B^{12})(1 - \theta_1 B - \theta_2 B^2)a_t$ in (C5.3.4). The second is an ARMA(1, 0)(0, 1)$_{12}$; in this instance substitute $n_t = [(1 - \theta_{12}B^{12})/(1 - \phi_1 B)]a_t$ in (C5.3.4).

C5.3.2 Estimation and Checking

The estimation results for model (C5.3.4) are quite similar for the two ARIMA disturbance models. However, the ARMA(1, 0)(0, 1)$_{12}$ alternative for n_t is superior for three reasons: (1) The DR model residual standard deviation $\hat{\sigma}_a$ is slightly smaller; (2) the absolute t value of the coefficient $\hat{\theta}_2$ in the other model is less than 1.6; and (3) the AR(1) choice for the nonseasonal component is more parsimonious that the MA(2) alternative. Therefore we will discuss in detail only the version of model (C5.3.4) with the ARMA(1, 0)(0, 1)$_{12}$ model for the differenced disturbance n_t.

The estimation results are shown in Table C5.6. All coefficients are significant at about the 5% level since all of their absolute t values are larger than 2.0. The estimated seasonal MA coefficient satisfies the invertibility condition $|\hat{\theta}_{12}| < 1.0$. The estimated AR coefficient satisfies the stationarity condition $|\hat{\phi}_1| < 1.0$. The residual standard deviation is slightly smaller than the one found for the ARIMA model (C5.1.3). There are several significant correlations among the estimated coefficients, but this is to be expected. The highest correlations are between coefficients appearing in the same transfer function ratio, which is not unusual. None of the absolute correlations is large enough (0.9 or larger) to suggest serious instability in the model.

The two residual CCFs in Figures C5.12 and C5.13 are fairly clean; this suggests that we have adequately captured the distributed lag response of the output to the inputs. All residual CCF coefficients are less than two

Table C5.6 Estimation Results for DR Model (C5.3.4)

Parameter	Estimate	t Value
$\omega_{1,0}$	0.0468	3.26
$\delta_{1,1}$	0.8001	9.02
$\omega_{2,0}$	0.0005	5.06
$\delta_{1,2}$	0.7444	10.54
θ_{12}	0.8050	26.35
ϕ_1	0.1910	3.75

$\hat{\sigma}_a = 0.125219$

Coefficient Correlation Matrix

	1	2	3	4	5	6
1	1.00					
2	−0.57[a]	1.00				
3	−0.28[a]	0.15[a]	1.00			
4	0.15[a]	−0.37[a]	−0.59[a]	1.00		
5	−0.02	0.01	0.03	−0.04	1.00	
6	−0.05	0.02	−0.03	0.03	−0.03	1.00

[a] Significant at 5% level.

standard errors from zero, and the approximate χ^2 statistics, computed using Equation (6.1.6), are insignificant. Figures C5.14 and C5.15 reproduce the sample $\hat{v}$ weights in Figures C5.10 and C5.11, respectively, along with solid lines that represent the $\hat{v}$ weights *implied by the rational form transfer functions* in model (C5.3.4). These implied $\hat{v}$ weights are found with Equation (4.3.11) in Chapter 4 using the $\hat{\omega}$ weights and $\hat{\delta}$ weights in (C5.3.4). In Figures C5.14 and C5.15, we see that the two Koyck-type rational form transfer functions in model (C5.3.4) each imply a set of $\hat{v}$ weights that approximately match the $\hat{v}$ weights we found with LTF model (C5.3.3). The match is not

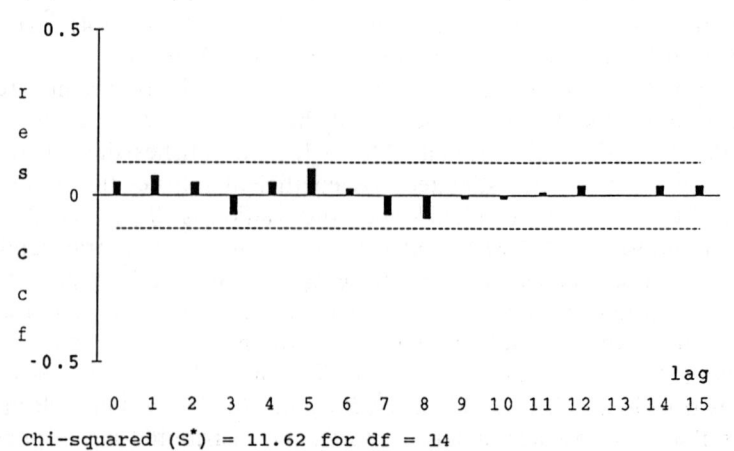

Chi-squared (S*) = 11.62 for df = 14

Figure C5.12 Residual CCF, $\hat{a}_t$ vs. $\hat{a}_{1,t-k}$.

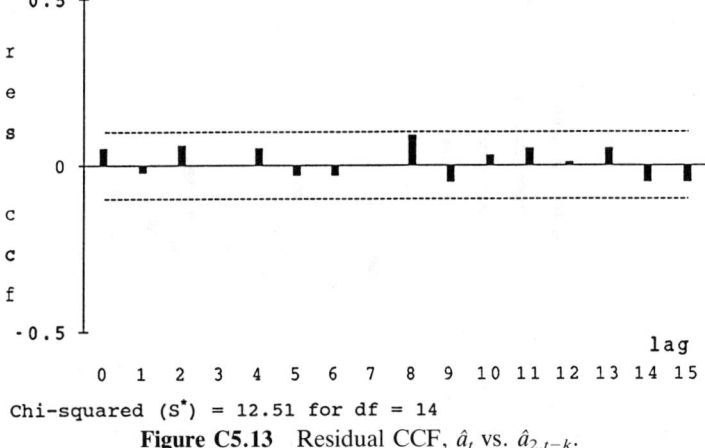

DR MODEL IDENTIFICATION, ESTIMATION, AND CHECKING   247

Chi-squared (s*) = 12.51 for df = 14

Figure C5.13   Residual CCF, $\hat{a}_t$ vs. $\hat{a}_{2,t-k}$.

perfect, of course; but if we allow for sampling variation, it seems that the rather simple rational form transfer functions in model (C5.3.4) provide a parsimonious and useful approximation to the $\hat{v}$-weight patterns.

The residual SACF in Figure C5.16 does not indicate any inadequacy in the disturbance ARIMA model. The normal probability plot of the residuals (Figure C5.17) is nearly a straight line except for two extreme values at each end. This suggests a possible problem with outliers. In fact, the plot of the standardized residuals (Figure C5.18) shows four values beyond the three standard deviation limits (two positive and two negative). It would be wise to study the data for misrecorded observations, or for special events that may have occurred at the times when the outliers appear. The outlier detection method in Chapter 8 might help us to characterize these outliers and choose an intervention transfer function to represent them in an expanded version of model (C5.3.4).

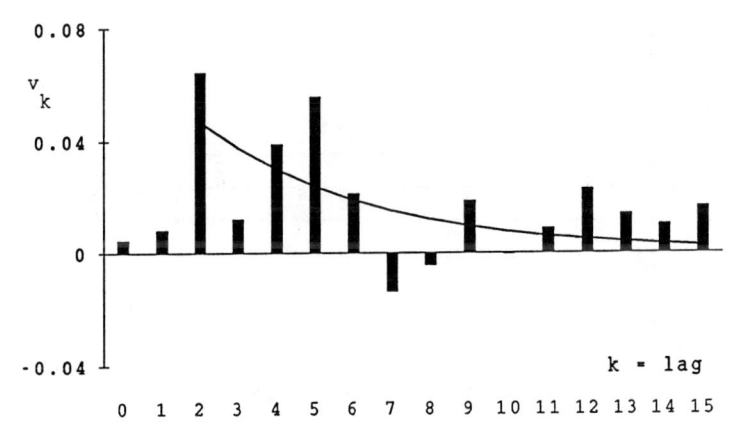

Figure C5.14   Sample impulse response function for $X'_{1,t}$ in Figure C5.10 with v weights implied by transfer function in model (C5.3.4).

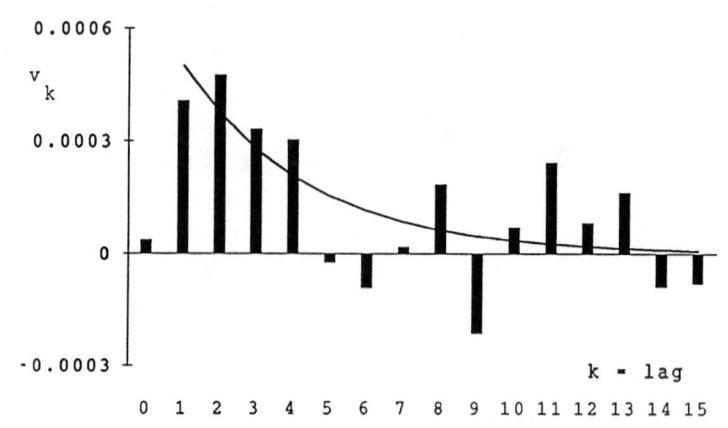

Figure C5.15 Sample impulse response function for $X_{2,t}$ in Figure C5.11 with v weights implied by transfer function in model (C5.3.4).

C5.4 FORECASTING

In this section we generate forecasts from DR model (C5.3.4) and compare their accuracy with forecasts from the ARIMA model for industrial production (C5.1.3). We will transform the forecasts back to the original metric to compare accuracy. As discussed in Chapter 9, the proper method of transformation back to the original metric depends on the nature of the forecast error loss function. For simplicity we use the naive retransformation $F = \exp(f)$, where F is the original metric forecast and f is the natural log metric forecast.

These data offer an interesting check on forecast accuracy. The Commerce Department has designated January 1980 ($t = 397$) as a peak in general business activity and July 1980 ($t = 403$) as a trough. Industrial production,

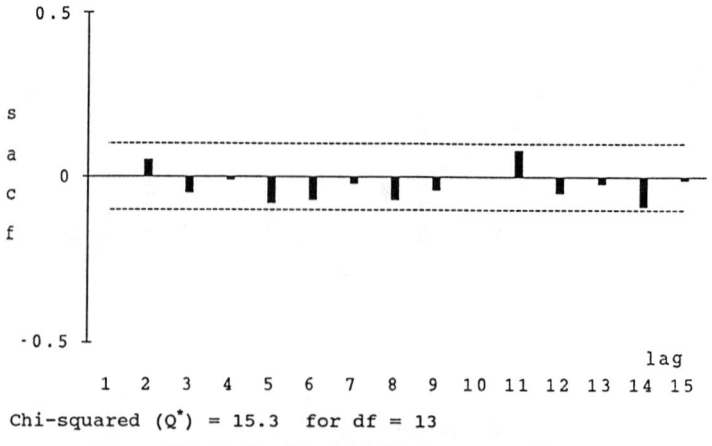

Chi-squared (Q^*) = 15.3 for df = 13

Figure C5.16 Residual SACF for model (C5.3.4).

FORECASTING

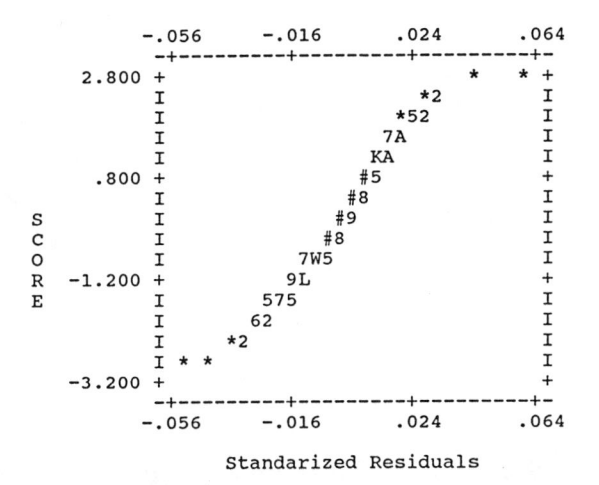

```
                     -.056       -.016        .024         .064
                    -+----------+------------+------------+-
            2.800 +                                 *    * +
                  I                                        I
                  I                             *2         I
                  I                            *52         I
                  I                            7A          I
                  I                            KA          I
             .800 +                            #5          +
                  I                            #8          I
          S       I                            #9          I
          C       I                            #8          I
          O       I                           7W5          I
          R    -1.200 +                        9L          +
          E       I                           575          I
                  I                            62          I
                  I          *2                            I
                  I  * *                                   I
            -3.200 +                                       +
                    -+----------+------------+------------+-
                     -.056       -.016        .024         .064
                            Standarized Residuals
```

Figure C5.17 Normal probability plot of residuals from model (C5.3.4).

one of several measures of current economic activity, reached a cyclical peak (according to Commerce Department seasonally adjusted data) in March 1980 ($t = 399$) rather than in January of that year. The seasonally adjusted series fell slightly in April ($t = 400$), then fell sharply through July 1980 ($t = 403$). The series rebounded in August and September of 1980 ($t = 404$ and 405), marking the start of a cyclical upturn. Thus our forecasts will indicate how well model (C5.3.4) is able to forecast both a cyclical downturn in industrial production (the sharp decline in the seasonally adjusted data starting with May 1980, $t = 401$) and an upturn (the recovery starting with August 1980, $t = 404$).

Figure C5.19 shows the five forecasts (solid line) from time origin $t = 400$ for periods $t = 401$ (lead time 1 month) through $t = 405$ (lead time 5 months). The last 10 available observed values (for $t = 391, 392, \ldots, 400$) and the

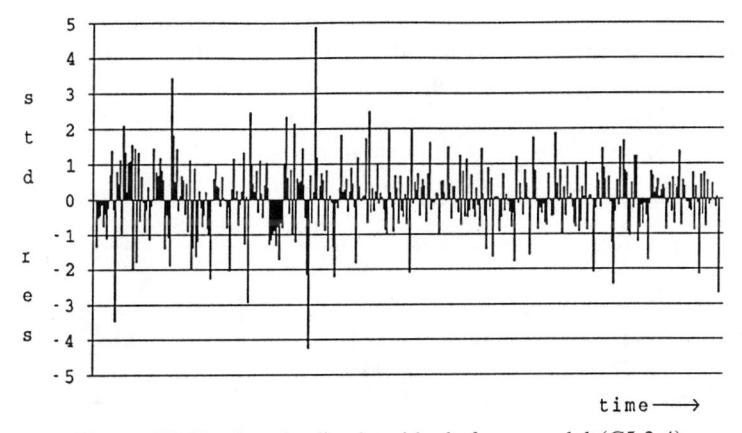

Figure C5.18 Standardized residuals from model (C5.3.4).

INDUSTRIAL PRODUCTION

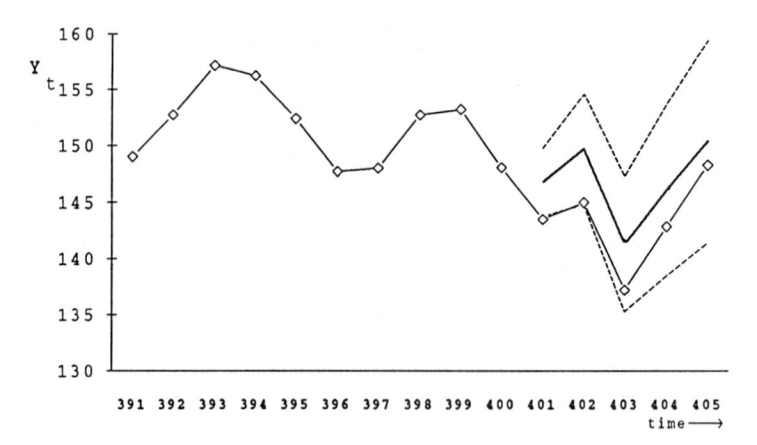

Figure C5.19 Forecasts from DR model (C5.3.4) for $t = 401$ through $t = 405$ from origin $t = 400$.

next 5 "future" values (for $t = 401, 402, \ldots, 405$) are represented by the diamond symbols. The dashed lines are 80% confidence intervals around the point forecasts. The forecasts are all a little too high, but they follow the month-to-month pattern of the actual data rather closely. Only the observed value for $t = 401$ falls outside the 80% confidence range. The seasonal pattern obscures somewhat the underlying business cycle pattern; but the forecasts do tend to capture the cyclical downtrend through $t = 403$ and the subsequent cyclical upswing, in addition to the seasonal pattern.

Forecasts from the DR model (C5.3.4) from time origins $t = 400$ through $t = 404$ (April 1980 through August 1980) are shown in Table C5.7. Observed values of Y_t are shown for comparison. Forecasts are generated through period $t = 405$ from each origin. Thus from origin $t = 400$ we have five forecasts (for $t = 401$ through $t = 405$); from origin $t = 401$ there are four forecasts (for $t = 402$ through $t = 405$); and so on. This gives five one-step ahead forecasts, four two-step ahead forecasts, and so forth.

Table C5.8 shows the percent forecast errors in the original metric for DR model (C5.3.4) and the mean absolute percent errors for the five differ-

Table C.5.7 Forecasts from Five Time Origins Using Model (C5.3.4)

		\multicolumn{5}{c}{Forecast Origin (time)}
Time	Observed	400	401	402	403	404
401	143.5	146.8	—	—	—	—
402	145.0	149.7	145.1	—	—	—
403	137.2	141.3	136.5	136.2	—	—
404	142.9	146.1	141.1	140.9	142.4	—
405	148.3	150.4	145.2	145.2	147.4	147.9

FORECASTING

Table C5.8 Percent Forecast Errors for Five Forecast Origins Using DR Model (C5.3.4)

	\multicolumn{5}{c}{Forecast Origin (time)}				
Time	400	401	402	403	404
401	−2.3%	—	—	—	—
402	−3.2	−0.1%	—	—	—
403	−3.0	0.5	0.7%	—	—
404	−2.2	1.3	1.4	0.3%	—
405	−1.4	2.1	2.1	0.6	0.3%

Mean Absolute Percent Errors

One-step ahead: 0.7%
Two-step ahead: 1.4
Three-step ahead: 2.1
Four-step ahead: 2.2
Five-step ahead: 1.4

ent lead times. These figures may be compared with the corresponding ones for the ARIMA model forecasts from model (C5.1.3) in Table C5.9. Clearly, the DR model forecasts are more accurate. This is consistent with the fact that $\hat{\sigma}_a$ for DR model (C5.3.4) is smaller than $\hat{\sigma}_a$ for ARIMA model (C5.1.3).

Not surprisingly, forecast accuracy tends to improve as the forecast origin

Table C5.9 Percent Forecast Errors for Five Forecast Origins Using ARIMA Model (C5.3.1)

	\multicolumn{5}{c}{Forecast Origin (time)}				
Time	400	401	402	403	404
401	−2.7%	—	—	—	—
402	−3.9	−0.3%	—	—	—
403	−3.9	0.4	0.8%	—	—
404	−3.3	1.3	1.7	0.8%	—
405	−2.3	2.4	2.9	1.8	0.7%

Mean Absolute Percent Errors

One-step ahead: 1.1%
Two-step ahead: 2.0
Three-step ahead: 2.7
Four-step ahead: 2.9
Five-step ahead: 2.3

advances. The mean absolute percent errors at the bottom of Table C5.8 summarize this tendency. For example, in Figure C5.7 the three-step ahead forecast from origin $t = 400$, for $t = 403$ (the cyclical trough), is 141.3; this is rather high compared to the observed value (137.2). But the one-step ahead forecast for $t = 403$ from origin $t = 402$ is 136.2; this is much closer to the observed value (137.2).

Forecasting Cyclical Patterns

The ARIMA model for the disturbance series in DR model (C5.3.4) no longer has the $AR(2)_{12}$ stochastic cycle component that characterized the ARIMA pattern for industrial production in model (C5.1.3). Apparently stock prices and vendor performance together account for the business cycle component, removing it from the DR model disturbance series.

QUESTIONS AND PROBLEMS

1. Adding or dropping an input in an LTF model can sometimes change drastically the estimated impulse response pattern(s) for the remaining included input(s). Try estimating an LTF model with $X'_{1,t}$ as the only input. Do you get a $\hat{v}_{1,k}$ pattern that is similar to the one in Figure C5.10?

2. Modify the orders of the transfer function for $X_{2,t}$ in (C5.3.4) to $(b_1, r_1, h_1) = (1, 0, 3)$. How many decay start-up terms appear in the numerator of this modified transfer function? Is this modified DR model superior to (C5.3.4)?

CHAPTER 7

Intervention Analysis

7.1 INTRODUCTION

We saw the use of deterministic inputs in DR models in Chapter 3 where the inputs are seasonal dummy variables, and in Case 2 where the inputs are trading day variables. Deterministic inputs can be used in DR models to represent identified events, called *interventions*. They may also be used to account for unexplained *outliers* (unusual observations) in a time series. In Chapter 8 we consider a method for *detecting* interventions and outliers. In this chapter we assume that the occurrence and timing of an intervention are known. Models of the type we will consider have a broad variety of applications. For example, Box and Tiao (1975) consider environmental and economic policy issues; Wichern and Jones (1977) consider marketing and management issues; and Zimring (1975) considers effects of gun control legislation.

Consider the saving rate data introduced in Chapter 1, reproduced in Figure 7.1. Congress passed a law granting a one-time tax rebate at time $t = 82$ (the second quarter of 1975, denoted 1975 II), indicated by the arrow in Figure 7.1. According to some economic theories of consumer behavior (e.g., the permanent income hypothesis), a temporary rise in disposable income will cause saving and the saving rate to rise. Inspection of Figure 7.1 suggests that the saving rate did, indeed, rise in 1975 II.

Suppose we want to estimate the effect of the 1975 tax rebate on the saving rate. We might be able to construct a quarterly *stochastic* variable reflecting the size of all tax rebates over the sample period. We could then use this variable as an input in a DR model for the saving rate. However, it might be difficult and time consuming to create such a variable. In addition, we might have difficulty separating the effect of the 1975 II rebate from the effects of other rebates. Alternatively, we might be able to adequately represent the event "one-time tax rebate, 1975 II" with a binary (0, 1) deterministic input variable. As discussed in Chapter 1, this variable could have the value $X_t = 1$ for $t = 1975$ II, and $X_t = 0$ for all other time periods. We will return to this example in Section 7.4.

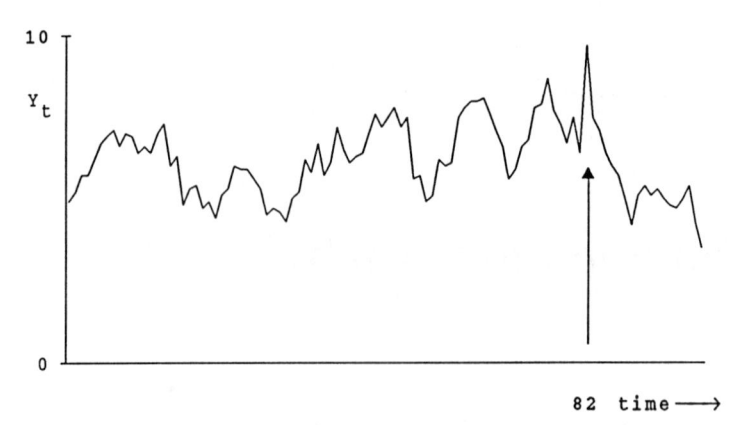

Figure 7.1 Quarterly U.S. saving rate, 1955 I–1979 IV.

The framework we use for evaluating an intervention effect represented by a single intervention variable is the rational form DR model that we first set out in Chapter 4:

$$Y_t = C + \frac{\omega(B)B^b}{\delta(B)} X_t + N_t \qquad (7.1.1)$$

where the definitions of all variables and operators are the same as in earlier chapters. The only difference is that X_t is a binary deterministic variable rather than a stochastic variable. As in earlier chapters, N_t may be described by an ARIMA process. Thus an intervention model is a special case of a DR (transfer function plus disturbance) model. We may have both stochastic and deterministic inputs in a DR model. In this chapter we restrict our attention to deterministic inputs.

In the next two sections we consider two main intervention types: *pulse* and *step*. Then in Section 7.4 we consider procedures for building DR models with intervention components. In Section 7.5 we consider how we can specify DR models that have multiple and compound intervention effects.

7.2 PULSE INTERVENTIONS

Time series variables might be interrupted by an infinite variety of interventions. We will focus on several simple types that are often useful. The first type is a *pulse* intervention. A pulse intervention is a *temporary* event that may affect the level of the output variable. Suppose the pulse intervention event occurs during period $t = i$. Then the pulse input variable X_t is defined as

PULSE INTERVENTIONS

$$X_t = 0 \quad t \neq i$$
$$= 1 \quad t = i$$

X_t is a binary (0 = "off," 1 = "on") variable. The temporary nature of a pulse intervention is reflected in X_t being "on" only during period i.

As a simple example of a pulse intervention, let Y_t be the number of students attending class each day at a multicampus university. If one campus is hit by an ice storm and is therefore shut down for a day, Y_t would drop on that day. If the effects of the storm disappear by the end of that day, Y_t might be expected to return to normal thereafter. In this case we can represent the effect of the storm on class attendance with a binary pulse input variable.

7.2.1 One-Period Temporary Response

The effect of a pulse intervention on series Y_t might appear as shown in Figure 7.2(*a*). In this example Y_t has a stationary mean except for the effect of the pulse intervention at $t = i$. (In all these examples we ignore stochastic fluctuations in Y_t so as to focus on the effect of the intervention on the level of Y_t.) During period $t = i$, Y_t shifts up (as illustrated) or down, away from the previously constant level. Immediately after $t = i$, the series *returns to its earlier level*.

Notice that this is a *one-period* response: Therefore Y_t has *no dynamic* (distributed lag) response to the intervention. The transfer function part of DR model (7.1.1) in this case is $\omega_0 X_t$, where ω_0 is the size of the displacement of Y_t during period i. If $\omega_0 = 100$, then Y_t temporarily rises by 100 units during period i. If $\omega_0 = -50$, then Y_t temporarily falls by 50 units during period $t = i$. In Figure 7.1 it appears that the saving rate series may have been interrupted by a pulse intervention of this type at period $t = i = 82$, with $\omega_0 > 0$.

For a series with a nonstationary mean, a pulse intervention might appear as shown in Figure 7.2(*b*). Series Y_t wanders (upward in this example) due to its nonstationary character until $t = i$. At period i the pulse intervention causes Y_t to shift down (as shown) or up. After period i the series *returns to the level determined by its nonstationary character*. This example is another *one-period* response. The transfer function part of (7.1.1) again is $\omega_0 X_t$. The nonstationary character of Y_t is reflected in the ARIMA structure of the disturbance N_t.

7.2.2 Multiperiod Temporary Response

Pulse interventions may also involve multiperiod (dynamic, or distributed lag) responses. That is, a pulse intervention at $t = i$ may cause a response not only during period i but also during later periods. Figure 7.2(*c*) shows

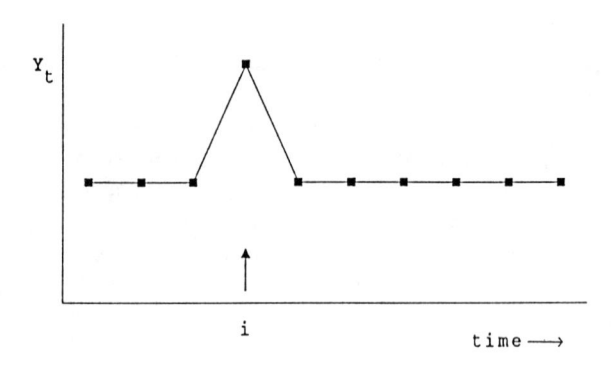

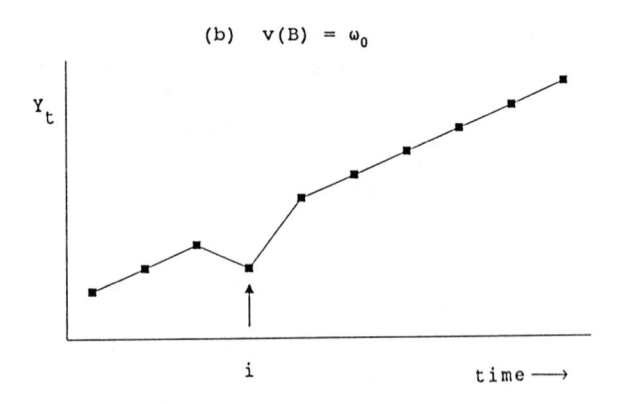

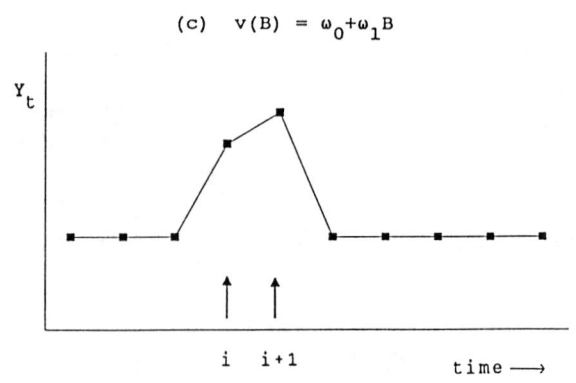

Figure 7.2 Examples of pulse interventions. (*a*) $v(B) = \omega_0$. (*b*) $v(B) = \omega_0$. (*c*) $v(B) = \omega_0 + \omega_1 B$. (*d*) $v(B) = \omega_0/(1 - \delta_1 B)$, $|\delta_1| < 1$. (*e*) $v(B) = \omega_0/(1 - \delta_1 B)$, $|\delta_1| < 1$. (*f*) $v(B) = \omega_0/(1 - \delta_1 B)$, $\delta_1 = 1$.

(d) $v(B) = \dfrac{\omega_0}{1 - \delta_1 B}$, $|\delta_1| < 1$

(e) $v(B) = \dfrac{\omega_0}{1 - \delta_1 B}$, $|\delta_1| < 1$

(f) $v(B) = \dfrac{\omega_0}{1 - \delta_1 B}$, $\delta_1 = 1$

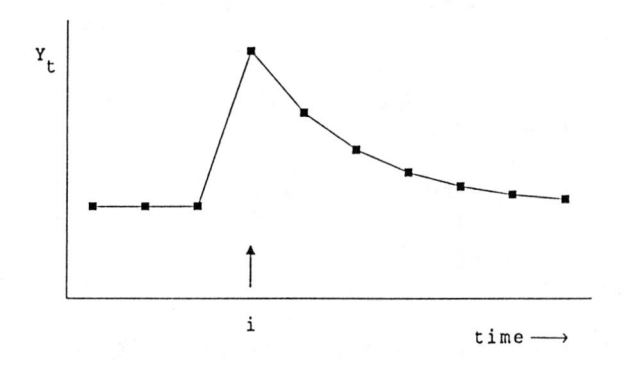

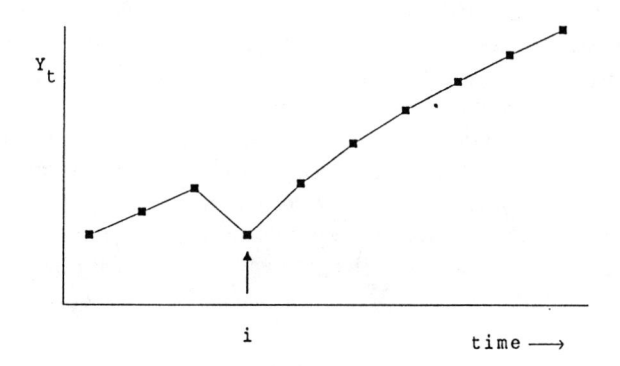

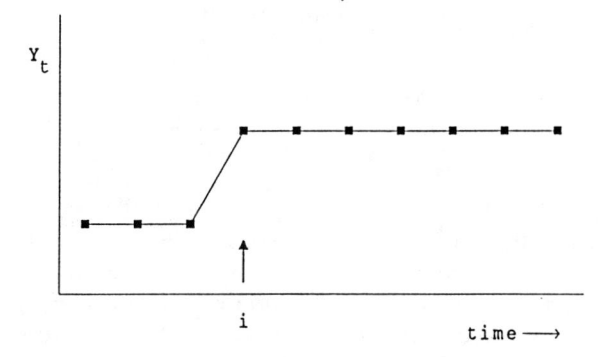

Figure 7.2 (*Continued*)

an example for a series that has a constant mean except for the effects of the intervention. This is the same as Figure 7.2(a), except that in 7.2(c) Y_t continues to react to the intervention during period $i + 1$. In 7.2(c) the response during $t = i + 1$ is greater than during $t = i$. In this case the transfer function part of (7.1.1) is $(\omega_0 + \omega_1 B)X_t$. Suppose $\omega_0 = 100$, and $\omega_1 = 150$. Then Y_t rises by 100 units during $t = i$, and it rises by 150 units during $t = i + 1$. Notice that these effects are *not cumulative*: Since $X_t = 0$ after $t = i$, therefore $\omega_0 X_t = 0$ after $t = i$, and $\omega_1 B X_t = 0$ after $t = i + 1$. This says that the initial response of $+100$ during period i is lost completely after period i, and the secondary response of $+150$ during period $i + 1$ is lost completely after period $i + 1$. Therefore Y_t is displaced during $t = i + 1$ only by the lagged response of $+150$ units, not by $100 + 150 = 250$ units. And Y_t is displaced after $t = i + 1$ by 0 units, not by $100 + 150 = 250$ units.

It is difficult to tell just from inspection, but the saving rate data in Figure 7.1 may involve a dynamic response. The observation at $t = 83$, for example, may be affected somewhat by the tax rebate that became effective during $t = 82$. We will examine these data further in Section 7.4.

The series in Figure 7.2(d) shows a *continuing dynamic response* following period i. In this example each response after $t = i$ is a constant fraction ($\delta_1 = 0.5$) of the response during the previous period. Notice that this graph has the same decaying appearance as an impulse response weight graph for a Koyck-type transfer function [see Figure 5.6(a)]. In fact, the specification of the transfer function part of (7.1.1) for this example is $[\omega_0/(1 - \delta_1 B)]X_t$, which is the Koyck model. ω_0 is the initial response at $t = i$ (the decay start-up term), and $\delta_1 = 0.5$ captures the exponential decay dynamic response for $t > i$. Suppose $\omega_0 = 100$, and $\delta_1 = 0.5$; then we have the following response pattern (displacements to Y_t):

t	Response
i	$\omega_0 = 100$
$i + 1$	$\delta_1 \omega_0 = (0.5)100 = 50$
$i + 2$	$\delta_1^2 \omega_0 = (0.5)^2 100 = 25$
$\vdots$	$\vdots$
$i + k$	$\delta_1^k \omega_0 = (0.5)^k 100$

Notice that coefficient δ_1 in this example satisfies the stability condition $|\delta_1| < 1$. Thus, in the limiting case as k goes to infinity, the response approaches zero for this example because the factor δ_1^k approaches zero. As with other pulse interventions, these responses are not cumulative. Since $X_t = 0$ after $t = i$, all preceding effects are lost. After $t = i$ we get only the distributed lag responses to the initial effect (ω_0), determined by the decay coefficient δ_1. It is tempting to compute the *gain* (g = total cumulated response) for the transfer function in this example, as discussed in Chapter 4.

STEP INTERVENTIONS

Setting $B = 1$ in the transfer function, we get $g = \omega_0/(1 - \delta_1)$, which is $100/0.5 = 200$ for our numerical example. This presents a problem, however: g is defined as the total change in Y_t when X_t rises *permanently* by one unit. But in this case X_t rises from 0 to 1 only temporarily, by definition of a pulse intervention. In the case of a pulse intervention, g is the *cumulation of the displacements to the level of Y_t*, but it is *not* the ultimate equilibrium change in the level of Y_t since the latter is necessarily zero in response to a pulse intervention.

Figure 7.2(*e*) is similar to Figure 7.2(*b*) for a series with a nonstationary mean, except that in 7.2(*e*) Y_t reacts dynamically to the pulse intervention after period $t = i$, with a decreasing response during periods $i + 1, i + 2, i + 3, \ldots$. This is the Koyck model, where $\omega(B)B^b/\delta(B) = \omega_0/(1 - \delta_1 B)$, with Y_t and X_t differenced appropriately to capture the nonstationary character of the disturbance. Once again, we have the restriction $|\delta_1| < 1$. Thus, as the response of Y_t to the pulse intervention fades away, Y_t *returns to the level determined by its nonstationary character*.

Finally, Figure 7.2(*f*) shows the special case of a Koyck-type pulse intervention with $\delta_1 = 1$. In this case the pulse intervention has the appearance of a *step* (a permanent change) rather than a pulse. The logic of this conclusion is seen as follows. Recall that δ_1 is the rate at which the response of Y_t decays following its initial response (ω_0) to the pulse event. For example, if $\delta_1 = 0.5$, each succeeding response of Y_t is equal to 0.5 times the previous period's response. Therefore, if $\delta_1 = 1$, then each succeeding response of Y_t is equal to 1.0 times the previous response; that is, there is no decay in the response. The implication is that the intervention effect is a *step* effect rather than a pulse when $\delta_1 = 1$. We now consider step interventions in more detail.

7.3 STEP INTERVENTIONS

The second major type of intervention we consider is a *step* intervention. This may be thought of as a *permanent* change in the level of Y_t. For a step intervention at time $t = i$, define

$$X_t = 0 \quad t < i$$
$$= 1 \quad t \geq i$$

The permanent nature of a step intervention is reflected in X_t being "on" continuously starting at period i. That is, X_t has the value 1 at $t = i$ *and permanently thereafter*.

We will use the word "permanent" as a helpful way to distinguish between pulse and step interventions. However, a step intervention need not be literally permanent. For example, instituting martial law in a country might have a step effect at time $t = i$ on $Y_t = $ volume of alcohol consumption, as illustrated in Figures 7.3(a) and 7.3(b). But if martial law is lifted at period $t = i + 3$, the step effect may cease, as shown in Figure 7.3(b). The intervention variable X_t for this example is defined as

$$X_t = 0 \quad i > t > i + 2$$
$$= 1 \quad t = i, \ i + 1, \ i + 2$$

In this case we may think of the intervention as having a permanent (step) effect on Y_t *until the policy action is reversed.* For simplicity, we will assume in the following examples that each step intervention is literally permanent.

7.3.1 One-Period Permanent Response

A stationary series with a constant mean may be affected by a step intervention as shown in Figure 7.3(a). Y_t is at a constant level until period i when it shifts up (shown here) or down. Since this shift is permanent, Y_t stays at its new level after period i. The transfer function representation in the DR model (7.1.1) for this type of intervention is $\omega_0 X_t$. Since $X_t = 1$ for $t \geq i$, the level of Y_t is higher by ω_0 units for $t \geq i$. If $\omega_0 = 100$, Y_t is permanently higher by 100 units. If $\omega_0 = -50$, Y_t is permanenly lower by 50 units.

Figure 7.3(c) illustrates a step intervention for a series with a nonstationary mean. The nonstationary character of the series causes it to wander upward (in this example) until period i, when the step intervention causes it to shift down (shown here) or up by ω_0 units. The nonstationary wandering continues after $t = i$, but *from a new permanently lower level* in this example. This is in contrast to Figure 7.2(b), where a *pulse* intervention affects the level of Y_t for only one period.

7.3.2 Multiperiod Permanent Response

Step interventions may also have dynamic effects. Figure 7.3(d) shows an example. Y_t has a stationary mean except for the step intervention at period i. The transfer function part of (7.1.1) for this example is $(\omega_0 + \omega_1 B)X_t$. ω_0 captures the rise (in this example) or fall in Y_t during period i, and ω_1 captures the rise (in this example) or fall during period $i + 1$.

Suppose $\omega_0 = 25$ and $\omega_1 = 75$. Then Y_t rises by $\omega_0 = 25$ units during period i, and by an *added* $\omega_1 = 75$ units during period $i + 1$. Since a step intervention is permanent, the effect from period i is not lost during later periods. Thus the total change in the level of Y_t during period $i + 1$, *compared to its level in periods before* $t = i$, is $+100$ units; this is found as $\omega_0 = 25$ units (the *continuing* period i response) *plus* $\omega_1 = 75$ units (the added permanent response at $t = i + 1$). This total (cumulated) change remains in place for all future periods since $X_t = 1$ for $t \geq i$. Since X_t rises permanently by one unit,

STEP INTERVENTIONS

(a) $v(B) = \omega_0$

(b) $v(B) = \omega_0$

(c) $v(B) = \omega_0$

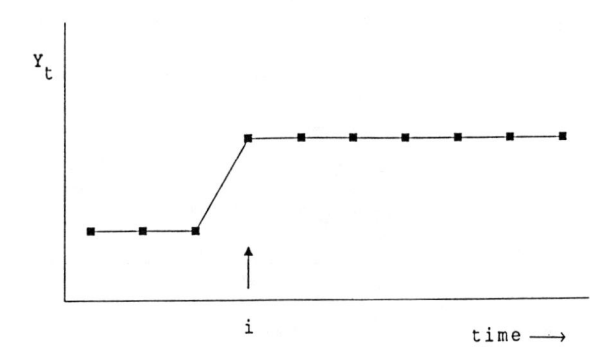

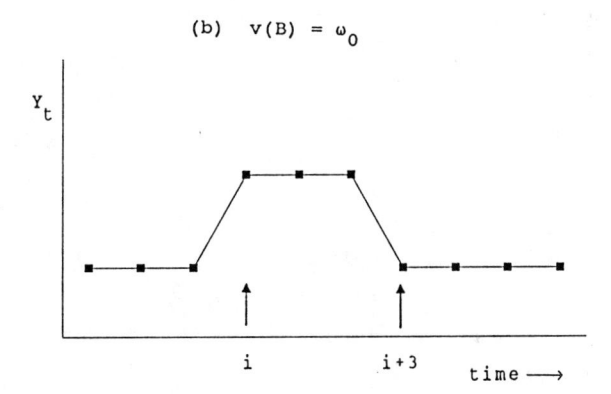

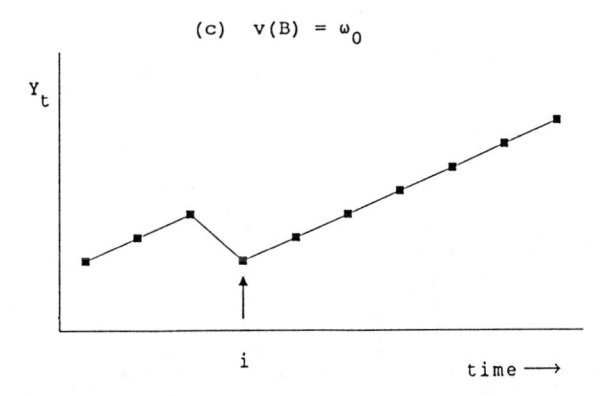

Figure 7.3 Examples of step interventions. (*a*) $v(B) = \omega_0$. (*b*) $v(B) = \omega_0$. (*c*) $v(B) = \omega_0$. (*d*) $v(B) = \omega_0 + \omega_1 B$. (*e*) $v(B) = \omega_0/(1 - \delta_1 B)$, $|\delta_1| < 1$. (*f*) $v(B) = \omega_0/(1 - \delta_1 B)$, $\delta_1 = 1$.

(d) $v(B) = \omega_0 + \omega_1 B$

(e) $v(B) = \dfrac{\omega_0}{1 - \delta_1 B}$, $|\delta_1| < 1$

(f) $v(B) = \dfrac{\omega_0}{1 - \delta_1 B}$, $\delta_1 = 1$

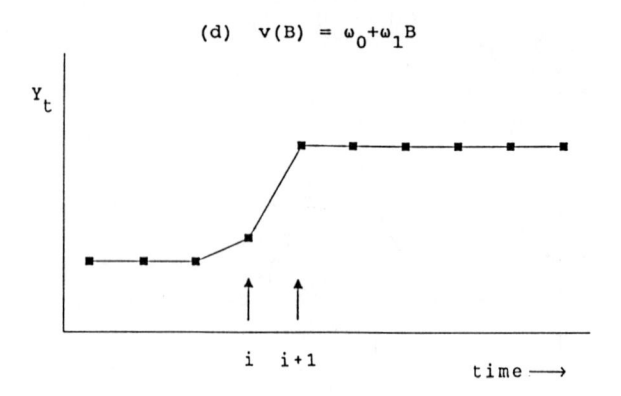

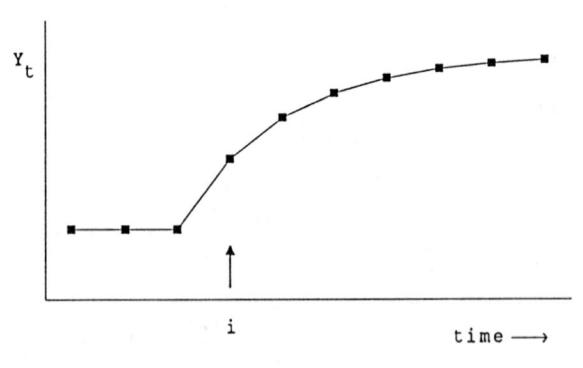

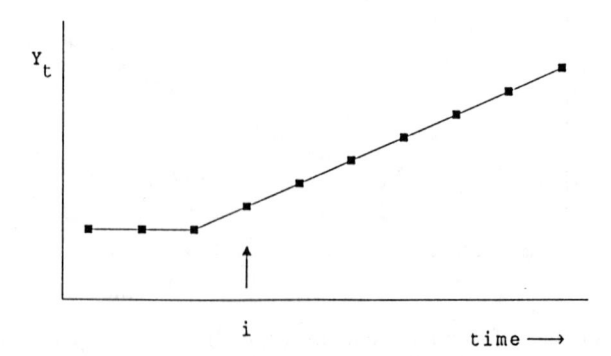

Figure 7.3 (*Continued*)

STEP INTERVENTIONS

we may compute the gain (g). Setting $B = 1$ in the transfer function, we have $g = \omega_0 + \omega_1$, or $g = 25 + 75 = 100$ for our numerical example.

Another dynamic response to a step intervention is shown in Figure 7.3(e). The transfer function is the Koyck model, $[\omega_0/(1 - \delta_1 B)]X_t$, with $|\delta_1| < 1$. Y_t responds initially to the step intervention at $t = 1$ by rising (in this example) or falling by $\omega_0 X_t = \omega_0$ units during period i. During period $i + 1$, Y_t rises *further* by a fraction (δ_1) of the period i response (ω_0) so the *added* response is $\delta_1 \omega_0$. Since the period i response is permanent (a step), the *cumulative* change through $t = i + 1$ is $\omega_0 + \delta_1 \omega_0 = \omega_0(1 + \delta_1)$ units. This response is also permanent. The *added* response during each subsequent period is a fraction (δ_1) of the previous period's *added* response, and each resulting cumulated response is permanent. Suppose $\omega_0 = 100$ and $\delta_1 = 0.5$; then we have the following response pattern:

t	Added Response	Cumulated Response
i	$\omega_0 = 100$	$\omega_0 = 100$
$i+1$	$\delta_1 \omega_0 = (0.5)100 = 50$	$\omega_0(1 + \delta_1) = 100(1 + 0.5) = 150$
$i+2$	$\delta_1^2 \omega_0 = (0.5)^2 100 = 25$	$\omega_0(1 + \delta_1 + \delta_1^2) = 100(1 + 0.5 + 0.5^2) = 175$
$\vdots$	$\vdots$	$\vdots$
$i+k$	$\delta_1^k \omega_0 = (0.5)^k 100$	$\omega_0(1 + \delta_1 + \delta_1^2 + \cdots + \delta_1^k) = 100(1 + 0.5 + 0.5^2 + \cdots + 0.5^k)$

How much is the ultimate cumulative change in Y_t (the gain, g) in response to this step intervention? Consider the cumulated response in the last line in the preceding table. In the limiting case as k goes to infinity, the second factor converges to $1/(1 - \delta_1)$ if $|\delta_1| < 1$. Therefore the gain is $\omega_0/(1 - \delta_1)$, which is $100/0.5 = 200$ for our numerical example. This gain may also be found as discussed in Chapter 4 by setting $B = 1$ in the transfer function.

Finally, Figure 7.3(f) shows a Koyck-type step intervention effect, with $\delta_1 = 1$. This has the appearance of a *ramp* rather than a step. The logic of this result is seen as follows. Recall that δ_1 is the rate at which the added response of Y_t decays following its initial response (ω_0) to the step event. For example, if $\delta_1 = 0.5$, then each succeeding added response of Y_t is equal to 0.5 times the previous period's added response. Therefore if $\delta_1 = 1$, each succeeding added response of Y_t is equal to 1.0 times the previous added response; that is, there is no decay when $\delta_1 = 1$. The implication is that the intervention effect is a ramp rather than a step. That is, with a step input the responses are cumulative. Therefore Y_t will first change by ω_0 units at time $t = i$, then by $\omega_0 + \omega_0 = 2\omega_0$ units at time $t = i + 1$, then by $\omega_0 + \omega_0 + \omega_0 = 3\omega_0$ units at time $t = i + 2$, and so forth.

Ramp effects are rather unusual. They may arise, for example, when data series are redefined. To capture a ramp effect that begins a time $t = i$, we define the following intervention variable:

$$X_t = 0 \quad t < 1$$
$$= t - i + 1 \quad t \leq i$$

Thus X_t increases linearly starting at $t = i$. If the transfer function is $\omega_0 X_t$, the ramp intervention is a ramp *up* in the data if $\omega_0 > 0$; it is a ramp *down* in the data if $\omega_0 < 0$.

7.4 BUILDING INTERVENTION MODELS

7.4.1 Transfer Function Identification by Hypothesis

One way to specify the transfer function part of (7.1.1) for an intervention component is *by hypothesis*. Identifying intervention transfer functions by hypothesis has clear advantages. It leads us to search for guidance from the relevant theory, perhaps from economics, psychology, physics, or some other discipline. Even if we find no relevant accepted theory, we are led to *think about the likely effects* of the possible intervention. No doubt, it's a good idea to think about what we are doing.

For example, consider the saving rate data in Figure 7.1. In this example we hypothesize that the one-time tax rebate during $t = 82$ will have a one-period temporary effect on the saving rate, with dead time $b = 0$. We are guided by economic theory (the permanent income hypothesis) that suggests that a temporary change in household income temporarily affects mainly saving (and the saving rate) rather than consumption spending. Therefore we create the pulse input

$$X_t = 1 \quad \text{for } t = i = 82 \quad \text{where } t = 82 \text{ is 1975 II}$$
$$= 0 \quad \text{for all other } t \tag{7.4.1}$$

The hypothesized transfer function is $\omega_0 X_t$, with X_t defined in (7.4.1). From (7.1.1) thus far our tentative DR model for this example is

$$Y_t = C + \omega_0 X_t + N_t \tag{7.4.2}$$

7.4.2 Disturbance ARIMA Model Identification

In Chapter 1 we estimated model (7.4.2) for the saving rate data using ordinary regression methods. In particular, we treated the disturbance N_t as zero-mean and normally distributed white noise. The resulting model (see Table 1.6, Example 2), with absolute t values in parentheses is

$$Y_t = 6.1343 + 3.5656 X_t + \hat{N}_t$$
$$(56.29) \quad (3.27) \tag{7.4.3}$$

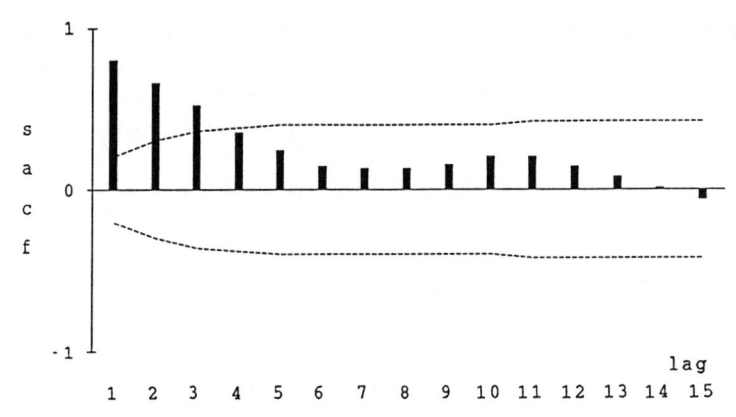

Figure 7.4 Residual SACF for model (7.4.3) applied to the saving rate data, with no ARIMA model for N_t.

with $\hat{\sigma}_N = 1.08427$. However, N_t might well be autocorrelated. As with any DR model, in building intervention models we are concerned about specifying an appropriate ARIMA model for the disturbance component.

The SACF for the estimated disturbance series ($\hat{N}_t = Y_t - 6.1343 - 3.5656 X_t$) for the saving rate intervention model (7.4.3) is shown in Figure 7.4. This series shows clear evidence of autocorrelation; therefore the ordinary regression estimates are inefficient, and the usual standard errors of the estimated coefficients and the associated t values are invalid. We will try to improve on these results by including an ARIMA model for N_t in (7.4.3).

Using the ARIMA Model for Y_t

One way to identify an ARIMA model for N_t in an intervention DR model is to *use the preliminary ARIMA model for Y_t*. After all, if we remove the intervention component from model (7.1.1), we have $Y_t = C + N_t$. Therefore the autocorrelation pattern of N_t should be the same as that of Y_t, aside from the intervention.

For the saving rate data example, using all $n = 100$ observations, the following ARIMA(1, 0, 2) model (with $\hat{\theta}_1 = 0$) for Y_t emerged from the usual identification, estimation, and diagnostic checking steps:

$$(1 - 0.7359 B) Y_t = 1.6138 + (1 + 0.3437 B^2) \hat{a}_t \quad (7.4.4)$$
$$(9.58) \quad\quad (3.35) \quad\quad (3.31)$$

with $\hat{\sigma}_a = 0.663716$; figures in parentheses are absolute t values. We could use this ARIMA(1, 0, 2) model as a tentative ARIMA model for N_t in the intervention DR model (7.4.3) for the saving rate series.

However, a problem can arise in using the preliminary ARIMA model for Y_t to identify an ARIMA model for N_t: The preliminary model for Y_t is built using a data set that *includes the effects of the intervention(s)*, and *interventions can change the autocorrelation patterns in a time series.* The effects may be slight or substantial, depending on the nature of the intervention, as discussed by Guttman and Tiao (1978) and Chang and Tiao (1983a). To avoid this problem we could identify an ARIMA model for Y_t (and therefore N_t) *using only data before (or after) the intervention* if enough data are available. In the saving rate example, the possible intervention occurs at $t = 82$. Therefore we have enough data if we use just the first 81 observations to identify an ARIMA model for Y_t. The following ARIMA(1, 0, 2) model (with $\hat{\theta}_1 = 0$) emerges from applying the usual Box–Jenkins modeling strategy to the first 81 observations:

$$(1 - 0.7243B)Y_t = 1.7624 + (1 + 0.3827B^2)\hat{a}_t \qquad (7.4.5)$$

with residual standard deviation $\hat{\sigma}_a = 0.560748$. This model is the same as model (7.4.3) based on all 100 observations, except for small differences in the values of the parameter estimates. Thus the possible intervention at $t = 82$ seems to have little effect on the autocorrelation pattern of Y_t in this example. Keep in mind, however, that interventions *will sometimes drastically change the autocorrelation pattern of a series.*

Using a Proxy AR Model for N_t

Instead of using a preliminary ARIMA model for Y_t to identify an ARIMA model for N_t in an intervention DR model, we could estimate a preliminary version of (7.1.1) that includes a *low-order AR proxy* for the disturbance ARIMA pattern. This is what we do in the LTF identification method discussed in Chapter 5.

Applying this method to the quarterly saving rate example, our preliminary intervention DR model is (7.4.2), with N_t specified as an AR(1)(1)$_4$. After estimating this preliminary model, we generate an estimate of the disturbance series in the usual way. In the saving rate example we compute $N_t = Y_t - \hat{C} - \hat{\omega}_0 X_t$. Then we study the SACF, SPACF, and ESACF of the sample disturbance series to identify a tentative ARIMA model for N_t.

7.4.3 DR Model Estimation, Checking, Reformulation, and Forecasting

Estimation, diagnostic checking, model reformulation, evaluation, and forecasting for intervention models are similar to the same steps discussed in earlier chapters and case studies. For the saving rate data example we have identified a tentative DR model in the form of (7.4.2), with X_t defined in (7.4.1). From the first 81 observations of Y_t, we identified an ARIMA(1, 0, 2) model (with $\hat{\theta}_1 = 0$) for N_t. Estimation of the DR model using the exact maximum likelihood option in the SCA System gives the following results (absolute t values in parentheses):

BUILDING INTERVENTION MODELS

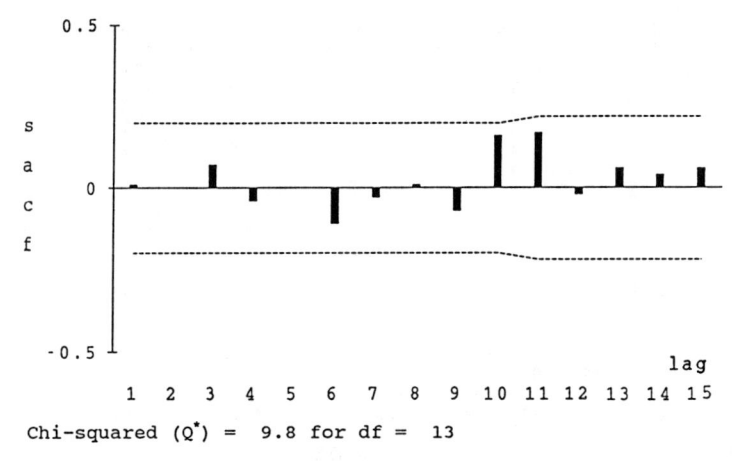

```
         0.5

    s

    a
    c          0
    f

        -0.5
                                                                    lag
             1  2  3  4  5  6  7  8  9 10 11 12 13 14 15
        Chi-squared (Q*) = 9.8 for df = 13
```

Figure 7.5 Residual SACF for model (7.4.6) applied to the saving rate data, with ARIMA(1, 0, 2) model for N_t.

$$Y_t = 6.0696 + 2.4784 X_t + \hat{N}_t$$
$$(15.8) \quad (5.44)$$

where

$$(1 - 0.8102B)\hat{N}_t = (1 + 0.2506B^2)\hat{a}_t$$
$$(11.76) \qquad\qquad (2.23)$$

$$\hat{\sigma}_a = 0.580589 \qquad\qquad (7.4.6)$$

All estimated coefficients in model (7.4.6) are significant (all absolute t values are greater than 2.0). The coefficient $\hat{\omega}_0$ says that the tax rebate produced a one-time temporary increase in the saving rate of almost 2.5 percentage points. The residual SACF in Figure 7.5 indicates that the ARIMA model for the disturbance is adequate. The normal probability plot of the residuals in Figure 7.6 is very nearly a straight line, suggesting that the shocks underlying model (7.4.6) are normally distributed. The correlation matrix of the estimated coefficients (not shown) has two significant correlations: The correlation between $\hat{\omega}_0$ and $\hat{\theta}_2$ is 0.28, and the correlation between $\hat{\theta}_2$ and $\hat{\phi}_1$ is 0.37. Neither of these correlations is large enough to raise concern about the stability of the estimated model. All things considered, model (7.4.6) is acceptable.

Because the intervention at $t = 82$ is temporary, and because the ARIMA model coefficients for the disturbance in model (7.4.6) are not very different from those for the ARIMA model for Y_t shown in (7.4.4), the forecasts from models (7.4.4) and (7.4.6) will be similar. However, other types of interventions could lead to forecasts from a DR intervention model that are quite different from those produced by an ARIMA model for Y_t.

In arriving at model (7.4.6) for this example, we used an ARIMA model

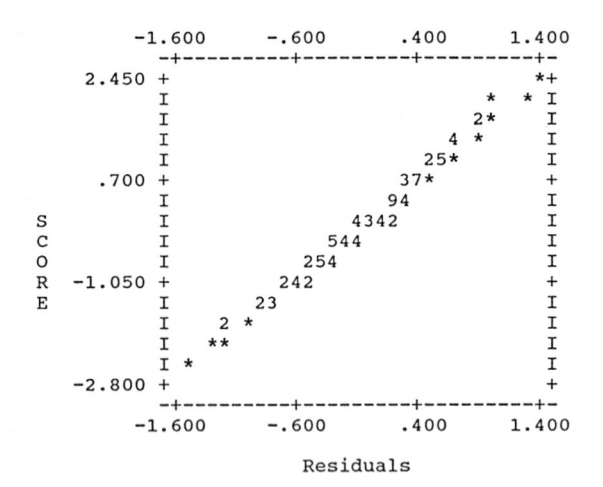

```
                    -1.600      -.600       .400       1.400
                    -+----------+----------+----------+-
             2.450 +                                   *+
                   I                              *  * I
                   I                             2*    I
                   I                         4  *      I
                   I                        25*        I
              .700 +                       37*         +
                   I                       94          I
         S         I                     4342          I
         C         I                     544           I
         O         I                    254            I
         R   -1.050+                   242             +
         E         I                  23               I
                   I           2   *                   I
                   I           **                      I
                   I   *                               I
             -2.800+                                   +
                    -+----------+----------+----------+-
                    -1.600      -.600       .400       1.400
                                Residuals
```

Figure 7.6 Normal probability plot of residuals from model (7.4.6) applied to the saving rate data, with ARIMA(1, 0, 2) model for N_t.

for Y_t (observations prior to the intervention) to identify a tentative ARIMA model for N_t. Using the AR proxy method instead to identify an ARIMA model for N_t eventually leads to the same DR model. The preliminary model, with an AR(1)(1)$_4$ proxy for N_t, produces an SACF for N_t (not shown) that is nearly identical to the one shown in Figure 7.4. The SPACF and ESACF for $\hat{N}_t$ (not shown) are consistent with an AR(1) model. Estimating (7.4.2) for the saving rate data, with N_t specified as an AR(1), leads to a residual SACF (not shown) that has a spike at lag 2 that is 1.7 times its standard error. Since the usual formula for the standard error of an autocorrelation coefficient can give an overstated value at the shorter lags in a *residual* SACF, it seems that an MA term at lag 2 may be called for. This leads directly to model (7.4.6), where we see that $\hat{\theta}_2$ is significant.

7.4.4 Intervention Model Identification: The LTF Method

We have suggested that we might try to identify the form of an intervention transfer function *by hypothesis*. Sometimes, however, there is little guidance from the relevant theory, or from trying to imagine the likely effects of an intervention using common sense. Even if we think it is likely that an intervention will raise or lower the level of Y_t, so that we expect either a positive or a negative sign on an estimated ω coefficient, we may know little about the exact form of the transfer function. (Is it a pulse or a step? Is it temporary or permanent?). Therefore in some cases we may take a more empirical approach to identifying an intervention transfer function. In this section we illustrate an application of the LTF method of Chapter 5 to the identification of an intervention model.

In the LTF method we estimate a free-form linear distributed lag regres-

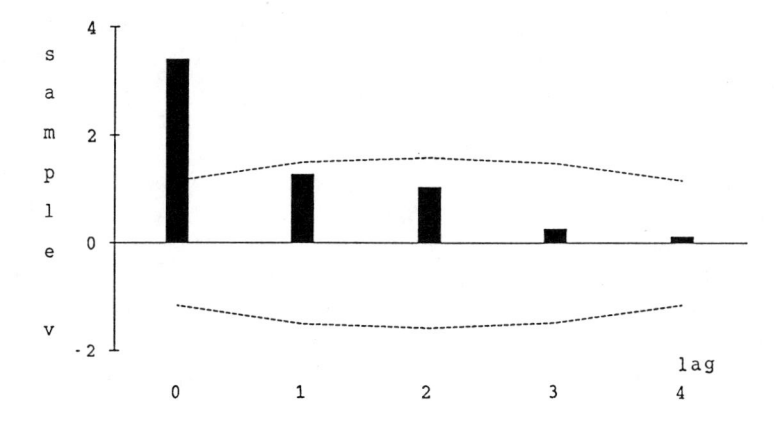

Figure 7.7 Estimated impulse response function from Equation (7.4.8).

sion (transfer function), which gives us a set of sample v weights. At the same time we specify a low-order AR proxy model for the disturbance ARIMA pattern; this is designed to give us more efficient estimates of the v weights by accounting for (we hope) most of the autocorrelation pattern in the disturbance.

Applying the LTF method to the saving rate data, we estimate the following preliminary model:

$$Y_t = C + (v_0 + v_1 B + v_2 B^2 + v_3 B^3 + v_4 B^4) X_t + N_t \qquad (7.4.7)$$

where N_t is specified as an AR(1)(1)$_4$. A maximum lag of 4 quarters on the sample v weights should be adequate since the tax rebate is expected to have a temporary effect on the saving rate. Estimation results (rounded) are as follows, with absolute t values in parentheses:

$$Y_t = \;\; 5.87 \;\; + \;\; (3.40 \; + \; 1.27B \; + \; 1.04B^2 \; + \; 0.26B^3 \; + \; 0.12B^4)X_t + \hat{N}_t$$
$$(11.46) \quad (5.86) \quad (1.70) \quad (1.32) \quad (0.35) \quad (0.20)$$

where

$$(1 - 0.89B)(1 + 0.13B^4)\hat{N}_t = \hat{a}_t$$
$$(14.77) \quad\quad (1.06) \qquad\qquad\qquad\qquad (7.4.8)$$

The estimated v weights are graphed in Figure 7.7. Only $\hat{v}_0$ is clearly significant at about the 5% level; all other $\hat{v}$ weights are less than twice their standard errors in absolute value. Therefore one reasonable conclusion is that the tax rebate at $t = 82$ had a one-period temporary effect on the saving rate, as we hypothesized in Section 7.4.1. Further identification of the disturbance series ARIMA pattern (not shown) leads to model (7.4.6). In this case the LTF method leads to the same model we found earlier.

There is an alternative interpretation of the $\hat{v}$-weight pattern in Equation (7.4.8) and Figure 7.7. Although the absolute t values of the $\hat{v}$ weights are less than 2.0 after lag 0, the $\hat{v}$ weights seem to decay from $\hat{v}_0$ in an exponential fashion. Since each coefficient is roughly 40% of the previous one ($\hat{v}_k = 0.4\hat{v}_{k-1}$, $k = 1, 2, 3, 4$), we could reasonably entertain a Koyck-type model, with δ_1 expected to equal about 0.4. There are two potential problems with this idea. First, it seems somewhat inconsistent with the economic theory (the permanent income hypothesis) that we used earlier for guidance in understanding the likely effects of the tax rebate. Second, the $\hat{v}$ weights in (7.4.8) are positively correlated as a group to a rather high degree. This is seen from the correlation matrix of the estimated coefficients, which is provided by the SCA System (a · entry indicates an insignificant value):

	1	2	3	4	5	6	7	8
1	1.00							
2	·	1.00						
3	·	0.64	1.00					
4	·	0.46	0.72	1.00				
5	·	0.35	0.55	0.72	1.00			
6	·	·	0.35	0.46	0.64	1.00		
7	−0.27	·	·	·	·	·	1.00	
8	·	·	·	·	·	·	−0.37	1.00

The rows and columns numbered 2 through 6 correspond to coefficients $\hat{v}_0$ through $\hat{v}_4$. All of the estimated coefficients are significantly positively correlated, except for $\hat{v}_0$ and $\hat{v}_4$. These correlations could be the cause of the decay pattern that we observe in Equation (7.4.8) and in Figure 7.7. The correlation among the $\hat{v}$ weights illustrated in this example *is not unusual when the LTF method is used to identify an intervention transfer function*. This is a weakness of the LTF method in identifying the dynamic effects of interventions, and the method must be used with care.

Despite the potential problems in the LTF identification method, let's proceed to the estimation stage on the assumption that the v weights decay. We have identified a possible intervention transfer function as $[\omega_0/(1 - \delta_1 B)]X_t$, where X_t is a pulse input defined as in (7.4.1). The disturbance series $\hat{N}_t$ estimated from the preliminary LTF model (7.4.8) has an SACF, SPACF and ESACF (not shown) that suggest an AR(1) model for N_t. Estimation of the identified DR model produces a set of residuals whose SACF suggests the need for an MA term at lag 2 in the disturbance ARIMA model. Subsequent estimation gives this model (absolute t values in parentheses):

BUILDING INTERVENTION MODELS 271

$$Y_t = 6.0386 + \frac{\underset{(5.57)}{2.9896}}{(1 - \underset{(2.41)}{0.4482}B)} X_t + \hat{N}_t$$
$$(16.00)$$

$$(1 - \underset{(11.62)}{0.8045}B)\hat{N}_t = (1 + \underset{(2.58)}{0.2874}B^2)\hat{a}_t \qquad (7.4.9)$$

$$\hat{\sigma}_a = 0.569211$$

The residual SACF (not shown) suggests no inadequacy in the disturbance ARIMA model. Most noteworthy in (7.4.9) is that $\hat{\delta}_1 = 0.4482$ is significant. This suggests that the tax rebate may have had a temporary but *multiperiod* effect on the saving rate, contrary to the model (7.4.6) that we obtained earlier. Can we rationalize model (7.4.9)? Maybe some people did not receive their rebates until after 1975 II. They may have moved, and their checks consequently may have been delayed in the mail. Or, the rebates may have been linked to the filing of income tax returns; people who filed for extensions may not have received their rebates until several quarters after 1975 II.

How shall we choose between model (7.4.6) and (7.4.9)? The answer is not clear. Purely on the grounds of statistical fit, (7.4.6) might be preferable since its residual standard deviation is only slightly larger, but it achieves this result with one less parameter. In terms of economic theory, either model could be acceptable. If our purpose is to *forecast* rather than to *assess the effect* of the tax rebate, it will make little difference which model we use in this example.

7.4.5 Transfer Function Identification Using Sequential Koyck Models

In this section we consider another empirical approach to identifying intervention transfer functions. We start with a Koyck model for a pulse input. That is, we first entertain the transfer function $[\omega_0/(1 - \delta_1 B)]X_t$, where X_t is a pulse input. Then we consider three possible outcomes:

1. If the estimated value $\hat{\delta}_1$ is not significant, we may conclude that the intervention involves a pulse with no decay.
2. If instead $\hat{\delta}_1$ is significant, with $0 < |\hat{\delta}_1| < 1$, then the intervention is a pulse with a decay.
3. Finally, if $\hat{\delta}_1 = 1$, we entertain the possibility that the intervention is a step rather than a pulse. The logic of this conclusion was stated in Section 7.2.2 with reference to Figure 7.2(*f*).

If we conclude that the intervention involves a step effect, then we entertain a Koyck model with a step input. That is, we try the transfer function

$[\omega_0/(1 - \delta_1 B)]X_t$, where X_t is a step input. Then we consider three possible outcomes:

1. If $\hat{\delta}_1 = 0$, we conclude that there is no decay in the added responses. Instead, the full amount of the step effect takes place immediately.
2. If instead $\hat{\delta}_1$ is significant, with $0 < |\hat{\delta}_1| < 1$, then we conclude that the step effect occurs with a decay in the added responses; that is, the full step effect is achieved only gradually, with the absolute added responses of Y_t to the intervention becoming smaller according to an exponential decay.
3. Finally, if $\hat{\delta}_1 = 1$, we then entertain the possibility that the intervention is a *ramp* rather than a step. The logic of this conclusion was stated in Section 7.3.2 with reference to Figure 7.3(f).

This method should not be used mechanically to identify intervention transfer functions. But it can sometimes provide useful information to prompt our thinking.

7.5 MULTIPLE AND COMPOUND INTERVENTIONS

7.5.1 Multiple Interventions

We can handle multiple interventions in a DR model using the framework for multiple inputs in Chapter 4. First we will consider multiple interventions occurring at *different* times. For example, suppose we have a pulse intervention at $t = i1$, a step intervention at $t = i2$, and another pulse intervention at $t = i3$. Figure 7.8 shows the hypothesized response of Y_t to these interventions. Since the first and third interventions are pulses, we define

$$X_{1,t} = 0 \quad t \neq i1$$
$$= 1 \quad t = i1$$

and

$$X_{3,t} = 0 \quad t \neq i3$$
$$= 1 \quad t = i3$$

The second intervention is represented as a step input:

$$X_{2,t} = 0 \quad t < i2$$
$$= 1 \quad t \geq i2$$

The first pulse intervention at $t = i1$ shown in Figure 7.8 has a one-period

MULTIPLE AND COMPOUND INTERVENTIONS 273

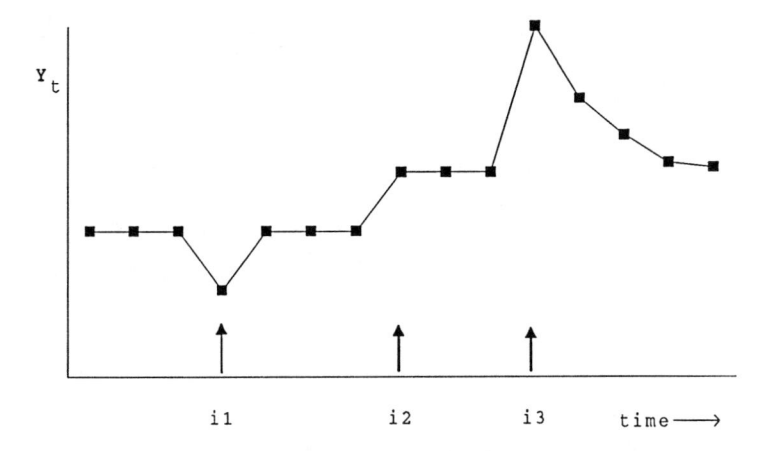

Figure 7.8 Example of multiple interventions.

effect, so its transfer function is $\omega_{1,0}X_{1,t}$. The step intervention at $t = i2$ has its full effect immediately, so its transfer function is $\omega_{2,0}X_{2,t}$. And the second pulse intervention at $t = i3$ has a dynamic effect that decays exponentially, so its transfer function is the Koyck model $[\omega_{3,0}/(1 - \delta_1 B)]X_{3,t}$. Combining these inputs gives the following three-input DR model:

$$Y_t = C + \omega_{1,0}X_{1,t} + \omega_{2,0}X_{2,t} + \frac{\omega_{3,0}}{1 - \delta_1 B}X_{3,t} + N_t \qquad (7.5.1)$$

Many other combinations of multiple interventions are possible. In each case, the procedure is the same: Specify a separate transfer function for each intervention, then combine the transfer functions as additive elements in a DR model.

7.5.2 Compound Interventions

A special type of multiple intervention is the *compound* intervention. This involves a *single* event with *several* distinct intervention effects. For example, consider a one-time sales promotion for Shampoo X (a special low price during one month). Suppose the customers are divided into three groups:

1. *New Permanent Customers.* These people have not bought Shampoo X before, but they are induced to try it because of its low price. They buy 30,000 units during month i. They like Shampoo X more than their current shampoo: in later months ($t > i$) they continue to buy 30,000 units of Shampoo X even though its price goes back up. This is a permanent response represented by a step intervention variable: $X_{1,t} = 1, t \geq i$, and $X_{1,t} = 0, t < i$. The transfer function for this effect is $\omega_{1,0}X_{1,t}$, where $\omega_{1,0} = 30,000$.

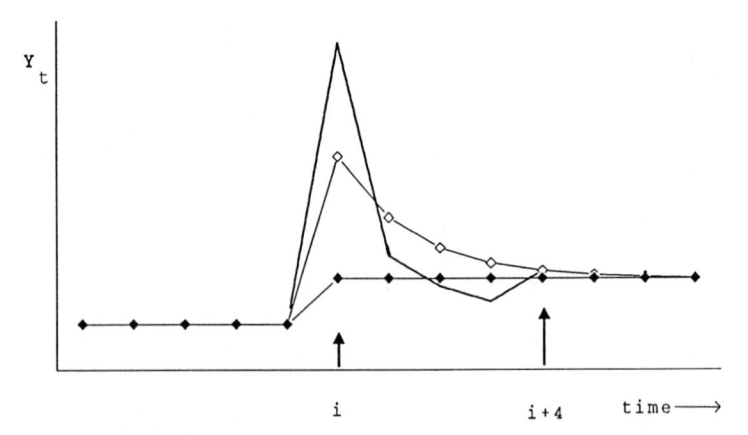

Figure 7.9 Example of a compound intervention.

2. *New Temporary Customers.* These people are also induced to try the product during $t = i$, in the amount $\omega_{2,0} = 80{,}000$ units. While some of these people continue to buy Shampoo X in later months, as a group they gradually move back to using their old shampoo. In each following month $t < i$, sales to this group are only 50% of the previous month's sales. This is a temporary response represented by a pulse intervention variable $X_{2,t} = 1$ for $t = i$, and $X_{2,t} = 0$ for $t \neq i$. The transfer function for this effect is a Koyck (exponential decay) model: $[\omega_{2,0}/(1 - \delta_1 B)]X_{2,t}$, with $\omega_{2,0} = 80{,}000$ and $\delta_1 = 0.5$.

3. *Existing Customers.* These buyers are induced by the lower price to stock up on the shampoo during the sale month at $t = i$. Their purchases are then lower for the next 3 months as they work down their excess inventory. Suppose they buy an extra 75,000 units during month i, and then buy 25,000 fewer units during each of the following 3 months. This is a temporary response, so it is also represented by the pulse intervention variable for response (2), $X_{2,t} = 1$ for $t = i$, and $X_{2,t} = 0$ otherwise. The transfer function for this effect is $(\omega_{3,0} + \omega_{3,1}B + \omega_{3,2}B^2 + \omega_{3,3}B^3)X_{2,t}$, where $\omega_{3,0} = 75{,}000$ and $\omega_{3,1} = \omega_{3,2} = \omega_{3,3} = -25{,}000$.

These responses are illustrated in Figure 7.9. Response (1) is the step response following the black diamonds. The top of the step is 30,000 units higher than the level before $t = i$. Response (2), which is in addition to response (1), is represented by the white diamonds. We may think of response (2) as beginning from the top of the step response during $t = i$, adding 80,000 units to that level, and then decaying toward the top of the black diamond step effect for $t > i$.

Response (3), which is in addition to responses (1) and (2), is represented by the solid line (no diamonds). We may think of response (3) as starting from the top of the white diamond at $t = i$, adding 75,000 to that level. At $t = i + 1$, the temporary addition of 75,000 from $t = i$ disappears, and there is a subsequent loss of 25,000 units due to consumers' inventory reduction. So the observed data series (ignoring stochastic variation for simplicity) does not follow the exponential decay of the white diamonds: It is 25,000 units lower than the white diamond at $t = i + 1$. At $t = i + 2$ there is another loss of 25,000 units, and at $t = i + 3$ there is a further loss of 25,000 units. Again, the observed series will not follow the white diamonds: will be 25,000 units below the white diamonds at $t = i + 2$ and $t = i + 3$. Finally at $t = i + 4$ the observed data will follow the decay pattern of the white diamonds, as response (3) is now entirely absent. Eventually the observed series converges to the top of the black diamond step function.

We must be careful in writing the full DR model for compound interventions. The recommended procedure is the same as for other multiple interventions: First specify a separate transfer function for each intervention component, then combine them as additive elements in the DR model. This may not be straightforward with compound interventions, as we will show for the present example. Adding the three transfer functions (ignoring the disturbance for convenience) gives this multiple input transfer function:

$$Y_t = C + \omega_{1,0} X_{1,t} + \frac{\omega_{2,0}}{1 - \delta_1 B} X_{2,t} + (\omega_{3,0} + \omega_{3,1} B + \omega_{3,2} B^2 + \omega_{3,3} B^3) X_{2,t} \tag{7.5.2}$$

There is a problem with (7.5.2): The second and third transfer function components have the *same* input variable $X_{2,t}$, and we want to estimate *two* current responses to this single variable ($\omega_{2,0}$ and $\omega_{3,0}$), *two* responses at lag 1 (the implicit Koyck response $\delta_1 \omega_{2,0}$, and $\omega_{3,1}$), and *two* additional responses at lags 2 and 3. Perfect colinearity (see Chapter 3) between the second input variable ($X_{2,t}$) and the third input variable (also $X_{2,t}$) makes estimation of (7.5.2) impossible in its present form. However, we can rewrite (7.5.2) to make estimation feasible. [Notice that there is no similar colinearity problem with the first intervention component, even though it also occurs at $t = i$: $X_{1,t}$ is defined differently from $X_{2,t}$. Notice also that we can't solve the colinearity problem in (7.5.2) by defining a new intervention variable $X_{3,t}$ for response (3). We could define such a variable, but $X_{3,t} = X_{2,t}$ so that the perfect colinearity remains.]

We may think of the last two transfer functions in (7.5.2) together as components of one transfer function. That is, by gathering terms the last two transfer functions may be written as

$$\left[\frac{\omega_{2,0}}{1 - \delta_1 B} + \omega_{3,0} + \omega_{3,1} B + \omega_{3,2} B^2 + \omega_{3,3} B^3 \right] X_{2,t} \tag{7.5.3}$$

Stating the terms within the square brackets with a common denominator allows us to rewrite (7.5.3) as

$$\frac{\omega_0 + \omega_1 B + \omega_2 B^2 + \omega_3 B^3 + \omega_4 B^4}{1 - \delta_1 B} X_{2,t} \qquad (7.5.4)$$

where

$$\begin{aligned}
\omega_0 &= \omega_{2,0} + \omega_{3,0} \\
\omega_1 &= \omega_{3,1} - \delta_1 \omega_{3,0} \\
\omega_2 &= \omega_{3,2} - \delta_1 \omega_{3,1} \qquad (7.5.5)\\
\omega_3 &= \omega_{3,3} - \delta_1 \omega_{3,2} \\
\omega_4 &= -\delta_1 \omega_{3,3}
\end{aligned}$$

The transfer function (7.5.4) replaces the last two transfer functions in (7.5.2). After estimating the parameters of this modified model, we may then find $\omega_{2,0}$ and the $\omega_{3,i}$ weights in (7.5.2) recursively, starting from the last equation in (7.5.5) and moving toward the first equation in (7.5.5).

Compound intervention models are somewhat complicated, but they can be useful. The lesson here is that they must be specified and estimated with special care. McCleary and Hay (1980) discuss the use of such models.

QUESTIONS AND PROBLEMS

1. In the Data Appendix is a series (Series 28) of monthly average calls for telephone directory assistance in Cincinnati, Ohio. These figures are the average number of calls per day during each month. For $t = 1$ to 146, there was no charge for assistance. Starting at $t = 147$, Cincinnati Bell began charging 20 cents for each request for assistance.
    a. Identify an ARIMA model for this series using the first 146 observations. (Analyze the data in the natural log metric.)
    b. Construct a binary (0, 1) step input variable (X_t) to represent the telephone company's policy regarding charges for assistance. What is the definition of X_t?
    c. Suppose instead that you define an input variable as "the charge per call for directory assistance, in cents." How does this input variable differ from X_t? Under what circumstances would it be preferable to use this input variable instead of the step input defined in part (b)?
    d. Based on your answers to parts (a) and (b), what DR model have you tentatively identified? Explain what sign(s) (+ or −) you expect to see associated with the estimated ω coefficient(s). Estimate and check the model.
    e. Use all 180 observations and the LTF method to identify a DR model, with X_t from part (b) as the input series.

QUESTIONS AND PROBLEMS 277

  f. Use the sequential Koyck model approach to identify a DR model for this data set. Does this approach confirm that imposition of the charge for directory assistance should be treated as an abrupt step input?
  g. Construct an ARIMA model for the entire data series, ignoring the intervention. Forecast with this ARIMA model and with your best DR model. Are the forecasts affected very much by the inclusion of an intervention component?

2. In the Data Appendix is a series of 120 daily observations (Series 29) of the score achieved by a schizophrenic patient on a test of perceptual speed. On the 61st day the patient began receiving a powerful tranquilizer (chlorpromazine) that could be expected to reduce perceptual speed. The drug regimen was continued throughout the rest of the sample period.
  a. Identify an ARIMA model for this series using the first 60 observations.
  b. Construct a binary (0, 1) step input variable (X_t) to represent the drug intervention. What is the definition of X_t?
  c. Suppose instead that you define an input variable as "the amount of chlorpromazine received per day, in milligrams." How does this input variable differ from X_t? Under what circumstances would it be preferable to use this input variable instead of the step input defined in part (b)?
  d. Based on your answers to parts (a) and (b), what DR model have you tentatively identified? Explain what sign(s) (+ or −) you expect to see associated with the estimated ω coefficient(s). Estimate and check the model.
  e. Use all 120 observations and the LTF method to identify a DR model, with X_t from part (b) as the input series. Do you identify a model that is different from the one you identified in part (d)?
  f. Use the sequential Koyck model approach to identify a DR model for this data set.
  g. Construct an ARIMA model for the entire data series, ignoring the intervention. Forecast with this ARIMA model and with your best DR model. Are the forecasts affected very much by the inclusion of an intervention component?

3. In the Data Appendix is a series of 144 monthly observations (Series 30) of the volume of freight, measured in ton-miles, carried by air carriers in the United States. Analyze the data in the log metric.
  a. Study the plot of the logs of these data. Do you notice any unusual observations? Study a plot of the first differences ($d = 1$). Now do you see any unusual observations? What about observations $t = 111$ and $t = 112$? Apply seasonal differencing ($D = 1$) in addition to $d = 1$ and

study the plot of the new differenced series. Which observations seem unusual?

b. Identify and estimate an ARIMA model for this series using the first 120 observations. Study a plot of the ARIMA model standardized residuals ($\hat{a}_t/\hat{\sigma}_a$). Which residuals stand out? How many standard deviations away from zero are these residuals? (This question anticipates some of the material in Chapter 8 on outlier detection.)

c. Identify and estimate an ARIMA model using the first 108 observations. Compare this model with the one you found in part (b). Do the unusual observations during 1978 seem to have an effect on the ARIMA structure of the data? On the values of the estimated coefficients?

d. Check the original source of the data to see if any observations in the Data Appendix are misrecorded. Might the original source contain misrecorded observations? Can you discover anything in the history of the air freight industry (a labor strike, for example) that could explain the unusual value(s) during 1978?

e. Construct a binary (0, 1) input variable (X_t) to represent an intervention that might have caused the unusual observation(s). What is the definition of X_t?

f. What DR model have you tentatively identified? Explain what sign(s) (+ or −) you expect to see associated with the estimated ω coefficient(s). Estimate and check the model using the first 120 observations.

g. Use the first 120 observations and the LTF method to identify a DR model, with X_t from part (e) as the input series. Do you identify a model that is different from the one you identified in part (f)?

h. Use the first 120 observations and the sequential Koyck model approach to identify a DR model for this data set.

i. Forecast with the ARIMA model that you constructed in part (b) and with your best DR model. Are the forecasts affected very much by the inclusion of an intervention component? Which model has more accurate forecasts?

CASE 6

Year-End Loading

C6.1 THE DATA

Figure C6.1 is a plot of an index of $n = 125$ monthly shipments of a consumer product from the manufacturer to distributors. The data are observed from January 1976 to May 1986; they are listed in the Data Appendix as Series 31. In 1983 the top-level managers began pressuring the marketing department to meet strict annual (calendar year) shipments targets. Near the end of 1983, 1984, and 1985, the marketing department found that shipments were going to fall short of the annual targets. They adopted a policy, called year-end loading, in which they pressured distributors to take extra shipments in December. The policy is controversial since it irritates distributors; the marketing manager also believes that extra shipments in December are offset by lower shipments in later months. In this example we use intervention analysis to assess the distributed lag effect on shipments of the year-end loading policy.

The variance of the shipments series in Figure C6.1 appears to increase slightly with the level of the data. However, this appearance is largely due to the perturbations to the data induced by the year-end loading policy. We will proceed with the original data; it may be necessary to use a transformation to stabilize the variance as more observations arise. At the end of this case study we will discuss the implications for assessing the year-end loading policy when the data are transformed.

The Inputs
The year-end loading policy is a sequence of three events whose effect on shipments (if any) is presumably temporary. Therefore these three events can be represented as three pulse intervention variables, defined as follows:

$X_{1,t} = 1$ if $t =$ December 1983 $X_{1,t} = 0$ otherwise

$X_{2,t} = 1$ if $t =$ December 1984 $X_{2,t} = 0$ otherwise

$X_{3,t} = 1$ if $t =$ December 1985 $X_{3,t} = 0$ otherwise

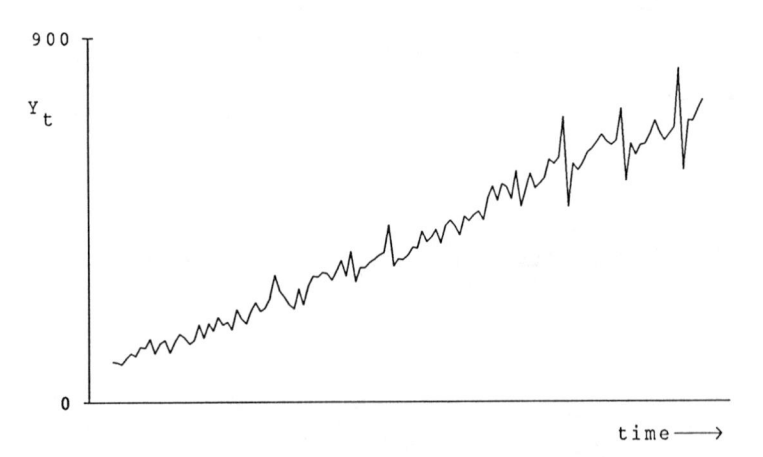

[Figure: line chart of Y_t vs time, ranging 0 to 900, showing an increasing trend with seasonal fluctuations]

Figure C6.1 Shipments, January 1976–May 1986.

If the year-end loading policy is effective, then December shipments for 1983, 1984, and 1985 will be higher than they would be otherwise. This would be indicated by a positive coefficient associated with each $X_{i,t}$ variable in a DR model. However, it is possible that higher December shipments (if any) are offset by lower shipments in later months as distributors work down their excess stocks of goods. To estimate these possible distributed lag effects, we will include time-lagged $X_{i,t}$ variables in our preliminary LTF model.

C6.2 DR MODEL IDENTIFICATION, ESTIMATION AND CHECKING

C6.2.1 Tentative Model 1

Using the LTF method, we begin by specifying a multiple input free-form distributed lag relationship between Y_t and the $X_{i,t}$ variables, with an $AR(1)(1)_{12}$ model as a proxy for the possible autocorrelation pattern in the disturbance series N_t:

$$Y_t = C + v_1(B)X_{1,t} + v_2(B)X_{2,t} + v_3(B)X_{3,t} + N_t \qquad (C6.1)$$

where $v_i(B) = v_{i,0} + v_{i,1}B + v_{i,2}B^2 + v_{i,3}B^3$ for $i = 1, 2, 3$, and $N_t = [1/(1 - \phi_{12}B^{12})(1 - \phi_1 B)]a_t$. We have chosen a maximum lag of three months for the distributed lag response to each intervention. This is a subjective judgment based on the opinion of the marketing manager that any offsetting reactions to increased December shipments are likely to occur within one or two months.

Estimation results for the preliminary LTF model (C6.1) appear in Table

DR MODEL IDENTIFICATION, ESTIMATION AND CHECKING   281

Table C6.1   Estimation Results for Preliminary LTF Model (C6.1)

Parameter	Estimate	Standard Error	t Value
C	811.7719	1303.3023	0.62
$v_{1,0}$	102.7996	26.7188	3.85
$v_{1,1}$	-115.1546	32.8777	-3.50
$v_{1,2}$	-6.0635	32.9905	-0.18
$v_{1,3}$	-19.4070	27.1027	-0.72
$v_{2,0}$	81.1649	26.7163	3.04
$v_{2,1}$	-94.6505	32.7202	-2.89
$v_{2,2}$	-0.4575	32.7202	-0.01
$v_{2,3}$	-26.2316	26.7165	-0.98
$v_{3,0}$	137.1003	26.7170	5.13
$v_{3,1}$	-121.7483	32.7212	-3.72
$v_{3,2}$	-8.5473	32.7212	-0.26
$v_{3,3}$	-19.2976	26.7169	-0.72
ϕ_1	0.9880	0.0171	57.85
ϕ_{12}	0.0099	0.1035	0.10

$\hat{\sigma}_a = 29.7429$

C6.1. As we noted in Chapter 7, the LTF method of identifying the dynamic effects of an intervention can produce highly correlated $\hat{v}$ weights; this may give a somewhat unreliable picture of the impulse response function. The following array shows the correlations among the $\hat{v}_{1,k}$ weights:

	$\hat{v}_{1,0}$	$\hat{v}_{1,1}$	$\hat{v}_{1,2}$
$\hat{v}_{1,1}$	0.61		
$\hat{v}_{1,2}$	0.40	0.65	
$\hat{v}_{1,3}$	0.25	0.42	0.58

(These values are available as part of the SCA System estimation results. The correlations among the $\hat{v}_{2,k}$ weights and $\hat{v}_{3,k}$ weights are similar to those above.) While these correlations are not extremely large, they are all significant at about the 5% level. This reinforces the point made in Chapter 7: We must be careful when using the LTF method to evaluate the dynamic effects of interventions. We will revisit this issue as we consider further versions of our DR model.

Our main concern in evaluating a preliminary LTF model is to decide if differencing is called for; that is, we want to decide if the disturbance series N_t is stationary in the mean. We can't observe N_t; but we can compute an estimate $\hat{N}_t$. As discussed in Chapters 5 and 7, $\hat{N}_t$ is found by using the $\hat{v}$ weights from the regression part of the preliminary LTF model. Solving (C6.1) for N_t and inserting the $\hat{v}$ weights and $\hat{C}$, we compute the estimated disturbance series

Table C6.2 Simplified ESACF for Disturbance Series in Model (C6.1)[a]

($q \rightarrow$)	0	1	2	3	4	5	6	7	8	9	10	11	12
($p' = 0$)	X	X	X	X	X	X	X	X	X	X	X	X	X
($p' = 1$)	X	0	0	0	0	0	0	0	0	0	0	0	0
($p' = 2$)	X	0	0	0	0	0	0	0	0	0	0	0	0
($p' = 3$)	0	0	0	0	0	0	0	0	0	0	0	0	0
($p' = 4$)	0	X	0	0	0	0	0	0	0	0	0	0	0
($p' = 5$)	0	X	0	0	0	0	0	0	0	0	0	0	0
($p' = 6$)	X	X	X	0	0	0	0	0	0	0	0	0	0

[a]X = 2 or more standard errors from zero.

$$\hat{N}_t = Y_t - \hat{C} - \hat{v}_1(B)X_{1,t} - \hat{v}_2(B)X_{2,t} - \hat{v}_3(B)X_{3,t}$$

This computation is performed automatically by the SCA System using appropriate commands. We may now study the SACF of $\hat{N}_t$ to see if differencing is needed. The SACF and ESACF for $\hat{N}_t$ are shown in Figure C6.2 and Table C6.2, respectively.

The slow decay of the SACF in Figure C6.2 suggests that N_t is nonstationary in the mean and that differencing ($d = 1$) is called for. The ESACF in Table C6.2 says that after differencing we should expect to find an MA(1) pattern for the disturbance. That is, we can locate a triangle of zeros in the lower right corner of the ESACF, with its upper left vertex in the ($p' = 1$, $q = 1$) cell; this implies an ARIMA(1, 0, 1) model for N_t. Recall from Chapter 2 that the EACF handles differencing as part of the AR order. In this case we have reason to think that the AR(1) operator for the disturbance pattern suggested by the ESACF can be written as $(1 - B)$, with $\phi_1 = 1.0$; this is the differencing operator with $d = 1$. Thus a tentative model for N_t is an ARIMA(0, 1, 1), where the (nonstationary) AR(1) suggested by the ESACF is represented by nonseasonal differencing ($d = 1$).

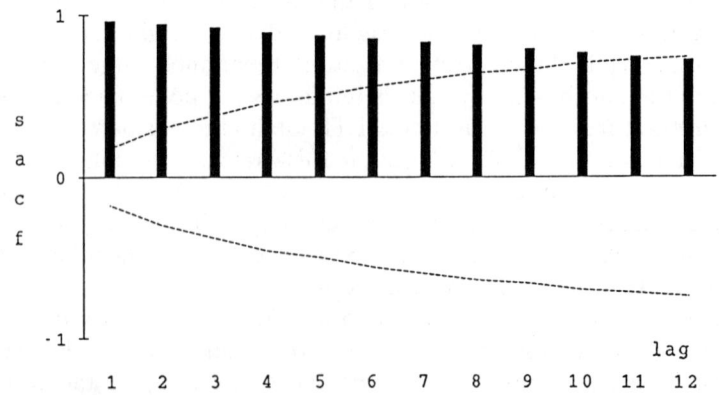

Figure C6.2 SACF for estimated disturbance series $\hat{N}_t$ in model (C6.1).

DR MODEL IDENTIFICATION, ESTIMATION AND CHECKING   283

Table C6.3 Estimation Results for Preliminary LTF Model (C6.2)

Parameter	Estimate	Standard Error	t Value
C	5.4518	1.6760	3.25
$v_{1,0}$	107.9799	23.2274	4.65
$v_{1,1}$	−117.5265	24.0067	−4.90
$v_{1,2}$	−7.2409	24.1840	−0.30
$v_{1,3}$	−29.4257	23.9885	−1.23
$v_{2,0}$	86.4176	23.2006	3.72
$v_{2,1}$	−93.9455	23.7627	−3.95
$v_{2,2}$	1.2555	23.7662	0.05
$v_{2,3}$	−27.4586	23.1887	−1.18
$v_{3,0}$	139.2227	23.1883	6.00
$v_{3,1}$	−123.4051	23.7608	−5.19
$v_{3,2}$	−8.9108	23.7615	−0.38
$v_{3,3}$	−25.1293	23.1849	−1.08
ϕ_1	−0.5169	0.0825	−6.26
ϕ_{12}	0.0294	0.1038	0.28

$\hat{\sigma}_a = 25.6372$

C6.2.2 Tentative Model 2

We could go directly to an ARIMA(0, 1, 1) model for the disturbance series. However, we will difference all variables ($d = 1$) in (C6.1) and continue with the LTF identification method to illustrate the results. Our new LTF model is:

$$\nabla Y_t = C + v_1(B)\nabla X_{1,t} + v_2(B)\nabla X_{2,t} + v_3(B)\nabla X_{3,t} + n_t \quad (C6.2)$$

where $n_t = \nabla N_t$ is represented by the proxy ARIMA model $n_t = [1/(1 - \phi_{12}B^{12})(1 - \phi_1 B)]a_t$. The values of C, ϕ_1, and ϕ_{12} are different in (C6.2) compared to (C6.1) since the data have been differenced.

Estimation results for (C6.2) are shown in Table C6.3. Our first concern is whether the differenced disturbance n_t has a stationary mean. The SACF of the estimated disturbance n_t is shown in Figure C6.3. $\hat{n}_t$ is computed by solving (C6.2) for n_t and substituting estimated coefficients for unknown parameters to get

$$\hat{n}_t = \nabla Y_t - \hat{C} - \hat{v}_1(B)\nabla X_{1,t} - \hat{v}_2(B)\nabla X_{2,t} - \hat{v}_3(B)\nabla X_{3,t}$$

The SACF of $\hat{n}_t$ is consistent with a stationary n_t series since it drops quickly to insignificant values. Further, the stationary disturbance appears to follow an MA(1) process, as suggested by the single large spike at lag 1 in the SACF. This conclusion is reinforced by the SPACF and ESACF (not shown) of $\hat{n}_t$. This also coincides with the conclusion we reached by studying the ESACF of the estimated disturbance $\hat{N}_t$ in model (C6.1).

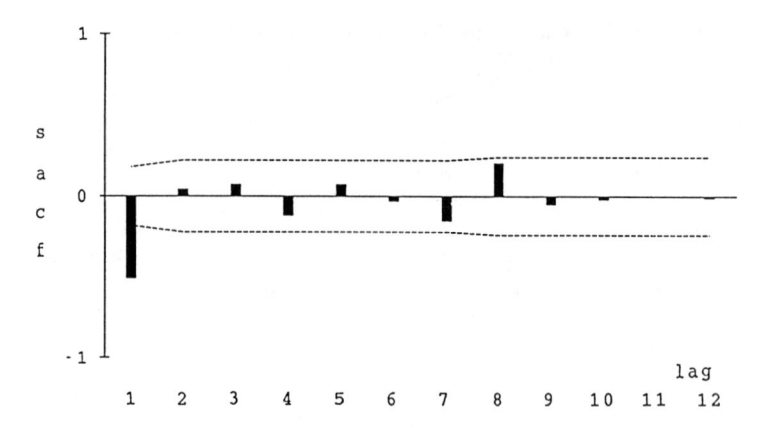

[Figure: bar chart of SACF with lags 1–12]

Figure C6.3 SACF for estimated disturbance series in model (C6.2).

In Table C6.3 we see that all sample v weights at lags 2 and 3 are insignificant; therefore we will drop these terms. This implies that the dead time for each intervention is $b_i = 0$, $i = 1, 2, 3$; that $r_i = 0$, $i = 1, 2, 3$; and $h_i = 1$, $i = 1, 2, 3$. We should be cautious in arriving at these conclusions since the $\hat{v}_{i,k}$ weights are still correlated. For example, the correlations among the $\hat{v}_{1,k}$ weights in the new preliminary LTF model (C6.2) are

	$\hat{v}_{1,0}$	$\hat{v}_{1,1}$	$\hat{v}_{1,2}$
$\hat{v}_{1,1}$	0.26		
$\hat{v}_{1,2}$	0.40	0.27	
$\hat{v}_{1,3}$	0.17	0.42	0.21

(The correlations among the $\hat{v}_{2,k}$ weights and $\hat{v}_{3,k}$ weights are similar.) All of these correlations except 0.17 are significant at the 5% level. Ideally we would have little or no significant correlation among the estimated $\hat{v}$ weights; this ideal is rarely attained, especially when we use the LTF method to identify intervention dynamics. These correlations seem small enough for us to have some confidence in the individual $\hat{v}$ weights and their t values. However, we should remember that our conclusions at this stage are necessarily more tentative than if the $\hat{v}$ weights were uncorrelated.

C6.2.3 A Final DR Intervention Model

We have arrived at what may be a final DR model:

$$\nabla Y_t = C + \omega_1(B)\nabla X_{1,t} + \omega_2(B)\nabla X_{2,t} + \omega_3(B)\nabla X_{3,t} + n_t \qquad (C6.3)$$

DR MODEL IDENTIFICATION, ESTIMATION AND CHECKING

Table C6.4   Estimation Results for DR Model (C6.3)

Parameter	Estimate	Standard Error	t Value
C	5.0394	0.3535	14.26
$\omega_{1,0}$	119.5462	22.3001	5.36
$\omega_{1,1}$	-106.0812	22.2624	-4.77
$\omega_{2,0}$	86.7808	22.4142	3.87
$\omega_{2,1}$	-93.9542	22.3464	-4.20
$\omega_{3,0}$	134.5518	22.4324	6.00
$\omega_{3,1}$	-121.0929	22.4785	-5.39
θ_1	0.8392	0.0508	16.51

$\hat{\sigma}_a = 23.0125$

Correlation Matrix of Parameter Estimates

	1	2	3	4	5	6	7	8
1	1.00							
2	-0.00	1.00						
3	-0.00	0.09	1.00					
4	-0.01	0.03	0.03	1.00				
5	-0.01	0.03	0.03	0.10	1.00			
6	-0.04	-0.01	-0.01	0.00	0.01	1.00		
7	-0.05	-0.01	-0.01	0.00	0.01	0.11	1.00	
8	-0.02	0.12	0.11	0.16	0.13	-0.09	-0.07	1.00

where $\omega_i(B) = \omega_{i,0} + \omega_{i,1}B$, $i = 1, 2, 3$. The ARIMA model for $n_t = \nabla N_t$ is now tentatively identified as an MA(1), $n_t = (1 - \theta_1 B)a_t$.

Estimation results for (C6.3) are shown in Table C6.4. This appears to be an acceptable final model. All estimated coefficients are significant at better than the 5% level since all absolute t values exceed 2.0. The estimate of θ_1 satisfies the invertibility condition $|\hat{\theta}_1| < 1.0$. The residual standard deviation is smaller than those for either of our two earlier preliminary LTF models. Interestingly, none of the coefficient correlations is significant. The residual SACF (Figure C6.4) is consistent with an independent random shock series when we allow for sampling variation. A few residual autocorrelations are moderately large (e.g., lags 4 and 24); but we are unlikely to improve the model much by adding terms at these lags.

The normal probability plot of the standardized residuals in Figure C6.5 is roughly a straight line, and the histogram of the standardized residuals in Figure C6.6 is only slightly skewed; these results do not seriously contradict the normality assumption about the random shocks. The standardized residuals are plotted in Figure C6.7. None of these values is highly unusual; only two lie in the neighborhood of three standard deviations away from zero. Research into the history of the series does not reveal any particular events that might have caused these two moderately large outliers. If we could identify such events, we might add more intervention components to our

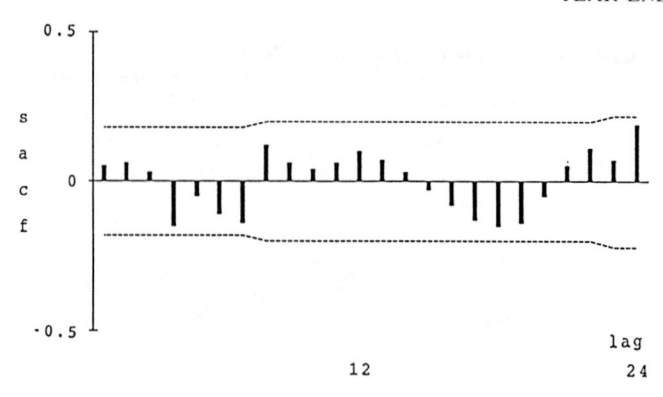

```
      0.5

s
a
c      0
f


     -0.5
                                                            lag
                         12                          24

Chi-squared = 31.9 for df = 23
```

Figure C6.4 Residual SACF for DR model (C6.3).

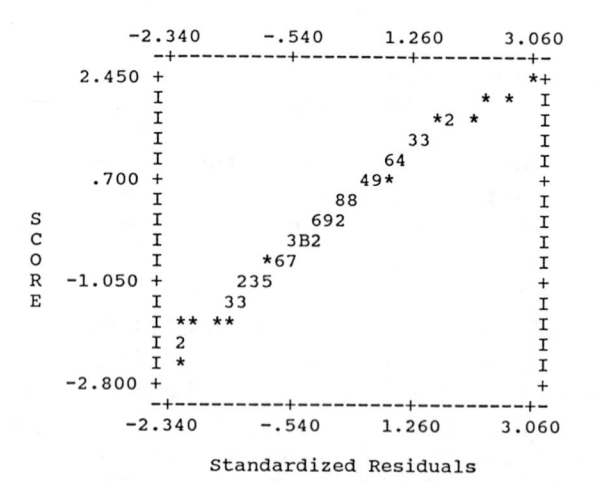

```
                -2.340      -.540     1.260     3.060
               -+---------+---------+---------+-
          2.450 +                             *+
                I                        * *  I
                I                     *2 *    I
                I                33            I
                I               64             I
           .700 +             49*              +
                I            88                I
      S         I           692                I
      C         I          3B2                 I
      O         I         *67                  I
      R  -1.050 +        235                   +
      E         I       33                     I
                I  ** **                       I
                I 2                            I
                I *                            I
         -2.800 +                              +
               -+---------+---------+---------+-
                -2.340      -.540     1.260     3.060
                         Standardized Residuals
```

Figure C6.5 Normal probability plot for residuals in DR model (C6.3).

```
LOWER    UPPER    FREQ-
BOUND    BOUND    UENCY  0    5   10   15   20   25   30   35
                        +----+----+----+----+----+----+----+
-3.600 - -3.000      0  I
-3.000 - -2.400      0  I
-2.400 - -1.800      5  I>>>>>
-1.800 - -1.200      9  I>>>>>>>>>
-1.200 - -0.600     20  I>>>>>>>>>>>>>>>>>>>>
-0.600 -  0.000     30  I>>>>>>>>>>>>>>>>>>>>>>>>>>>>>>
 0.000 -  0.600     25  I>>>>>>>>>>>>>>>>>>>>>>>>>
 0.600 -  1.200     21  I>>>>>>>>>>>>>>>>>>>>>
 1.200 -  1.800      8  I>>>>>>>>
 1.800 -  2.400      3  I>>>
 2.400 -  3.000      1  I>
 3.000 -  3.600      1  I>
                        +----+----+----+----+----+----+----+
                        0    5   10   15   20   25   30   35
```

Figure C6.6 Histogram of standardized residuals for DR model (C6.3).

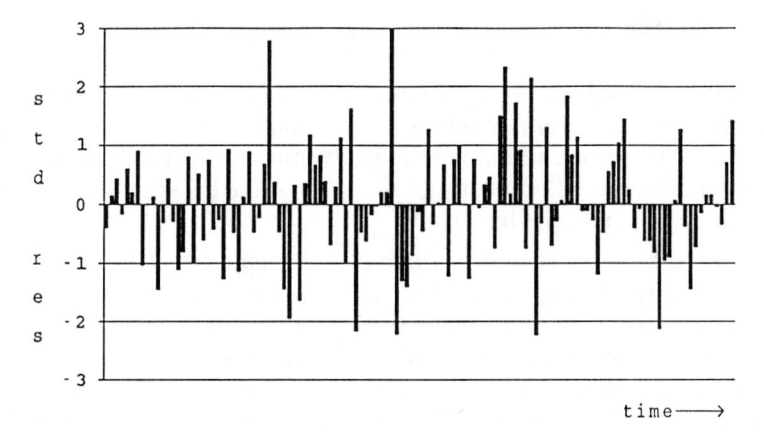

Figure C6.7 Standardized residuals from model (C6.3).

DR model. A formal method for detecting and identifying outliers and interventions is presented in Chapter 8.

C6.3 INTERPRETATION OF THE DR MODEL

How may we understand model (C6.3)? It appears that the year-end loading policy has had some short-term success. For all three episodes the estimates of the coefficients associated with December (the $\hat{\omega}_{i,0}$ coefficients) are positive and significant. However, it also seems that the marketing manager's point about future offsetting responses (lower future shipments) is valid: The estimates of the coefficients associated with each January following a year-end loading of shipments (the $\hat{\omega}_{i,1}$ coefficients) are all negative and significant. A comparison of the December and the January coefficients suggests that there is no net gain of shipments due to year-end loading. For the first episode the December gain is estimated at about 120 index units; the January loss is about 106 units. Though this implies a gain of 14 index units over these 2 months, the difference between the two coefficients is not significant since each coefficient has a standard error of more than 22. A similar conclusion holds for the other two episodes.

The main issue in this case involves evaluating the year-end loading policy. However, model (C6.3) can be further interpreted in two ways. First, if we remove the intervention components we are left with an ARIMA(0, 1, 1) model; as discussed in Chapter 2, this is a nonseasonal EWMA forecast model. Second, the model contains a deterministic trend component. As discussed in Chapter 2, after differencing a constant term represents a deterministic trend. In this case we estimate that shipments have tended to grow linearly each month by the fixed amount $\hat{C} = 5.0394$ index units.

C6.4 COMMENTS

1. In this case we are mainly interested in evaluating the historical effects of a policy. However, our conclusions have implications for forecasting. In particular, if the year-end loading policy continues, we would forecast that it will tend to increase shipments in December at the expense of an offsetting reduction of shipments in January.

2. If we use model (C6.3) to forecast, we might prefer to leave out the deterministic trend component. While the constant term (constant monthly growth component) is a statistically significant feature of the historical data, it might be unwise to project a similar growth into the future. Any future deviation of the data from this historical growth path could lead to poor forecasts, especially if model (C6.3) is used to forecast more than a few months ahead. It would be easier to accept a fixed growth component in the model if the data were generated by a physical mechanism that we could expect to remain unchanged in the future. But these data have no such underlying physical mechanism.

3. At the start of this case we pointed out that the data show a slight tendency toward a rising variance at higher levels. We decided to work with the original data, but we suggested that as more observations arise we might see a more clear need to transform the data to achieve a constant variance. Analyzing the data in a transformed metric would complicate the interpretation of the intervention coefficients. To see this, suppose we apply a square root transformation and arrive at this simple model:

$$Y'_t = 60 + 10X_t - 10X_{t-1} + a_t \qquad (C6.4)$$

where $Y'_t = Y_t^{1/2}$, and where X_t is a pulse intervention defined as $X_t = 1$ at time $t = i$, and $X_t = 0$ otherwise. One-step ahead predictions of Y'_t (denoted $\hat{Y}'_t$) from this model for $t = i - 1, i, i + 1$, and $i + 2$, along with the corresponding predictions in the original metric, (Y'_t), are

t	$\hat{Y}'_t$	$\hat{Y}_t$
$i - 1$	$60 + 10(0) - 10(0) = 60$	$60^2 = 3600$
i	$60 + 10(1) - 10(0) = 70$	$70^2 = 4900$
$i + 1$	$60 + 10(0) - 10(1) = 50$	$50^2 = 2500$
$i + 2$	$60 + 10(0) - 10(0) = 60$	$60^2 = 3600$

The predicted change from the constant level 3600 in the original metric for $t = i$ is $4900 - 3600 = +1300$ units, while the predicted change from the constant level 3600 for $t = i + 1$ is $2500 - 3600 = -1100$ units. Thus, while the coefficients of X_t and X_{t-1} in (C6.4) are equal, they do not imply exactly offsetting responses to the intervention in the original metric. [In this illlustration we used the simple retransformation $\hat{Y}_t = (\hat{Y}'_t)^2$. Obtaining forecasts in the original metric can be more complicated, as discussed in Section 9.2.3.]

CHAPTER 8

Intervention and Outlier Detection and Treatment

In Chapter 7 we considered intervention models, with external events (interventions) represented by deterministic binary (0 or 1) input variables. We assumed that the *occurrence* and *timing* of the events were *known*. Our task was to estimate the dynamic response of an output variable to one or more external events using a rational form transfer function. However, a time series may be interrupted by an event whose *existence* is initially *unknown* to the analyst. In fact, analysts often discover possible external events only when studying a model's residual series for outliers (unusual values). Even if it is known that an external event has occurred, the timing of the event may be unknown. And even if the existence and timing are known, the dynamic *response type* (pulse, step, abrupt, gradual, etc.) may be unknown.

In this chapter we consider a formal method for *detecting* the existence, timing, and type of certain kinds of external events in time series. After detecting such an event, we may represent it by an intervention-type transfer function component in a DR model. We focus on three types of external events:

1. One-period response pulse events as discussed in Chapter 7, also called *additive outliers* (denoted AO).
2. Step interventions as discussed in Chapter 7, also called permanent *level shifts* (denoted LS).
3. *Innovational outliers* (denoted IO), which are additions to the random shock series a_t; an IO affects the output series Y_t through the ARIMA *structure of the disturbance series* in the model for Y_t.

8.1 THE RATIONALE FOR INTERVENTION AND OUTLIER DETECTION

8.1.1 Better Understanding of the Data

Why bother with intervention and outlier detection? First, it may lead to a better understanding of the series under study. Ideally, we can explain and interpret all identified external events in a time series. The methods discussed in this chapter are designed only to *detect* external events; they do not provide explanations. However, research into the history of a time series may lead to a plausible explanation of a detected event. For example, a labor strike may be found to coincide with a detected temporary decline in a product shipment series. This knowledge gives the analyst a better understanding of the history of the series. We will not gain this understanding if we apply intervention and outlier detection methods in a purely mechanical fashion.

In some cases no clear explanation of a detected possible event may be found. For example, outliers may arise due to untraceable recording errors in the data. And, of course, some detected events might reflect nothing more than chance movements of the data. How should we treat such unexplained but detected possible events? Some analysts prefer to include automatically a component in a DR model to represent any detected event. This may be done using a rule based on the statistical significance of the detected event, as discussed in Section 8.4. Other analysts prefer to make a judgment about whether a detected event is sufficiently plausible to warrant its presence in a DR model. In either case, the analyst at least has gained knowledge about certain observations that deserve scrutiny.

8.1.2 Better Model Specification and Estimation

Interventions and outliers can seriously alter the correlation patterns within a time series. The possible effects range from mild to severe, depending on the nature of the intervention or outlier. Some of these effects are discussed by Guttman and Tiao (1978) and Chang and Tiao (1983a, 1983b). By adjusting for external events, we can improve the overall specification of ARIMA or DR models and obtain better estimates of the parameters. An early study of the problem of estimating model parameters in the presence of outliers appears in Denby and Martin (1979). This problem has been studied more recently by Chang et al. (1988), and Chen et al. (1990).

8.1.3 Better Forecasts

External events can affect the accuracy of forecasts from time series models. Some possible effects of external events on forecast accuracy are discussed by Ledolter (1988). The effects are often slight, but they can be substantial,

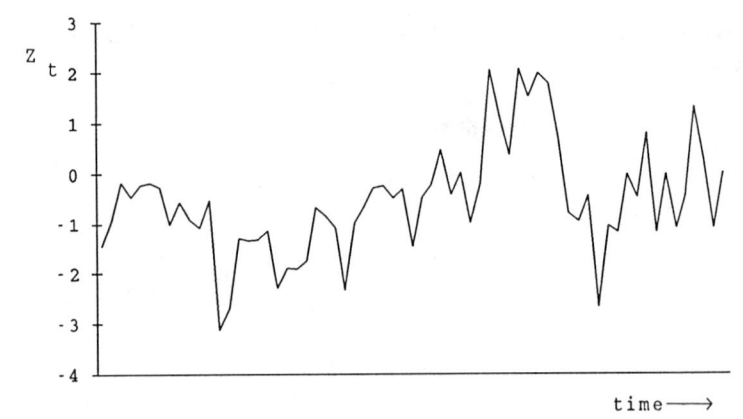

Figure 8.1 Simulated series z_t based on AR(1) with $\phi_1 = 0.7$ and $\sigma_a^2 = 1.0$.

depending on the timing and nature of the event. For example, suppose some external event has caused a large level shift (LS) over the last few observations of a data series. If we are able to detect this intervention, our forecasts should be more accurate if we can build the LS effect into our forecast function.

If a detected event with an identifiable cause might recur, we can use a DR model to produce forecasts that are conditional on that event happening again. For example, a manager may want an answer to the question, "What if another labor strike of the same magnitude occurs in the second quarter of next year?" If the effect of the historical strike is estimated as an intervention component in a DR model, the model's forecasts can be modified to include the "what-if" assumption that a similar strike occurs at a future date.

8.2 MODELS FOR INTERVENTION AND OUTLIER DETECTION

Consider a time series z_t whose behavior is described by an AR(1) process with a mean of zero, $(1 - \phi_1 B)z_t = a_t$, or

$$z_t = \frac{1}{1 - \phi_1 B} a_t \qquad (8.2.1)$$

We will illustrate the main ideas of outlier and intervention detection throughout this section using this simple AR(1). Figure 8.1 is a plot of 65 simulated observations generated by process (8.2.1), with $\phi_1 = 0.7$ and $\sigma_a^2 = 1.0$. Later we will introduce different types of outliers and interventions into this simulated series to see their effects on its behavior.

Of course, the AR(1) in (8.2.1) is a special case of an ARIMA $(p, d, q)(P, D, Q)_s$ process:

MODELS FOR INTERVENTION AND OUTLIER DETECTION   293

$$z_t = \frac{\theta(B^s)\theta(B)}{\phi(B^s)\phi(B)\nabla_s^D \nabla^d} a_t \qquad (8.2.2)$$

Processes (8.2.1) and (8.2.2) are assumed to be stationary (after suitable differencing) and invertible. These processes might also have a constant term; we ignore the constant term for now to simplify the notation.

We will simplify the presentation by focusing on the AR(1) in (8.2.1). However, all of our results can be extended to series described by the more general ARIMA$(p, d, q)(P, D, Q)_s$ in (8.2.2). The ideas presented here can also be applied when the underlying mechanism for z_t is a DR mechanism rather than an ARIMA process.

Now suppose that instead of observing z_t we actually observe a "contaminated" series Y_t composed of the original series z_t plus a contamination term $f(t)$:

$$Y_t = f(t) + z_t \qquad (8.2.3)$$

where $f(t)$ is a function describing an external event such as an additive or innovational outlier (AO or IO) or a level shift (LS) intervention.

In practice the contamination term $f(t)$ could have an infinite variety of forms. To simplify, we will suppose that $f(t)$ has the form of a rational distributed lag transfer function of the type discussed in Chapters 4 through 7:

$$f(t) = \frac{\omega(B)}{\delta(B)} X_t \qquad (8.2.4)$$

where X_t is a deterministic binary pulse variable representing an intervention to series z_t at time i. Thus $X_t = 1$ at $t = i$, and $X_t = 0$ otherwise; $\omega(B)$ is the usual rational transfer function numerator operator of degree h; and $\delta(B)$ is the usual denominator operator of degree r. Now let's consider three special cases for $f(t)$.

8.2.1 Additive Outlier

Consider a very simple form of $f(t)$ in (8.2.4): Suppose that all coefficients in $\omega(B)$ and $\delta(B)$ are zero, except for ω_0 (written here as ω_A). Then from (8.2.4), $f(t) = \omega_A X_t$. Substitute this form of (8.2.4) into (8.2.3) to get

$$Y_t = \omega_A X_t + z_t \qquad (8.2.5)$$

Equation (8.2.5) shows that, for this simple form of $f(t)$, the contaminated series Y_t is identical to the original series z_t *except for one observation*: Y_t is shifted up (if $\omega_A > 0$) or down (if $\omega_A < 0$) by ω_A units at time $t = i$. Such an

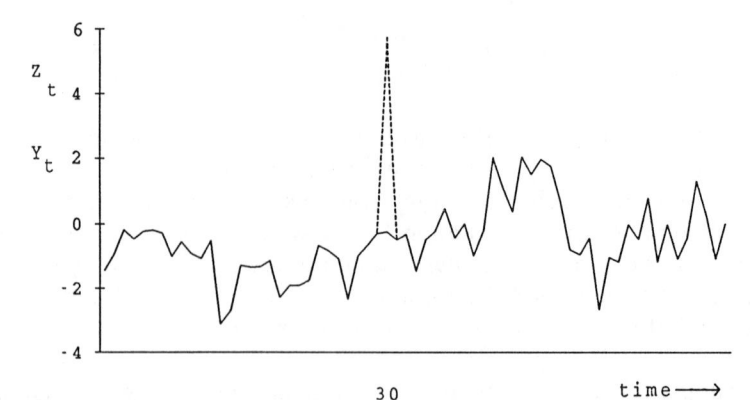

Figure 8.2 Simulated series z_t, and series Y_t with AO effect at $t = 30$ superimposed.

external event is called an *additive outlier* (AO) event. A common cause of AOs is data recording errors.

To illustrate an AO visually, we introduce an AO into series z_t (shown in Figure 8.1) at $t = i = 30$, with $\omega_A = 6.0$. The resulting contaminated series Y_t is shown in Figure 8.2; it is superimposed on the original series z_t. Notice that series Y_t is identical to series z_t *except at $t = i = 30$ when the AO occurs*. The solid line in Figure 8.2 is series z_t; it is also series Y_t for all t except $t = 30$. The dashed line shows the increase in series Y_t at $t = 30$.

If we know when an AO occurs ($t = i$ is known), or if we have tentatively identified the time ($t = i$) when one occurs, we can estimate the size (ω_A) of the AO effect within a DR model. To see this, substitute the AR(1) process for z_t in (8.2.1) into (8.2.5) to get

$$Y_t = \omega_A X_t + \frac{1}{1 - \phi_1 B} a_t \qquad (8.2.6)$$

This is the one-period response pulse intervention DR model discussed in Chapter 7. The first term in (8.2.6) is a regression-type intervention term with coefficient ω_A and input variable X_t; the second term is a disturbance with the same AR(1) process as z_t. Thus, AO effects can be handled within the DR framework developed in this book. Of course, we often don't know when an AO occurs; we will consider how we can identify the timing of AO events as we proceed.

We can generalize our results so far to handle the more general ARIMA $(p, d, q)(P, D, Q)_s$ process shown in (8.2.2). If z_t is described by that process, we substitute (8.2.2) into (8.2.5) to get

$$Y_t = \omega_A X_t + \frac{\theta(B^s)\theta(B)}{\phi(B^s)\phi(B)\nabla_s^D \nabla^d} a_t \qquad (8.2.7)$$

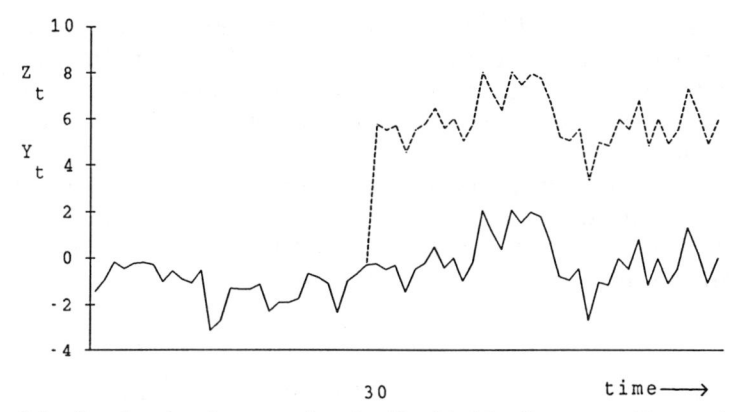

MODELS FOR INTERVENTION AND OUTLIER DETECTION   295

[Figure 8.3: plot showing series z_t and Y_t with LS effect at $t=30$ superimposed; y-axis labeled Z_t, Y_t ranging from -4 to 10; x-axis marked at 30 labeled time]

Figure 8.3 Simulated series z_t, and series Y_t with LS effect at $t = 30$ superimposed.

8.2.2 Level Shift

Consider again $f(t)$ in (8.2.4). Now suppose that only ω_0 (written here as ω_S) and δ_1 are nonzero, with $\delta_1 = 1$. Then $1 - \delta_1 B = 1 - B = \nabla$, and $f(t) = (\omega_S/\nabla)X_t$, where again $X_t = 0$ for all t except at $t = i$ when $X_t = 1$. In Section 7.2.2 we saw that, with this definition of X_t, the expression $(\omega_S/\nabla)X_t$ is one way to write a step, or permanent *level shift* (LS) intervention term, where the onset of the shift is abrupt. (We also saw in Section 7.3 that another way to write this LS intervention term, without ∇ in the denominator, is to define $X_t = 0$ for $t < i$, $X_t = 1$ for $t \geq i$, and write $\delta(B) = 1$; with this definition of X_t, $f(t)$ in (8.2.4) is $\omega_S X_t$.) We will work with X_t defined as a pulse intervention, so $f(t) = (\omega_S/\nabla)X_t$; substitute this form of (8.2.4) into (8.2.3) to get

$$Y_t = \frac{\omega_S}{\nabla} X_t + z_t \qquad (8.2.8)$$

Equation (8.2.8) shows that, for this LS form of $f(t)$, the new contaminated series Y_t is identical to the original series z_t until $t = i$; then Y_t is shifted up (if $\omega_S > 0$) or down (if $\omega_S < 0$) by ω_S units *for all $t \geq i$.*

To illustrate an LS effect visually, we introduce an LS effect into the simulated series z_t (Figure 8.1), at $t = i = 30$, with $\omega_S = 6.0$. The resulting contaminated series Y_t is shown in Figure 8.3; it is superimposed on the original zeries z_t. Series Y_t is identical to series z_t until $t = i = 30$ *when the LS occurs*; for $t \geq 30$, Y_t is higher than z_t by $\omega_S = 6$ units. The solid line in Figure 8.3 is series z_t. For $t = 1, 2, \ldots, 29$, this solid line is also series Y_t, which is identical to z_t until the LS occurs. The dashed line is series Y_t for $t \geq 30$. All Y_t observations for $t \geq 30$ are higher than the original series z_t by $\omega_S = 6$ units.

This LS effect, and the AO effect in Figure 8.2, are easy to see despite the presence of the AR(1) pattern of z_t and the presence of the shock term

a_t. In other situations an LS or AO effect may be less easy to see: ω_S or ω_A may be rather small; or the variance (σ_a^2) of the shock term may be quite large; or the ARIMA pattern of z_t may serve to obscure the LS or AO effect. Thus, it is useful to have a formal detection method that does not rely on visual inspection of the data.

If we know when an LS effect occurs ($t = i$ is known), or if we have tentatively identified the time ($t = i$) when one occurs, we can estimate the size (ω_S) of the LS effect within a DR model. To see this, substitute the AR(1) process for z_t in (8.2.1) into (8.2.8) to get

$$Y_t = \frac{\omega_S}{\nabla} X_t + \frac{1}{1 - \phi_1 B} a_t \tag{8.2.9}$$

This is the permanent step intervention DR model discussed in Chapter 7. The first term in (8.2.9) is an intervention term, and the second term is a disturbance with the same AR(1) process as z_t. Thus, LS effects can be handled within a DR framework. Of course, we often don't know when an LS occurs; we will consider how we can identify the timing of LS events as we proceed.

Once again we can generalize our results to deal with the more general ARIMA$(p, d, q)(P, D, Q)_s$ process for z_t shown in (8.2.2). If z_t is described by that process, we substitute (8.2.2) into (8.2.8) to get

$$Y_t = \frac{\omega_S}{\nabla} X_t + \frac{\theta(B^s)\theta(B)}{\phi(B^s)\phi(B)\nabla_s^D \nabla^d} a_t \tag{8.2.10}$$

8.2.3 Innovational Outlier

Consider again the contamination function $f(t)$ in (8.2.4). Suppose now that in the numerator only ω_0 (written here as ω_I) is nonzero. And suppose that in the denominator only δ_1 is nonzero *and that δ_1 is equal to ϕ_1*. Thus, $f(t) = [\omega_1/(1 - \phi_1 B)]X_t$. Once again $X_t = 0$ for all t except at $t = i$ when $X_t = 1$. From our discussion in Chapter 7 you should recognize this form of $f(t)$ as a temporary pulse intervention term, with a Koyck-type decay response; the decay rate is given by $\delta_1 = \phi_1$. This type of outlier effect is called an *innovational outlier* (IO); we will see the reason for this terminology shortly. To see the effect of an IO on a time series, first substitute this IO form of (8.2.4) into (8.2.3) to get

$$Y_t = \frac{\omega_I}{1 - \phi_1 B} X_t + z_t \tag{8.2.11}$$

Equation (8.2.11) shows that, for this IO form of $f(t)$, the contaminated series Y_t is identical to the original series z_t until $t = i$; then Y_t shifts up (if

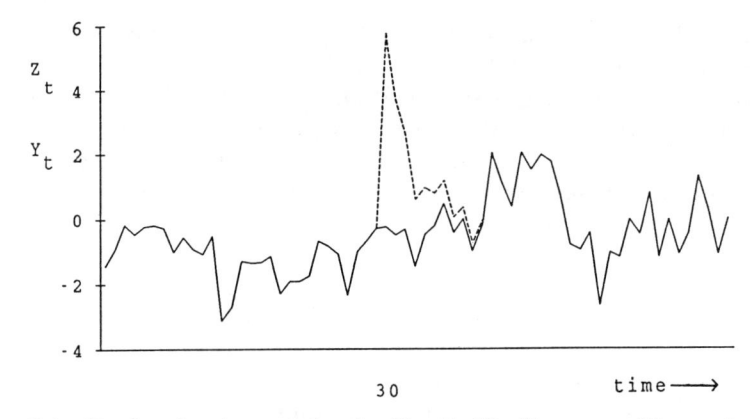

MODELS FOR INTERVENTION AND OUTLIER DETECTION   297

Figure 8.4 Simulated series z_t and series Y_t with IO effect at $t = 30$ superimposed.

$\omega_I > 0$) or down (if $\omega_I < 0$) by ω_I units at $t = i$; after $t = i$, this effect fades exponentially at a rate determined by the decay coefficient $\delta_1 = \phi_1$.

To illustrate an IO effect visually, we introduce an IO effect into the simulated series z_t (Figure 8.1) at $t = i = 30$, with $\omega_I = 6.0$ and $\delta_1 = \phi_1 = 0.7$. The resulting contaminated series Y_t is shown in Figure 8.4; it is superimposed on the original series z_t. Series Y_t is identical to z_t until $t = i = 30$ when the IO occurs. The solid line in Figure 8.4 is series z_t. For $t = 1, 2, \ldots, 29$, this solid line is also series Y_t, which is identical to z_t until the IO occurs. The dashed line is series Y_t for $t \geq 30$.

For $t \geq i = 30$, Y_t is higher than z_t by $\phi_1^{t-i} \omega_I$ units. This represents a Koyck-type decay as discussed in Chapter 4. Thus in our example Y_t is higher than z_t by $0.7^0(6) = 6$ units at $t = i = 30$; by $0.7(6) = 4.2$ units at $t = 31$; by $0.7^2(6) = 2.94$ units at $t = 32$; and so forth. The effect of the IO fades until eventually the contaminated series Y_t is indistinguishable from the original series z_t: In Figure 8.4 the dashed line (Y_t) gradually gets closer to the solid line (z_t) as we move further past $t = i = 30$.

If we know when an IO effect occurs ($t = i$ is known), or if we have tentatively identified the time ($t = i$) when one occurs, we can estimate the size (ω_I) of the IO effect within a DR model. To see this, substitute the AR(1) process for z_t in (8.2.1) into (8.2.11) to get

$$Y_t = \frac{\omega_I}{1 - \phi_1 B} X_t + \frac{1}{1 - \phi_1 B} a_t \qquad (8.2.12)$$

This is a temporary pulse intervention model with a Koyck-type decay response, as discussed in Chapter 7. The first term in (8.2.12) is an intervention term with decay start-up coefficient ω_I, decay coefficient $\delta_1 = \phi_1$, and pulse input X_t; the second term is a disturbance with the same AR(1) process as z_t. Thus, IO effects can be handled within a DR framework. Of course, we often don't know when an IO occurs; it remains for us to consider how we can identify the timing of IO events.

Using (8.2.12) we can explain why the effects we have been discussing are called "innovational" outliers. The random shock series (a_t) is sometimes called the innovation series. The idea of an IO is that the event produces *an addition to the random shock (innovation) series* a_t at time i. This is seen more clearly by gathering terms and rewriting (8.2.12) as

$$Y_t = \frac{1}{1 - \phi_1 B}(\omega_I X_t + a_t) \qquad (8.2.13)$$

Equation (8.2.13) says that the effect of event X_t is added to a_t at $t = i$ in the amount ω_I. The initial effect at $t = i$ on the observed series Y_t is also in the amount ω_I. But, unlike an AO, this initial effect of an IO then *propagates through the subsequent Y_t values*, since Y_t may be thought of as depending on past a_t values, as described by (8.2.13). The exact nature of this propagation depends on the nature of the ARIMA structure of the disturbance series. We have used an AR(1) disturbance pattern as an example; thus we find in (8.2.12) and (8.2.13) that an IO effect propagates through Y_t after $t = i = 30$ via an AR(1) mechanism.

Once again we can extend our results for the IO to deal with the more general ARIMA$(p, d, q)(P, D, Q)_s$ process for z_t shown in (8.2.2). If z_t is described by that process instead of by an AR(1), we replace $1/(1 - \phi_1 B)$ in (8.2.13) by the ratio of MA, AR, and differencing elements in (8.2.2). Factoring the resulting expression gives the following DR model:

$$Y_t = \frac{\theta(B^s)\theta(B)\omega_I}{\phi(B^s)\phi(B)\nabla_s^D \nabla^d} X_t + \frac{\theta(B^s)\theta(B)}{\phi(B^s)\phi(B)\nabla_s^D \nabla^d} a_t \qquad (8.2.14)$$

This expression is interpreted in the same way that we interpreted the simpler DR model result in (8.2.12). The second term in (8.2.14) is our usual expression for the ARIMA process of the disturbance in a DR model. The first term in (8.2.14) is an intervention term with input X_t, which produces a distributed lag response in Y_t; this response is described by the rational transfer function by which X_t is multiplied. As discussed in Chapters 4 and 5, the numerator of this ratio captures unpatterned and decay start-up terms; the denominator captures the decay pattern. Of course, this is a rather special version of a rational transfer function: Here the so-called unpatterned terms, decay start-up terms, and decay rate terms are all (except for ω_I) related to the coefficients in the disturbance term ARIMA process.

In this chapter we focus on three types of interventions and outliers (AO, LS, and IO). However the framework established here can be extended to deal with many other types of external events. (See Questions 3 and 4 at the end of the chapter.) Tsay (1988) also discusses the application of the framework presented here to the detection of variance changes in time series.

So far the discussion in this chapter involves merely adapting the intervention model framework in Chapter 7 to show how certain kinds of outliers may be represented within a DR model. We now begin consideration of certain ideas that are important for detecting outliers and interventions.

8.3 LIKELIHOOD RATIO CRITERIA

In this section we establish some ideas that will lead to a practical procedure for detecting AO, LS, and IO events. The practical procedure is given in Section 8.4. The material in these sections is based on the work of Chang et al. (1988), who give additional references. Tsay (1986, 1988) takes a similar approach. The ideas presented here have been implemented as part of several computing packages, including the SCA System.

To fix the underlying ideas, we assume in this section that the ARIMA coefficients and σ_a^2 for process (8.2.2) are known. In Section 8.4 we consider the case where these parameters are estimated. Two main ideas are developed here:

1. The initial effect of an AO, LS, or IO (coefficients ω_A, ω_S, and ω_I) can be *estimated by least-squares regression methods* using a set of constructed "residuals."
2. The hypothesis that an AO, LS, or IO has occurred can be *tested with a likelihood ratio statistic*, which is similar to a t statistic.

Before developing these ideas in detail, we first consider two useful operators, $\pi(B)$ and $c(B)$, where B is the backshift operator. Then we consider a set of model "residuals" that can be studied for evidence of interventions and outliers.

8.3.1 The π-Weight and c-Weight Operators

$\pi(B)$ Operator

As discussed in Chapter 2, any invertible ARIMA process can be written in equivalent AR form: We can write the model so that z_t is a function of past z's only (no MA terms). In effect we replace the AR and MA terms with a set of AR-like terms. To obtain this form of process (8.2.2), divide both sides of that expression by the MA operators to get

$$\frac{\phi(B^s)\phi(B)\nabla_s^D\nabla^d}{\theta(B^s)\theta(B)} z_t = a_t \qquad (8.3.1)$$

The ratio on the left side of (8.3.1) operates on z_t only, not on a_t. In other words we are stating z_t as function of past z's only (AR-like terms), not past a's.

Instead of representing these AR-like terms in ratio form, we define an operator in B that is equal to the ratio in (8.3.1):

$$\pi(B) = 1 - \pi_1 B - \pi_2 B^2 - \cdots = \frac{\phi(B^s)\phi(B)\nabla_s^D \nabla^d}{\theta(B^s)\theta(B)}$$

This is the π-weight function for the ARIMA model in (8.2.2), as discussed by Box and Jenkins (1976). Substituting $\pi(B)$ for the ratio of operators in (8.3.1), we can write (8.2.2) in π-weight form as

$$\pi(B)z_t = a_t$$

The order of $\pi(B)$ is not finite if any MA terms are present. However, if the process is invertible then the π_j values tend to become small fairly quickly as j increases.

The π weights are found by multiplying both sides of the definition of $\pi(B)$ by $\theta(B^s)\theta(B)$ to get $\theta(B^s)\theta(B)(1 - \pi_1 B - \pi_2 B^2 - \cdots) = \phi(B^s)\phi(B)\nabla_s^D \nabla^d$. Then we equate coefficients of like powers of B on either side of this expression. Some software packages compute the π weights for a given model, but it can be done by hand. For example, for the AR(1) process in (8.2.1) we have $\pi(B) = 1 - \phi_1 B$. Equating coefficients of like powers of B on either side of this expression gives $\pi_1 = \phi_1$, and $\pi_j = 0$ for $j \geq 2$. Thus the π-weight operator for this simple example is $\pi(B) = 1 - \pi_1 B$.

As another example, consider an ARIMA(1, 0, 1) model. To find the π weights, equate coefficients of like powers of B in this expression: $(1 - \theta_1 B)(1 - \pi_1 B - \pi_2 B^2 - \cdots) = 1 - \phi_1 B$. Expand the left side of this expression to get

$$1 - (\pi_1 + \theta_1)B - (\pi_2 - \theta_1\pi_1)B^2 - (\pi_3 - \theta_1\pi_2)B^3 - \cdots = 1 - \phi_1 B \quad (8.3.2)$$

Equating coefficients of B^1 on either side of (8.3.2) gives $\pi_1 - \theta_1 = \phi_1$, or $\pi_1 = \phi_1 - \theta_1$. Equating coefficients of B^2 gives $\pi_2 - \theta_1\pi_1 = 0$, or $\pi_2 = \theta_1\pi_1$. A similar procedure for higher powers of B for this example shows $\pi_j = \theta_1\pi_{j-1}$ (or $\pi_j = \theta_1^{j-1}\pi_1$) for $j > 1$. Because $\theta_1 \neq 0$, the number of nonzero π weights for this model is not finite. However, if $|\theta_1| < 1$ so that the process is invertible, then the absolute values of the π weights tend to become small fairly quickly. Clearly the π-weight form may not be a parsimonious way to write an ARIMA model; but this form will be useful in developing a procedure for detecting outliers and interventions.

LIKELIHOOD RATIO CRITERIA

$c(B)$ Operator

Our second operator is defined as $c(B) = 1 - c_1 B - c_2 B^2 - \cdots = \pi(B) \nabla^{-1}$. As we will see, $c(B)$ is useful for detecting LS events. The c weights are found by equating coefficients of like powers of B in the expression $\nabla c(B) = \pi(B)$. The results are $c_1 = \pi_1 - 1$ and $c_j = c_{j-1} + \pi_j$ for $j > 1$.

8.3.2 The Residual Series e_t

Now suppose we filter Y_t through the $\pi(B)$ operator to produce the "residual" series e_t:

$$\begin{aligned} e_t &= \pi(B) Y_t \\ &= (1 - \pi_1 B - \pi_2 B^2 - \pi_3 B^3 - \cdots) Y_t \\ &= Y_t - \pi_1 Y_{t-1} - \pi_2 Y_{t-2} - \pi_3 Y_{t-3} - \cdots \end{aligned} \quad (8.3.3)$$

In (8.3.3) we are just computing the difference between the observed value of Y_t (the first term on the right side) and the one-step ahead predicted value of Y_t (the remaining terms on the right side); the predicted value is computed using the π-weight form of the ARIMA process for Y_t. For example, for the AR(1) in (8.2.1) we found $\pi(B) = 1 - \pi_1 B$, with $\pi_1 = \phi_1$. For this example we have $e_t = (1 - \pi_1 B) Y_t = Y_t - \pi_1 Y_{t-1}$. Y_t is the observed value, and $\pi_1 Y_{t-1}$ is the one-step ahead predicted value. The residual series e_t is the difference between these two values. Thus e_t may also be interpreted as a set of one-step ahead prediction errors.

Suppose there are n observations available for the original data series. Estimating an ARIMA$(p, d, q)(P, D, Q)_s$ model entails the loss of $n_1 = p + Ps + d + Ds$ observations, starting with z_1. Thus the residual series will start at time $t = 1 + n_1$ and contain only $m = n - n_1$ instead of n values. For example, for the AR(1) in (8.2.1) we cannot compute $e_1 = Y_1 - \pi_1 Y_0$ because Y_0 is not observed; therefore $n_1 = p = 1$. Series e_t would start at time $t = 1 + n_1 = 1 + 1 = 2$, and the number of e_t values would be only $m = n - 1$.

In practice our first set of e_t values is given by the model residuals that are available as a by-product of model estimation. Equation (8.3.3) will be especially helpful in *modifying* the e_t series as needed. *We will use (and modify) this residual series to detect interventions and outliers since different types of interventions and outliers leave characteristic "footprints" in the residual e_t series.*

Now let's see how we can make use of the residual series e_t. From (8.3.3) we can write $Y_t = \pi^{-1}(B) e_t$. Substitute this into (8.2.7), (8.2.10), and (8.2.14), use the definitions of $\pi(B)$ and $c(B)$, and rearrange to obtain, respectively,

(AO) $$e_t = \omega_A \pi(B) X_t + a_t \quad (8.3.4)$$

(LS) $$e_t = (\omega_S/\nabla)\pi(B)X_t + a_t$$
$$= \omega_S c(B)X_t + a_t \quad (8.3.5)$$

(IO) $$e_t = \omega_I X_t + a_t \quad (8.3.6)$$

From these three equations we see that the residual series e_t is identical to the random shock series a_t when there are no AO, LS, or IO effects (i.e., when the ω weights are zero). According to (8.3.4) and (8.3.5), an AO or LS at time $t = i$ affects a *string* of residual e_t values, $e_i, e_{i+1}, \ldots$. But from (8.3.6), an IO at $t = i$ affects only the residual e_i. While both an AO and an LS affect strings of residuals, the *pattern* in the e_t series produced by an AO is different from the pattern produced by an LS. Thus, if we have a correct ARIMA model but have not detected certain interventions and outliers, we can check the e_t series for telltale footprints that characterize AO, LS, and IO events.

AO Example

We will now consider some simple examples to clarify the behavior of the residual series e_t. We continue with the AR(1) example for z_t in (8.2.1). Suppose again that an external event occurs at $t = i = 30$, so that $X_t = 1$ at $t = 30$ and $X_t = 0$ for all other t. With no differencing and no seasonal AR terms, and with $p = 1$, we have $n_1 = p + Ps + d + Ds = 1 + 0 + 0 + 0 = 1$; therefore in practice the model residuals and e_t can be computed starting at $t = 1 + n_1 = 1 + 1 = 2$.

Suppose the external event is an AO. Following the procedure for finding the π weights discussed earlier, for an AR(1) we find $\pi_1 = \phi_1$, and $\pi_j = 0$ for $j > 1$. Therefore we can write (8.3.4) as $e_t = \omega_A(1 - \pi_1 B)X_t + a_t$, or

$$e_t = \omega_A X_t - \omega_A \pi_1 X_{t-1} + a_t \quad (8.3.7)$$

According to (8.3.7), if z_t follows an AR(1) and is contaminated by an AO at $t = i$, then the constructed residual series e_t is equal to a_t except at $t = i$ and $t = i + 1$. Let's illustrate this result visually. Figure 8.1 shows series z_t, which was simulated by an AR(1) process with $\phi_1 = 0.7$ and $\sigma_a^2 = 1$. Figure 8.2 shows series Y_t, which is equal to z_t plus an AO effect at $t = 30$, with $\omega_A = 6$. Figure 8.5 shows series e_t for this example as computed from (8.3.7). The values at $t = i = 30$ and $t = i + 1 = 31$ stand out. These are the telltale footprints in series e_t of the AO at $t = 30$ for this example.

Of course, these AO footprints in Figure 8.5 are embedded within a random shock series. According to (8.3.7), we can *isolate the pure effect* of the AO on the residual series e_t if we subtract a_t from e_t. (In practice we never observe series a_t directly; but since z_t in this example is a *simulated* series, we happen to know the values of a_t.) Figure 8.6 shows the series e_t in Figure 8.5 minus the random shock component (a_t) of simulated series

LIKELIHOOD RATIO CRITERIA

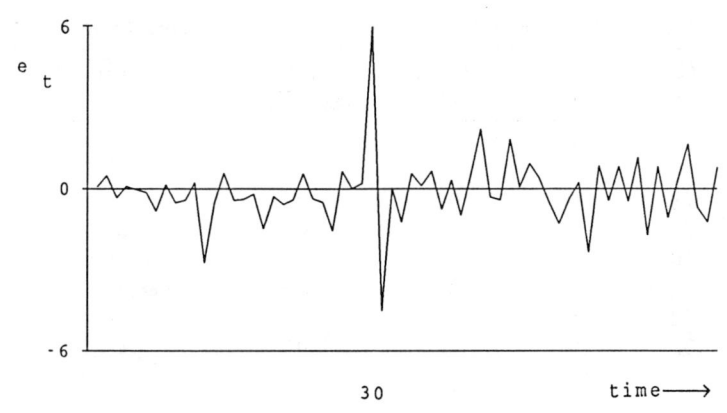

Figure 8.5 Residual series e_t computed from (8.3.7) applied to series Y_t (with AO) in Figure 8.2.

z_t. Figure 8.6 shows that $e_t - a_t$ for this example is zero everywhere except at $t = i = 30$ and $t = i + 1 = 31$.

Some tabular computations will clarify the effects of this AO on the residual series e_t. Table 8.1 shows the behavior of e_t under the assumptions for this example. Column (1) is the time period. Column (2) gives the values of X_t: all zeros, except $X_{30} = 1$. Columns (3) and (4) show the first two terms on the right side of (8.3.7), disregarding signs. Column (5) shows the value of e_t, which is found by subtracting column (4) from column (3), and adding a_t; we subtract column (4) to account for the negative sign in (8.3.7). Column (5) in Table 8.1 tells us the following: If z_t follows an AR(1), and if z_t is contaminated by an AO at $t = 30$, then the computed residual sequence e_t is zero-mean and normally distributed white noise (a_t) except at $t = 30$ and $t = 31$. This is the result illustrated in Figure 8.5.

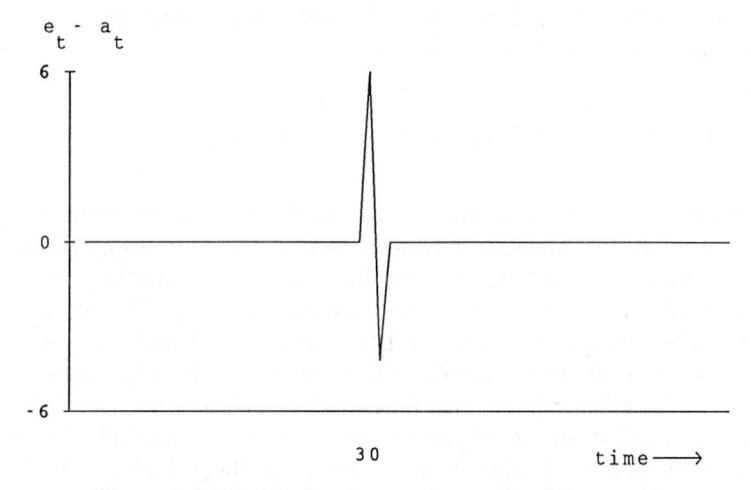

Figure 8.6 Residual series e_t minus a_t for AO example.

Table 8.1 Example of Effect of AO at $t = 5$ on Residual Series e_t

(1) t	(2) X_t	(3) $\omega_A X_t$	(4) $\omega_A \pi_1 X_{t-1}$	(5) e_t
⋮	⋮	⋮	⋮	⋮
28	0	0	0	a_{28}
29	0	0	0	a_{29}
30	1	ω_A	0	$\omega_A + a_{30}$
31	0	0	$\omega_A \pi_1$	$-\omega_A \pi_1 + a_{31}$
32	0	0	0	a_{32}
33	0	0	0	a_{33}
⋮	⋮	⋮	⋮	⋮

In this example e_t is affected at $t = i = 30$ *and later*. How long the AO continues to affect e_t after $t = i$ depends on the π weights. For this example the effect of the AO on e_t stops after $t = 31$. When the ARIMA process for z_t has only AR terms, the order of $\pi(B)$ is finite. If there are any MA terms in the ARIMA process for z_t, the order of $\pi(B)$ is not finite and the effect of the AO on series e_t would continue indefinitely. However, if the ARIMA process is invertible, then the π weights tend to decline in absolute value fairly quickly, and the effect of the AO on e_t tends to fade fairly quickly.

LS Example
Now suppose the external event at $t = i$ is an LS rather that an AO. As seen in (8.3.5), we now must find the c weights after finding the π weights. Using the result in Section 8.3.1, for an AR(1) we have $c_j = \pi_1 - 1, j \geq 1$. Therefore we can write Equation (8.3.5) for this example as $e_t = \omega_S(1 - c_1 B - c_1 B^2 - c_1 B^3 - \cdots)X_t + a_t$, or

$$e_t = \omega_S X_t - \omega_S c_1 X_{t-1} - \omega_S c_1 X_{t-2} - \omega_S c_1 X_{t-3} - \cdots + a_t \qquad (8.3.8)$$

According to (8.3.8), if z_t follows an AR(1) and is contaminated by an LS at $t = i$, then the constructed residual series e_t is equal to a_t through $t = i - 1$; thereafter e_t is affected by the LS event. Let's illustrate this result visually. Again, Figure 8.1 shows the simulated series z_t. Figure 8.3 shows series Y_t, which is equal to z_t plus an LS effect at $t = 30$, with $\omega_S = 6$. Figure 8.7 shows series e_t for this example as computed from (8.3.8). The values of e_t at $t = i = 30$ and thereafter are all shifted away from the overall level of the previous random shock values; the value at $t = 30$ especially stands out. These are the telltale footprints in series e_t of the LS at $t = 30$ for this example.

LIKELIHOOD RATIO CRITERIA

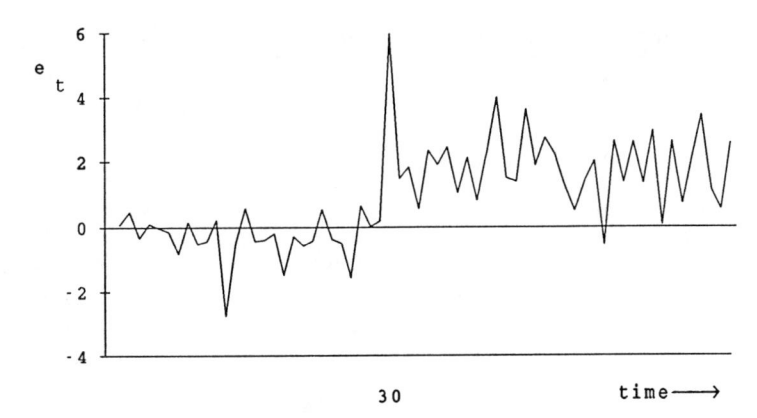

Figure 8.7 Residual series e_t computed from (8.3.8) applied to series Y_t (with LS) in Figure 8.3.

The LS footprints in Figure 8.7 are embedded within a random shock series. According to (8.3.8), we can isolate the pure effect of the LS on the residual series e_t if we subtract a_t from e_t. (Since z_t in this example is a *simulated* series, we know the values of a_t.) Figure 8.8 shows the series e_t in Figure 8.7 *minus* the random shock component (a_t) of simulated series z_t. Figure 8.8 shows that $e_t - a_t$ for this example is zero through $t = i - 1 = 29$; at $t = i = 30$ the LS effect causes e_t to rise by $\omega_S = 6$ units; thereafter the e_t series drops back toward zero (the overall level of the random shock series), but it doesn't drop all the way to zero. From (8.3.8) we see that $e_t - a_t$ is $\omega_S c_1 = -6(-0.3) = 1.8$ units above zero for $t > 30$ for this example.

Table 8.2 shows the behavior of part of the residual series e_t under the assumptions for this example. Columns (1) and (2) are identical to those in Table 8.1. Columns (3), (4), (5), and (6) show the first four terms on the

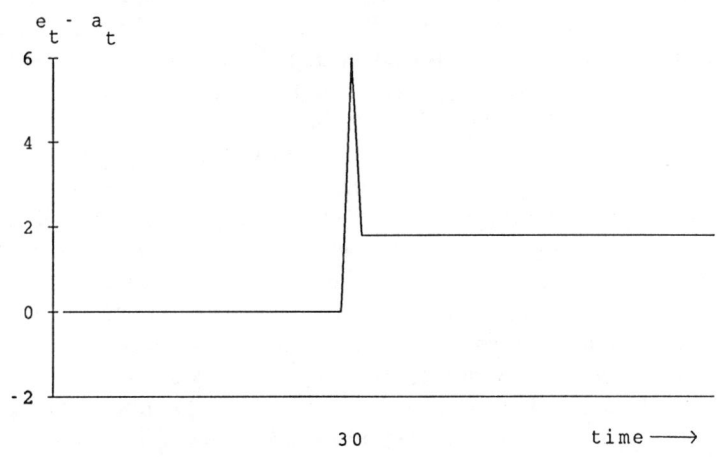

Figure 8.8 Residual series e_t minus a_t for LS example.

Table 8.2 Example of Effect of LS at $t = 30$ on Residual Series e_t

(1) t	(2) X_t	(3) $\omega_S X_t$	(4) $\omega_S c_1 X_{t-1}$	(5) $\omega_S c_1 X_{t-2}$	(6) $\omega_S c_1 X_{t-3}$	(7) e_t
⋮	⋮	⋮	⋮	⋮	⋮	⋮
28	0	0	0	0	0	a_{28}
29	0	0	0	0	0	a_{29}
30	1	ω_S	0	0	0	$\omega_S + a_{30}$
31	0	0	$\omega_S c_1$	0	0	$-\omega_S c_1 + a_{31}$
32	0	0	0	$\omega_S c_1$	0	$-\omega_S c_1 + a_{32}$
33	0	0	0	0	$\omega_S c_1$	$-\omega_S c_1 + a_{33}$
⋮						⋮
$t > 30$		⋅	⋅	⋅		$-\omega_S c_1 + a_t$

right side of (8.3.8), disregarding signs. Column (7) is e_t, which is found by subtracting columns (4), (5), and (6) from column (3), and adding a_t; we subtract to account for the negative signs on the right side of (8.3.8). The ellipses (...) after $t = 33$ in Table 8.2 indicate that the LS *continues to affect* e_t *indefinitely*, unlike the AO illustrated in Table 8.1 where the AO effects stop after $t = 31$. This is a general feature of LS events. While an AO affects e_t after $t = i$, its effect on e_t eventually ceases if $\pi(B)$ has a finite number of terms (as in the AO example). However, the effect of an LS on e_t always continues indefinitely since $c(B)$ is never of finite order.

IO Example
Next suppose the external event has the form of an IO. Equation (8.3.6) says that an IO at $t = i$ affects series e_t at $t = i$ only. Let's illustrate this result visually. Figure 8.4 shows series Y_t, which is equal to the simulated series z_t (shown in Figure 8.1) plus an IO effect at $t = 30$, with $\omega_I = 6$. Figure 8.9 shows series e_t for this example as computed from (8.3.6). The value at $t = i = 30$ stands out clearly from the rest of the series. This is the telltale footprint in series e_t of the IO at $t = 30$ for this example.

The IO footprint in Figure 8.9 is embedded within a random shock series. According to (8.3.6), we can isolate the pure effect of the IO on the residual series e_t if we subtract a_t from e_t. (Again, since z_t in this example is a simulated series, we happen to know the values of a_t.) Figure 8.10 shows the series e_t in Figure 8.9 *minus* the random shock component (a_t) of simulated series z_t. Figure 8.10 shows that $e_t - a_t$ for this example is zero for all t except $t = 30$, when the IO effect causes e_t to rise by $\omega_I = 6$ units.

Table 8.3 shows the behavior of the residual series e_t under the assumptions for this example. Columns (1) and (2) are identical to those in Tables

LIKELIHOOD RATIO CRITERIA

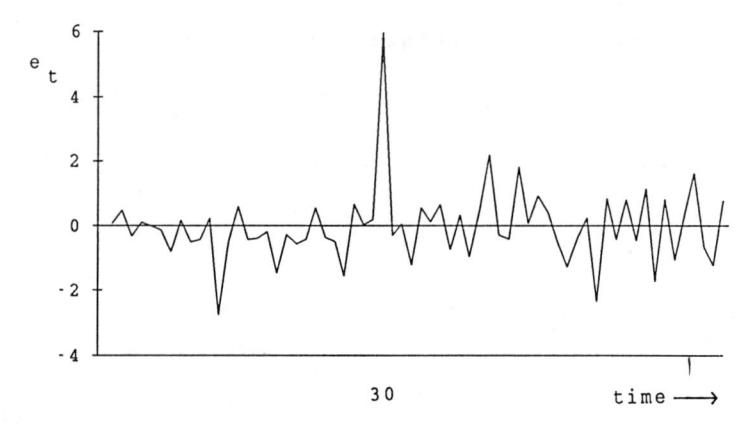

Figure 8.9 Residual series e_t computed from (8.3.6) applied to series Y_t (with IO) in Figure 8.4.

8.1 and 8.2. Column (3) shows the first term on the right side of (8.3.6). Column (4) is e_t, which is found by adding a_t to column (3). Column (4) confirms what we see in Figures 8.9 and 8.10: The IO event at time $t = 30$ affects series e_t at time $t = 30$ only. In fact, *any* IO at $t = i$ *affects only e_i*, unlike an AO event at $t = i$, which *may* affect e_t also after period $t = i$, and unlike an LS event at $t = i$, which *always* affects e_t also after period $t = i$.

We have seen that AO, LS, and IO events affect the residual series e_t. Further, these events affect e_t in different ways. Thus, by studying the e_t series from a given model for telltale footprints, we may be able to detect the possible existence, timing, and type of these three external events.

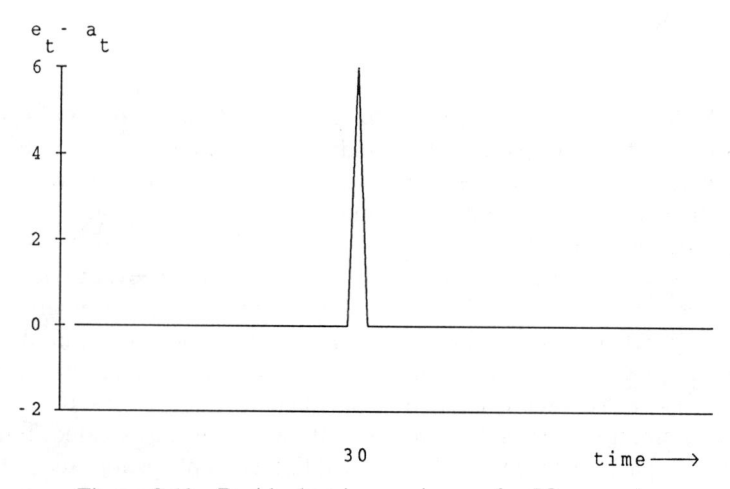

Figure 8.10 Residual series e_t minus a_t for IO example.

Table 8.3 Example of Effect of IO at $t = 30$ on Residual Series e_t

(1) t	(2) X_t	(3) $\omega_I X_t$	(4) e_t
⋮	⋮	⋮	⋮
28	0	0	a_{28}
29	0	0	a_{29}
30	1	ω_I	$\omega_I + a_{30}$
31	0	0	a_{31}
32	0	0	a_{32}
⋮	⋮	⋮	⋮

8.3.3 Least-Squares Regression Estimates of ω's

The ω coefficients in (8.3.4), (8.3.5), and (8.3.6) can be estimated with ordinary least squares-regression methods. The least-squares estimators (denoted ω^*) of the ω_A, ω_S, and ω_I parameters for an AO, LS, or IO event at time $t = i$ are

$$\omega_A^* = k_A \pi(F) e_i$$
$$= k_A (1 - \pi_1 F - \pi_2 F^2 - \cdots - \pi_{n-i} F^{n-i}) e_i \quad (8.3.9)$$
$$\omega_S^* = k_S c(F) e_i$$
$$= k_S (1 - c_1 F - c_2 F^2 - \cdots - c_{n-i} F^{n-i}) e_i \quad (8.3.10)$$
$$\omega_I^* = e_i \quad (8.3.11)$$

where F is the *forward shift operator* defined such that $F^j e_t = e_{t+j}$; $k_A = (1 + \pi_1^2 + \pi_2^2 + \cdots + \pi_{n-i}^2)^{-1}$; and $k_S = (1 + c_1^2 + c_2^2 + \cdots + c_{n-i}^2)^{-1}$. The derivation of these expressions is given later. In practice we use the sample residual series $\hat{e}_t$ to obtain estimates $\hat{\omega}_A$, $\hat{\omega}_S$, and $\hat{\omega}_I$.

In practice we don't know at which time $t = i$, if any, an AO, LS, or IO occurs. Thus we consider the possibility that each type of event could occur at every time $t = i = 1 + n_1, 2 + n_1, \ldots, n$. Computing the estimate ω_I^* for each t in (8.3.11) is quite simple: Each model residual computed at the estimation stage is an initial estimate of e_t, so we immediately have the estimate $\omega_I^* = e_i$ for each t, as seen in (8.3.11).

The computations for (8.3.9) and (8.3.10) are more complicated, but both ω_A^* and ω_S^* can be computed recursively. For example, suppose $n_1 = 0$ so that computations can start at $t = 1$. Coefficient ω_A^* is found recursively for $t = 1, 2, \ldots, n$ as follows:

LIKELIHOOD RATIO CRITERIA   309

1. Set $t = i = 1$. Compute k_A for period $i = 1$ as $k_A = (1 + \pi_1^2 + \pi_2^2 + \cdots + \pi_{n-1}^2)^{-1}$. Then compute $\pi(F)e_i$ in (8.3.9) for $i = 1$ as $e_1 - \pi_1 e_2 - \pi_2 e_3 - \cdots - \pi_{n-1} e_n$. Multiply this result by k_A to get ω_A^* for $t = i = 1$.

2. Recursively for $t = i = 2, 3, \ldots, n$, denote k_A^{-1} for $t - 1 = i - 1$ as k_{A*}^{-1} and compute a new $k_A^{-1} = k_{A*}^{-1} - \pi_{n-i+1}^2$. Then compute the new value of $\pi(F)e_i$ in (8.3.9) as $e_i - \pi_1 e_{i+1} - \pi_2 e_{i+2} - \cdots - \pi_{n-i} e_n$. Multiply this new value of $\pi(F)e_i$ by the new k_A to get the new ω_A^* for each $t = i$.

A similar recursive procedure applies to the computation of ω_S^* for $t = i = 1 + n_1, 2 + n_1, \ldots, n$.

To see the derivation of Equations (8.3.9) to (8.3.11), first define W_t as

$$W_t = \begin{cases} \pi(B)X_t & \text{in Equation (8.3.4)} \quad \text{(AO)} \\ c(B)X_t & \text{in Equation (8.3.5)} \quad \text{(LS)} \\ X_t & \text{in Equation (8.3.6)} \quad \text{(IO)} \end{cases} \tag{8.3.12}$$

Now we can write Equations (8.3.4), (8.3.5), and (8.3.6) as

$$e_t = \omega_0 W_t + a_t \tag{8.3.13}$$

where W_t is defined in (8.3.12), and ω_0 equals ω_A, ω_S, or ω_I, depending on the type of event considered. Equation (8.3.13) is a simple *linear regression model through the origin*; therefore the least-squares estimator of ω_0 is

$$\omega_0^* = \frac{\sum_{t=1+n_1}^{n} W_t e_t}{\sum_{t=1+n_1}^{n} W_t^2} \tag{8.3.14}$$

The result (8.3.14) for linear regression through the origin is derived in many textbooks; for example, see Abraham and Ledolter (1983, p. 13). Equation (8.3.14) leads to Equations (8.3.9), (8.3.10), and (8.3.11), respectively, for the three definitions of W_t shown in (8.3.12).

It is easiest to see the result of using (8.3.14) in the case of an IO, leading to Equation (8.3.11). From (8.3.12), $W_t = X_t$ for an IO; therefore we substitute X_t for W_t in (8.3.14). The computations that produce Equation

Table 8.4 Application of Least-Squares Estimator (8.3.14) to Equation (8.3.6) to Estimate ω_I

(1) t	(2) $W_t = X_t$	(3) e_t	(4) $W_t e_t$	(5) W_t^2
1	0	e_1	0	0
2	0	e_2	0	0
3	0	e_3	0	0
$\vdots$	$\vdots$	$\vdots$	$\vdots$	$\vdots$
$i-1$	0	e_{i-1}	0	0
i	1	e_i	e_i	1
$i+1$	0	e_{i+1}	0	0
$\vdots$	$\vdots$	$\vdots$	$\vdots$	$\vdots$
n	0	e_n	0	0
Column sums			e_i	1

(8.3.11) are shown in Table 8.4, assuming $n_1 = 0$. The values of X_t and e_t are shown in columns (2) and (3), respectively. Column (4) shows the product $W_t e_t$ ($= X_t e_t$) for each t. The only product $W_t e_t$ that is nonzero is at $t = i$, when it is equal to e_i. ($W_t = X_t$ is zero for all t except $t = i$, when $W_t = X_t = 1$.) The sum of the products $W_t e_t$ is shown at the bottom of column (4) of Table 8.4. This sum is the numerator in Equation (8.3.14), and it is equal to e_i for the case of an IO. Column (5) in Table 8.4 shows the squared values of W_t (equal to the squared values of X_t); these sum to 1.0 as shown at the bottom of Table 8.4. This sum is the denominator in Equation (8.3.14), and it is equal to 1.0 for the case of an IO. Thus the least-squares estimator for an IO at $t = i$ is $\omega_I^* = e_i$, as shown in Equation (8.3.11).

Next we will use (8.3.14) to derive (8.3.9). [The derivation of (8.3.10) is left as an exercise at the end of the chapter.] Table 8.5 gives the information needed to show how (8.3.14) leads to (8.3.9) for an AO, again assuming $n_1 = 0$. Columns (1) and (2) in Table 8.5 are the same as columns (1) and (2) in Table 8.4: An AO occurs at time $t = i$, so $X_t = 1$ at that time. But now $W_t = \pi(B)X_t = (1 - \pi_1 B - \pi_2 B^2 - \cdots)X_t$ for an AO, as shown in column (3). At $t = 1$ there are no earlier values of X_t (or we may assume $X_t = 0$ for all $t < 1$), so only the first term in $\pi(B)$ is relevant: All other terms involve the backshift operator to some power greater than zero, but there are no earlier X_t values. The first term in $\pi(B)$ is 1; multiply by $X_1 = 0$ to get zero. This is the first entry in column (3).

At $t = 2$ there is just one previous value for X_t (at $t = 1$), so only the first two terms in $\pi(B)$ are relevant. The first term in $\pi(B)$ is 1; multiply by $X_2 = 0$ to get zero. The second term in $\pi(B)$ is $-\pi_1 B$; multiply by X_2 to get $-\pi_1 X_1$, which equals zero because $X_1 = 0$. The sum of these two terms is

LIKELIHOOD RATIO CRITERIA

Table 8.5 Application of Least-Squares Estimator (8.3.14) to Equation (8.3.4) to Estimate ω_A

(1) t	(2) X_t	(3) $W_t = \pi(B)X_t$	(4) e_t	(5) $W_t e_t$	(6) W_t^2
1	0	0	e_1	0	0
2	0	0	e_2	0	0
3	0	0	e_3	0	0
⋮	⋮	⋮	⋮	⋮	⋮
$i-1$	0	0	e_{i-1}	0	0
i	1	1	e_i	e_i	1
$i+1$	0	$-\pi_1$	e_{i+1}	$-\pi_1 e_{i+1}$	π_1^2
$i+2$	0	$-\pi_2$	e_{i+2}	$-\pi_2 e_{i+2}$	π_2^2
⋮	⋮	⋮	⋮	⋮	⋮
n	0	$-\pi_{n-i}$	e_n	$-\pi_{n-i} e_n$	π_{n-i}^2

zero, and that is the second entry in column (3). Similar reasoning leads to the remaining values in column (3).

Column (4) in Table 8.5 is e_t, just like column (3) in Table 8.4. Column (5) is the product of W_t in column (3) and e_t in column (4); we sum these products to find the numerator in Equation (8.3.14). Summing these terms gives $e_i - \pi_1 e_{i+1} - \cdots - \pi_{n-i} e_n$; using the forward shift operator, this may be written as $\pi(F)e_i = (1 - \pi_1 F - \cdots - \pi_{n-i} F^{n-i})e_i$, which is how the numerator in Equation (8.3.14) is expressed in (8.3.9).

Finally, column (6) in Table 8.5 is the square of each W_t value in column (3). The sum of these squared terms gives the denominator in Equation (8.3.14). Their sum is $1 + \pi_1^2 + \pi_2^2 + \cdots + \pi_{n-i}^2$. Define k_A^{-1} as this sum. In (8.3.14) we must divide by this sum, or multiply by its inverse (k_A) as shown in Equation (8.3.9). This completes the derivation of Equation (8.3.9).

8.3.4 Variances of the Estimators

To perform statistical tests in our outlier detection procedure, we need the variances of the estimators in (8.3.9), (8.3.10), and (8.3.11). These variances are

$$\text{var}(\omega_A^*) = k_A \sigma_a^2 \qquad (8.3.15)$$

$$\text{var}(\omega_S^*) = k_S \sigma_a^2 \qquad (8.3.16)$$

$$\text{var}(\omega_I^*) = \sigma_a^2 \qquad (8.3.17)$$

respectively. These variances are found using a standard result for the least-squares regression estimator through the origin: The variance of this least-squares estimator ω_0^* is

$$\text{var}(\omega_0^*) = \frac{\sigma_a^2}{\sum_{t=1+n_1}^{n} W_t^2} \tag{8.3.18}$$

For the case of an IO, we saw in Table 8.4 that the denominator in (8.3.18) is equal to 1.0. This leads immediately to the result in (8.3.17). For the case of an AO, we saw in Table 8.5 that the denominator in (8.3.18) is k_A^{-1}. Dividing σ_a^2 by k_A^{-1} as shown in (8.3.18) gives (8.3.15) as the variance of ω_A^*. The derivation of (8.3.16) using (8.3.18) is left as an exercise at the end of the chapter.

8.3.5 Likelihood Ratio Tests

In practice we don't know if an AO, LS, or IO event has occurred at any time t. We use a hypothesis testing procedure to decide if such events have occurred. Let H_0 denote the null hypothesis that no AO, LS, or IO occurs at time i; that is, the ω coefficient in (8.3.4), (8.3.5), or (8.3.6) is zero at $t = i$. Let H_A denote the alternate hypothesis, $\omega_A \neq 0$; let H_S denote the alternate, $\omega_S \neq 0$; and let H_I denote the alternate, $\omega_I \neq 0$. Tests may be performed with the following likelihood ratio statistics (denoted as L):

$$H_0 \text{ vs. } H_A: L_{A,i} = \omega_A^* k_A^{-1/2} / \sigma_a \tag{8.3.19}$$

$$H_0 \text{ vs. } H_S: L_{S,i} = \omega_S^* k_S^{-1/2} / \sigma_a \tag{8.3.20}$$

$$H_0 \text{ vs. } H_I: L_{I,i} = \omega_I^* / \sigma_a \tag{8.3.21}$$

In (8.3.19), (8.3.20), and (8.3.21) we are just dividing each estimated ω^* coefficient by its corresponding standard error [the square root of the variance given by (8.3.15), (8.3.16), or (8.3.17)]. Under the null hypothesis H_0, and assuming that both time i and the parameters of the ARIMA model in (8.2.1) are known, the statistics L_A, L_S, and L_I are normally distributed with mean zero and variance 1.0.

In practice we don't know the parameters of the ARIMA model in (8.2.2). Therefore we don't know the e_t values, the π-weight values in $\pi(B)$, or the c-weight values in $c(B)$ that are needed to apply least-squares regression to Equations (8.3.4), (8.3.5), and (8.3.6) to estimate the ω weights. Further, we don't know σ_a^2. However, *under the null hypothesis of no AO, LS, or IO events*, the maximum-likelihood estimates of the parameters in (8.2.2) are consistent. Therefore we estimate model (8.2.2) in the usual way assuming that no external events are present. Then we may use the following tests, which are asymptotically equivalent (see Chang, 1982) to (8.3.19), (8.3.20), and (8.3.21):

$$H_0 \text{ vs. } H_A: \hat{L}_{A,i} = \hat{\omega}_A \hat{k}_A^{-1/2}/\hat{\sigma}_a \qquad (8.3.22)$$

$$H_0 \text{ vs. } H_S: \hat{L}_{S,i} = \hat{\omega}_S \hat{k}_S^{-1/2}/\hat{\sigma}_a \qquad (8.3.23)$$

$$H_0 \text{ vs. } H_I: \hat{L}_{I,i} = \hat{\omega}_I/\hat{\sigma}_a \qquad (8.3.24)$$

where the circumflex denotes a sample value.

8.4 AN ITERATIVE DETECTION PROCEDURE

In this section we set out an iterative detection procedure based on the ideas developed in Section 8.3. This procedure is the one suggested by Chang et al. (1988) and used in the OUTLIER option of the SCA System (Liu and Hudak, 1986). A BASIC computer program for this procedure is given in Appendix 8A.

Suppose there is an unknown number of AO, LS, and IO events in a time series Y_t, occurring at unknown times $t = i1, i2, \ldots$. A detection procedure is as follows:

1. Identify and estimate an ARIMA model (or DR model) for series Y_t assuming that no AO, LS, or IO events are present.
2. Compute the model residuals $(\hat{e}_t)$ and estimate σ_a^2 as

$$\hat{\sigma}_a^2 = m^{-1} \sum_{t=1+n_1}^{n} \hat{e}_t^2$$

where m is the number of residuals available $(m = n - n_1)$.

3. Compute the likelihood ratios in (8.3.22), (8.3.23), and (8.3.24). Set $\hat{L}_{0,t}$ equal to the largest of these statistics; that is, $\hat{L}_{0,t} = \max\{|\hat{L}_{A,t}|, |\hat{L}_{S,t}|, |\hat{L}_{I,t}|\}$ for the m time periods $t = 1 + n_1, 2 + n_1, \ldots, n$.

4. Find $\hat{L} = \max\{\hat{L}_{0,t}\}$. Compare $\hat{L}$ with a predetermined critical value C_d (discussed later). If $\hat{L} \leq C_d$, stop the procedure. If $\hat{L} > C_d$, then a possible AO, LS, or IO is detected. The time $(t = i)$, type (AO, LS, or IO), and estimated ω coefficient of the identified possible event are those associated with $\hat{L}$.

    a. If a possible AO is detected, its size is estimated by $\hat{\omega}_A$ using (8.3.9). Remove this AO effect from the residual series by replacing each $\hat{e}_t$ with $\hat{e}_t - \hat{\omega}_A \hat{\pi}(B) X_t$ for $t \geq i$. Reestimate σ_a^2 using the new $\hat{e}_t$ series; use this new estimate to recompute $\hat{L}_{A,t}$.

    b. If a possible LS is detected, its size is estimated by $\hat{\omega}_S$ in (8.3.10).

Remove this LS effect from the residual series by replacing each $\hat{e}_t$ with $\hat{e}_t - \hat{\omega}_S \hat{c}(B) X_t$ for $t \geq i$. Reestimate σ_a^2 using the new $\hat{e}_t$ series; use this new estimate to recompute $\hat{L}_{S,t}$ in (8.3.23).

c. If a possible IO is detected, its effect is estimated by $\hat{\omega}_I$ according to (8.3.11). Remove this IO effect from the residual series by replacing $\hat{e}_t$ at time $t = i$ with $\hat{e}_t - \hat{\omega}_I = 0$. Reestimate σ_a^2 using the new $\hat{e}_t$ series; use this new estimate to recompute $\hat{L}_{I,t}$ in (8.3.24).

5. If a possible AO, LS, or IO is detected in step 4, repeat steps 3 and 4 using the same initial estimates of the time series model parameters in (8.2.2), but using the new estimate of σ_a^2 and the new residual series. Repeat steps 3 and 4 until no further possible AO, LS, or IO events are found. Each time a new possible AO, LS, or IO event is detected at step 4, alter (1) the latest residual series, (2) the estimate of σ_a^2, and (3) the latest $\hat{L}_t$ statistic as discussed in step 4.

6. Suppose T possible AO, LS, or IO effects are found at times $i1, i2, \ldots, iT$. Treat these times as known and estimate the ω coefficients for each effect simultaneously within a DR model. For example, suppose we find $T = 3$ effects, with a possible AO detected at time $t = i1$, a possible LS at time $t = i2$, and a possible IO at time $t = i3$. Then we estimate the model

$$Y_t = \omega_A X_{1,t} + \omega_S X_{2,t} + \frac{\theta(B^s)\theta(B)}{\phi(B^s)\phi(B)\nabla_s^D \nabla^d}(\omega_I X_{3,t} + a_t) \tag{8.4.1}$$

where $X_{1,t} = 1$ at $t = i1$ and $X_{1,t} = 0$ otherwise; $X_{2,t} = 0$ for $t < i2$ and $X_{2,t} = 1$ for $t \geq i2$; $X_{3,t} = 1$ at $t = i3$ and $X_{3,t} = 0$ otherwise. The model may also call for a constant term. Diagnostic checking may lead to us to modify (8.4.1).

In this detection procedure we use the same ARIMA model coefficients throughout. However, external events can distort the autocorrelation patterns in a data series. Thus we might be using a misspecified ARIMA model in the detection procedure. Chen et al. (1990) discuss an iterative procedure designed to deal with this problem. This iterative procedure is more complicated than steps 1 to 6; it is implemented as part of the OESTIM and OFORECAST options in the SCA System.

Choice of C_d

The critical value C_d is similar to a critical standard normal value or t value. However, it is not feasible to determine the exact repeated sampling distributions of the likelihood ratios in (8.3.22), (8.3.23), and (8.3.24). Therefore we don't know the exact significance level (α) associated with various values of C_d. Some simulation experiments are reported by Chang et al. (1988) for AO and IO events. Those results are based on sample sizes ranging from $n = 50$ to $n = 150$, with either an AR(1) or an MA(1) mechanism for

APPLICATION 315

Table 8.6 The First 12 Estimated π Weights for Saving Rate ARIMA(1, 0, 2) Model

j	Estimated π_j
1	$\hat{\phi}_1 =$ 0.7359
2	$-\hat{\theta}_2 =$ 0.3437
3	$\hat{\theta}_2 \hat{\pi}_1 = -0.2529$
4	$\hat{\theta}_2 \hat{\pi}_2 = -0.1181$
5	$\hat{\theta}_2 \hat{\pi}_3 =$ 0.0869
6	$\hat{\theta}_2 \hat{\pi}_4 =$ 0.0406
7	$\hat{\theta}_2 \hat{\pi}_5 = -0.0299$
8	$\hat{\theta}_2 \hat{\pi}_6 = -0.0140$
9	$\hat{\theta}_2 \hat{\pi}_7 =$ 0.0103
10	$\hat{\theta}_2 \hat{\pi}_8 =$ 0.0048
11	$\hat{\theta}_2 \hat{\pi}_9 = -0.0035$
12	$\hat{\theta}_2 \hat{\pi}_{10} = -0.0016$

(8.2.2), and with $\phi_1 = \theta_1 = 0.6$. According to those results, setting $C_d = 4.0$ corresponds roughly to $\alpha = 0.01$; $C_d = 3.5$ corresponds roughly to $\alpha = 0.05$; and $C_d = 3.25$ corresponds roughly to $\alpha = 0.10$. Thus $C_d = 4.0$ is recommended for low sensitivity to possible external events, $C_d = 3.5$ for medium sensitivity, and $C_d = 3.0$ for high sensitivity.

8.5 APPLICATION

In this section we apply the ideas developed in this chapter to the saving rate data ($n = 100$ observations) presented in Chapters 1 and 7.

8.5.1 $\hat{\pi}$ Weights and $\hat{c}$ Weights

In Chapter 7 we identified an ARIMA(1, 0, 2) model (with $\hat{\theta}_1 = 0$) for the saving rate data, with $\hat{\phi}_1 = 0.7359$, $\hat{\theta}_2 = -0.3437$, $\hat{C} = 1.6138$, and with residual standard deviation $\hat{\sigma}_a = 0.663716$. For now we assume that this model is correct and that there are no interventions or outliers present. The SCA System can provide the needed $\hat{\pi}$ weights; computations for the first 12 $\hat{\pi}$ weights are shown in Table 8.6. The estimated model is invertible; therefore the $\hat{\pi}_j$ weights decline in absolute value as j increases. In this example the decline is rather quick: All estimated π weights are less than 0.005 in absolute value after $j = 9$.

We also need the $\hat{c}$ weights to test for possible LS events. Table 8.7 shows the computation of the first 12 $\hat{c}$ weights. Unlike the $\hat{\pi}$ weights, the $\hat{c}$ weights do not gradually approach zero. From the definition of the c weights, we have $c_j = (\Sigma_{i=1}^{j} \pi_i) - 1$ for $j \geq 1$. Setting $B = 1$ in the expression defining

Table 8.7 The First 12 Estimated c Weights for the Saving Rate ARIMA(1, 0, 2) Model

j	Estimated c_j
1	$\hat{\pi}_1 - 1 = -0.2641$
2	$\hat{c}_1 + \hat{\pi}_2 = 0.0796$
3	$\hat{c}_2 + \hat{\pi}_3 = -0.1733$
4	$\hat{c}_3 + \hat{\pi}_4 = -0.2914$
5	$\hat{c}_4 + \hat{\pi}_5 = -0.2045$
6	$\hat{c}_5 + \hat{\pi}_6 = -0.1639$
7	$\hat{c}_6 + \hat{\pi}_7 = -0.1938$
8	$\hat{c}_7 + \hat{\pi}_8 = -0.2077$
9	$\hat{c}_8 + \hat{\pi}_9 = -0.1975$
10	$\hat{c}_9 + \hat{\pi}_{10} = -0.1927$
11	$\hat{c}_{10} + \hat{\pi}_{11} = -0.1962$
12	$\hat{c}_{11} + \hat{\pi}_{12} = -0.1978$

the π weights for the saving rate ARIMA model, it can be seen that $\Sigma_{i=1}^{j} \pi_i$ converges to $1 - (1 - \phi_1)/(1 - \theta_2)$ as j goes to infinity, since the model is invertible; therefore c_j converges to $-(1 - \phi_1)/(1 - \theta_2)$. Inserting the estimated ARIMA model coefficients into this expression gives $-(1 - 0.7359)/(1 + 0.3437) = -0.196546\ldots$, which is close to the $\hat{c}_j$ values starting at about $j = 7$ in Table 8.7.

8.5.2 Residuals and C_d

We also need the estimated residual variance of the saving rate ARIMA model, and the value of C_d (the critical value for the likelihood ratio tests). The model residuals were computed when the ARIMA(1, 0, 2) model was estimated; they are plotted in Figure 8.11. Since there is a first-order AR

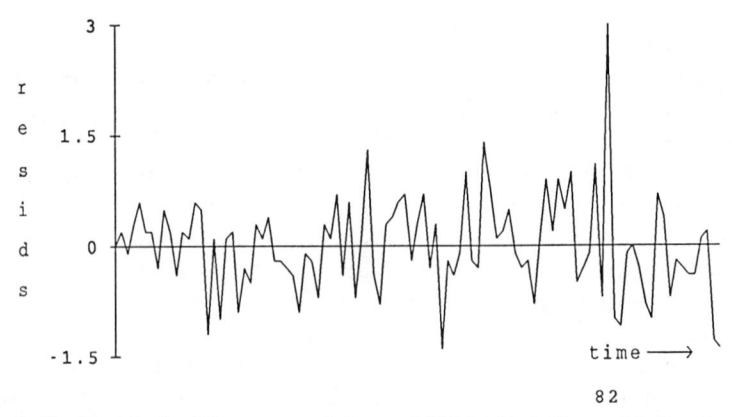

Figure 8.11 Residuals ($\hat{a}_t$) computed from ARIMA(1, 0, 2) model for the saving rate data.

APPLICATION

Table 8.8 Intervention Detection Results for Saving Rate Data with $C_d = 2.5$

t	Type Detected	Estimated ω Coefficient	$|\hat{L}$ Ratio$|$
82	AO	2.37	5.37
99	LS	−1.56	2.88
43	AO	1.13	2.76
62	IO	1.45	2.78
55	IO	−1.44	2.88
89	AO	−1.06	2.92

component in the ARIMA model ($p = 1$), we lose $n_1 = p + Ps + d + Ds = 1 + 0 + 0 + 0 = 1$ observation at the start of the series. Therefore there are only $m = n - n_1 = 100 - 1 = 99$ residuals. These residuals are used to compute the initial residual variance as indicated in Section 8.4 at step 2 of the detection procedure.

The value of C_d must be assigned. We choose $C_d = 2.5$; this is a rather small value for C_d, but it will lead to the detection of multiple interventions and outliers in this example, permitting us to see how all three types (AO, LS, and IO) are estimated. As discussed in the previous section, in general a better choice for C_d may be between 3.0 and 4.0.

8.5.3 Detection Results and Full Model Estimation

Table 8.8 shows the results of applying the detection procedure to the saving rate residuals. The first and most significant possible event detected is the AO at $t = 82$, which is the second quarter of 1975. In Chapter 7 we estimated an intervention model with a pulse event at $t = 82$, and we found that a simple one-period response pulse input provided an adequate fit to data. This is exactly what the detection routine has suggested to us, since an AO is a one-period response pulse event in the original data. Two other possible AO events (at $t = 43$ and $t = 89$) are indicated in Table 8.8. There is also a possible LS event at $t = 99$, and two possible IO events, at $t = 62$ and $t = 55$.

Now we want to estimate a complete model like (8.4.1), with all the possible events we have detected, along with the coefficients of the tentative disturbance ARIMA model. Estimation of a complete model is important; this model represents our best specification of the behavior of series Y_t, and the estimated coefficients can be quite different in the full model compared to the preliminary estimates.

Specifying AO and LS events in a DR model is straightforward. An AO is just a one-period pulse event, with $X_t = 1$ at $t = i$ and $X_t = 0$ otherwise, as discussed in Chapter 7. An LS is a step event, with $X_t = 0$ for $t < i$, and $X_t = 1$ for $t \geq i$.

Table 8.9 Estimation Results for Saving Rate DR Model (8.5.1) with Detected Possible AO, LS, and IO Events Included

Parameter	Estimate	$\|t\text{ Value}\|$
C	6.1635	18.32
ω_{82}	2.3346	6.25
ω_{99}	−1.5114	3.17
ω_{43}	1.1378	3.26
ω_{62}	1.4574	3.01
ω_{55}	−1.4915	3.05
ω_{89}	−1.0702	3.03
θ_2	−0.3762	3.68
ϕ_1	0.7976	12.07
$\hat{\sigma}_a = 0.480503$		

Specifying an IO event is more complicated. An IO at time $t = i$ is represented by a binary pulse variable, $X_t = 1$ at $t = i$ and $X_t = 0$ otherwise. In (8.2.14) and (8.4.1) we see that the coefficients in the disturbance ARIMA model are *also part of the transfer function for the IO event input*. The coefficients that are shared by the disturbance ARIMA part and the IO part of the model must be restricted to have identical values. The way this restriction is achieved in practice depends on the software used. Appendix 8B shows how this restriction is achieved when the SCA System is applied to the saving rate example. [The SCA System also has an option to automatically detect and adjust for outliers; with this option it is not necessary to specify and estimate a model like (8.4.1) in a separate step.]

The full model for the saving rate example is

$$Y_t = C + \omega_{82}X_{1,t} + \omega_{99}X_{2,t} + \omega_{43}X_{3,t} + \frac{(1 - \theta_2 B^2)\omega_{62}}{1 - \phi_1 B}X_{4,t}$$

$$+ \frac{(1 - \theta_2 B^2)\omega_{55}}{1 - \phi_1 B}X_{5,t} + \omega_{89}X_{6,t} + \frac{1 - \theta_2 B^2}{1 - \phi_1 B}a_t \qquad (8.5.1)$$

where $X_{1,t}$, $X_{3,t}$, $X_{4,t}$, $X_{5,t}$, and $X_{6,t}$ are binary pulse (AO or IO) input variables equal to 1.0 at $t = 82, 43, 62, 55$, and 89, respectively, and equal to zero elsewhere; $X_{2,t}$ is a binary step (LS) input variable equal to zero for $t < 99$, and equal to 1.0 for $t \geq 99$; the ω coefficient subscripts correspond to the time period of the possible event.

Table 8.9 shows the estimation results for model (8.5.1). All estimated coefficients have approximate t values larger than 3.0 in absolute value. In this example the new ARIMA coefficients ($\hat{\phi}_1$ and $\hat{\theta}_2$) are quite close to their values in the original ARIMA model with no intervention or outlier components included. However, estimated ARIMA coefficients can change drastically when intervention and outlier components are added to the model. In this example the outlier and intervention event coefficients are all close to the preliminary estimates (in Table 8.8) obtained from the detection procedure. The residual SACF for this model (not shown) does not indicate any model inadequacy.

In this example we set C_d at a low level (2.5) to detect a variety of effects, mainly for pedagogical reasons. Setting C_d at such a low level may have led to the detection of possible outliers or interventions that reflect nothing more than the ordinary stochastic variation in the data. Setting C_d at a higher level, between 3.0 and 4.0 as suggested in Section 8.4, will lead to the detection of fewer possible events; this reduces the chance that we are mistakenly modeling as special events effects that are really just ordinary chance occurrences in the data.

It is wise to study any detected possible event to see if it might reasonably correspond to some associated cause. In the case of the saving rate data, the AO at $t = 82$ corresponds to the one-time tax rebate, and the rise in the saving rate at that time can be explained.

8.6 DETECTED EVENTS NEAR THE END OF A SERIES

Dealing with possible outliers and interventions near the end of a data series requires some thought. For example, consider the possible LS at $t = 99$ that we detected for the saving rate data. From (8.3.9) and (8.3.10) we see that the e_t values *after* $t = i$ are important in computing the ω^* coefficients and the likelihood ratios for AO and LS effects. Thus, at $t = i = 99$ in the saving rate case we have only one later e_t value (e_{100}) to help identify both the occurrence and type of any possible outlier effect. Thus the detection at $t = 99$ is based on rather little information. Or consider the situation for $t = i = n$, the last observation. If an outlier is detected there, we can't distinguish between an AO, LS, or IO at all. This is shown by Equations (8.3.9), (8.3.10), and (8.3.11) and by (8.3.19), (8.3.20), and (8.3.21); these equations show that the estimated coefficients and variances for the three possible effects (AO, LS, and IO) are identical when $i = n$.

Because of these difficulties we may be skeptical about whether possible outliers and interventions near the end of a data series are correctly identified. *When our goal is to forecast*, it is important to decide whether or not to include detected end-of-series events in the model, and their type. Observations near the end of a data series tend to be important in the forecast function. The choice we make in modeling end-of-series events can thus have a large effect on the forecasts. Our forecasts of the saving rate series, for example, will be substantially affected if we include an LS effect at $t = 99$.

When an outlier is detected at the last observation ($t = i = n$), perhaps we will know something about the possible cause of the outlier (e.g., a strike; a change in a law); this may help us decide how to treat such an outlier. Or we may want the model forecasts to reflect a what-if assumption based on our choice of outlier type. Another possibility is to ignore the outlier. However, keep in mind that ignoring an outlier at $t = n$ is equivalent to *treating it as an IO for forecasting purposes* (see Hillmer, 1984). Perhaps the best way to try to resolve this dilemma, when possible, is to seek information about the potential causes of detected possible events. This is highly recommended in any case, no matter where the detected possible events occur within a series. We may then be able to make a wiser choice about whether, and how, to incorporate possible outliers and interventions that arise near the end of a data series.

QUESTIONS AND PROBLEMS

1. Suppose $d = D = 0$, and that all coefficients in the ARIMA process in (8.2.2) are zero. Find the first five π weights and c weights.

2. Using least-squares principles, derive the results given in Equation (8.3.10) for estimating the size of an LS event and in Equation (8.3.16) for finding the associated variance.

3. Tsay (1988) discusses the transient change (TC) event. Writing ω_0 as ω_T, we define $f(t) = [\omega_T/(1 - \delta_1 B)]X_t$, $0 < \delta_1 < 1$. The expression corresponding to (8.3.12) for this event is $W_t = [\pi(B)/(1 - \delta_1 B)]X_t$, with $X_t = 1$ at $t = i$, and $X_t = 0$ for all other t. This is a one-period pulse event, where the effect on Y_t fades with simple exponential decay. Derive expressions for the least-squares estimator ω_T^*, the corresponding variance, and the corresponding likelihood ratio. Note that δ_1 must have a preassigned value when using this result in a detection procedure.

4. A gradual level shift (GS) event may be defined by writing ω_0 as ω_G, and defining $f(t) = [\omega_G/\nabla(1 - \delta_1 B)]X_t$, with $0 < \delta_1 < 1$. The expression corresponding to (8.3.12) for this event is $W_t = [c(B)/(1 - \delta_1 B)]X_t$, with $X_t = 1$ at $t = i$, and $X_t = 0$ for all other t. This is a multiperiod gradual level shift event, where Y_t moves gradually to a permanent new level. Derive the least-squares estimator ω_G^*, the corresponding variance, and the corresponding likelihood ratio. As with the TC event in Question 3, the value of δ_1 must be preassigned when using this result in a detection procedure.

5. See Questions 1, 2, and 3 at the end of Chapter 7. What outliers and interventions do you detect in Series 28, Series 29, and Series 30 using the detection method outlined in the present chapter?

APPENDIX 8A. BASIC PROGRAM TO DETECT AO, LS, AND IO EVENTS

At the end of this appendix is a BASIC computer program to detect AO, LS, and IO events. The program requires the $\hat{\pi}$ weights as input, as shown at line 1100 in the code; the $\hat{\pi}$ weights are stored in array p(.). It also requires the model residuals as input, stored in array r(.), as shown at line 1200 in the code. The $\hat{c}$ weights are computed recursively from the $\hat{\pi}$ weights at line 1300; they are stored in array c(.).

Line 1400 starts the computations of the likelihood ratios. The variables sum1 and sum2 in the code are preliminary values of $\hat{k}_A^{-1}$ and $\hat{k}_S^{-1}$, denoted in the code as kA__1 and kS__1, respectively, with values assigned at line 1450. Adjustments are made to kA__1 and kS__1 at lines 1510 and 1520 to give the correct initial values. Assigning preliminary values to kA__1 and kS__1 at 1450 and then adjusting permits us to compute them recursively after assignment of the preliminary values.

Line 1500 starts the loop in which the likelihood ratios are computed. The largest absolute likelihood ratio (maxL in the code) among the three likelihood ratios (L__ao, L__ls, and L__io in the code) is found for each time period. Information about the time at which this value occurs (T in the code) is recorded, and the corresponding coefficient estimate, variance, and likelihood value (w, v, and L, respectively, in the code) are recorded. Then the same computations are performed for the next period. If a larger value of maxL is found for any time period, maxL is set equal to this value, and the time (T), type (AO, LS, IO), estimated initial effect (w), and associated variance (v) of the possible event are updated.

After $t = n$ (line 1590), if the largest absolute likelihood ratio (the value of maxL) is less than or equal to C_d (variable Crit in the code), the search is stopped. Otherwise, the residual sequence $\hat{e}_t$ [array r(.) in the code] is adjusted to remove the estimated effect, a new estimated residual variance (variable va) is computed from the new residual series, and the coefficient variance and likelihood ratio for the latest detected event are adjusted to reflect this new value of va. The results of each pass through the data are printed; the search continues until no value of maxL greater than variable Crit is found.

```
1000 dim p (999)    \   dim r (999)    \   dim c (999)
1050 input "# observations original data"; n
     input "# residuals & pi-weights     "; m
1100 input "input file of pi-weights      "; i$
     open i$ for input as file 1
     input #1, p(i) for i = 1 to m    \   close 1
1200 input "input file of residuals       "; i$
     open i$ for input as file 1
     input #1, r(i) for i = 1 to m    \   close 1
```

```
1220 input "Critical |L|            "; Crit
     input "output file of results   "; o$
     open o$ for output as file 1
1250 va = 0   \   va = va + r(i)**2 for i = 1 to m   \   va = va/m
1300 c(1) = p(1) − 1   \   c(i) = c(i − 1) + p(i) for i = 2 to m
1350 print #1, "Initial Std Err Residuals = "; sqr(va)
     print #1, "Time", "Type", " Coeff", "L-ratio"
1400 sum1 = 1   \   sum1 = sum1 + p(i)**2 for i = 1 to m
     sum2 = 1   \   sum2 = sum2 + c(i)**2 for i = 1 to m
1450 kA__1 = sum1   \   kS__1 = sum2   \   maxL = 0   \   Ti = 0   \   T$ = ""
1500 for i = 1 to m
1505 sum = r(i)   \   sum = sum − p(j − i)*r(j) for j = i + 1 to m
1510 kA__1 = kA__1 − p(m − i + 1)**2   \   kA = 1/kA__1
     w__ao = kA* sum   \   v__ao = kA*va   \   L__ao = w__ao/sqr(v__ao)
     sum = r(i)   \   sum = sum − c(j − i)*r(j) for j = i + 1 to m
1520 kS__1 = kS__1 − c(m − i + 1)**2   \   kS = 1/kS__1   \   w__ls = kS*sum
     v__ls = kS*va   \   L__ls = w__ls/sqr(v__ls)   \   w__io = r(i)
     v__io = va   \   L__io = r(i)/sqr(va)
1540 if abs(L__ao) > maxL then maxL = abs(L__ao)   \   T = i
     w = w__ao   \   v = v__ao   \   L = L__ao   \   T$ = "AO" end if
1550 if abs(L__ls) > maxL then maxL = abs(L__ls)   \   T = i
     w = w__ls   \   v = v__ls   \   L = L__ls   \   T$ = "LS" end if
1560 if abs(L__io) > maxL then maxL = abs(L__io)   \   T = i
     w = w__io   \   v = v__io   \   L = L__io   \   T$ = "IO" end if
1570 next i
1590 if maxL <= Crit then print #1
     print #1, "Final Std Err Residuals = "; sqr (va)
     goto 9000   end if   ! *** EXIT PROGRAM *** !
1600 if T$ = "AO" then r(T) = r(T) − w
     r(i) = r(i) + w * p(i − T) for i = T + 1 to m end if
1610 if T$ = "LS" then r(T) = r(T) − w
     r(i) = r(i) + w * c(i − T) for i = T + 1 to m end if
1620 if T$ = "IO" then r(T) = 0 end if
1630 val = 0   \   val = val + r(i)**2 for i = 1 to m
     val = val/m   \   L = w/sqr(v*val/va)   \   va = val
1700 print #1, T + (n − m), T$, w, L   \   goto 1450   ! NEXT ITERATION
9000 close 1   \   end
```

APPENDIX 8B. SPECIFYING IO EVENTS IN THE SCA SYSTEM

The SCA System has an option to estimate coefficients of outlier effects in a full DR model automatically. However, these coefficients may also be specified and estimated explicitly. The following SCA commands show how to specify explicitly model (8.5.1) for the saving rate example:

APPENDIX 8B. SPECIFYING IO EVENTS IN THE SCA SYSTEM

```
UTS NAME IS AL. MODEL IS SR = knst          @
    + (w01)Puls82(BIN)                      @
    + (w02)Step99(BIN)                      @
    + (w03)Puls43(BIN)                      @
    + (f)(w04)(f+th2*B**2)/(1−ph1*B)Puls62(BIN) @
    + (f)(w05)(f+th2*B**2)/(1−ph1*B)Puls55(BIN) @
    + (w06)Puls89(BIN)                      @
    + (2; th2)/(1; ph1)NOISE. FIX f.

f = −1.0
```

The saving rate data have been stored under variable name SR. The binary input variables have been stored under the names Puls82, Step99, Puls43, Puls62, Puls55, and Puls89.

The key point is the two lines in the UTS paragraph that specify the IO components of the model. In the SCA System a numerator term in a rational transfer function must be written with a + sign. However, an IO component requires negative signs on MA terms (such as $\theta_2 B^2$ in the saving rate example) since that is how they appear in the disturbance ARIMA model. The correct specification is obtained by fixing coefficient f at $f = -1$. If you multiply out the IO components in the UTS paragraph you will see that the numerator coefficients in the IO terms carry the correct signs.

We may state the same coefficient name in several places in an SCA System model. Such a coefficient is constrained to have the same value in its various locations. Coefficient th2, for example, will have the same value in both IO components as it has in the ARIMA component.

The sentence "FIX f" in the UTS paragraph holds variable f constant during estimation. The assignment $f = -1.0$ must be made *before model estimation* to ensure that f is fixed at that value during estimation.

CHAPTER 9

Estimation and Forecasting

9.1 DR MODEL ESTIMATION

At the identification stage we tentatively specify a rational form transfer function model of orders (b, r, h) and a disturbance series ARIMA model of orders (p, d, q) $(P, D, Q)_s$. For the sales and leading indicator data presented in Chapter 1 and discussed in Chapters 5 and 6, we identified the following DR model:

$$\nabla Y_t = C + \frac{\omega_0 B^3}{1 - \delta_1 B} \nabla X_t + (1 - \theta_1 B) a_t \qquad (9.1.1)$$

The tentative transfer function model has orders $(b, r, h) = (3, 1, 0)$. The tentative disturbance series ARIMA model is ARIMA(0, 1, 1).

At the second stage of our modeling strategy we *estimate* the parameters of the identified DR model using the available data. There are a variety of estimation methods that give approximate maximum-likelihood (ML) estimates. Some software packages offer several estimation options. Unfortunately, some of these options can lead to serious bias in the estimation of moving average (MA) coefficients in the disturbance series ARIMA component.

All estimation results in this book were found using the SCA System (Liu and Hudak, 1986). The SCA System provides two estimation options. One provides approximate ML estimates based on the conditional likelihood function, following Box and Jenkins (1976). This involves choosing coefficients that minimize the sum of squared residuals ($\Sigma \hat{a}_t^2$). This option involves less computing time than more exact ML algorithms, but it can lead to badly biased estimates of MA coefficients, especially when these coefficients are near the invertibility boundary. This option is useful in the early phase of model building or if no MA terms are present. The SCA System also provides a more exact approximation to ML estimates using a procedure

DR MODEL ESTIMATION

due to Hillmer and Tiao (1979). This option requires more computing time, but it produces better estimates of MA coefficients.

We will not present all aspects of DR model estimation in detail; we will set out the main sequence of steps and provide references to more detailed literature. We will point out two key issues that have to do with the *starting values* of certain computed series.

9.1.1 A Two-Input DR Model

To simplify the notation we will not put circumflexes (^) over estimated coefficients in the following discussion. Consider the following DR model with $M = 2$ inputs, where the differenced output and input series (denoted y_t, $x_{1,t}$, and $x_{2,t}$) are now stationary in the mean:

$$y_t = \frac{\omega_1(B) B^{b_1}}{\delta_1(B)} x_{1,t} + \frac{\omega_2(B) B^{b_2}}{\delta_2(B)} x_{2,t} + \frac{\theta(B)}{\phi(B)} a_t \qquad (9.1.2)$$

where the orders of $\omega_1(B)$, $\omega_2(B)$, $\delta_1(B)$, $\delta_2(B)$, $\theta(B)$, and $\phi(B)$ are h_1, h_2, r_1, r_2, q, and p, respectively, and where the dead times for inputs $x_{1,t}$ and $x_{2,t}$ are b_1 and b_2, respectively. To simplify the presentation, we ignore multiplicative seasonal terms in the disturbance ARIMA model.

9.1.2 Initial Values for Coefficients

To estimate (9.1.2) we must first assign some *initial values* to the coefficients. The terminology here can be confusing. We will use "initial values" to refer to coefficient values. Later we will use "starting values" to refer to values computed for certain time series variables. Initial coefficient values can often be found from identification stage information. For example, estimates of the coefficients in $\omega_1(B)$ and $\delta_1(B)$ can be found from the preliminary LTF estimates of the v weights $(v_{1,0}, v_{1,1}, \ldots, v_{1,K_1})$, where K_i denotes the maximum lag length on $x_{i,t}$ in the preliminary LTF estimation. This is done using a procedure similar to the one discussed in Chapter 4 for finding v weights from v weights and δ weights. An example is given by Box and Jenkins (1976, p. 383). Often the initial value 0.1 works well for ARIMA coefficients if other better values are not available.

9.1.3 Transfer Function Outputs

Next, using the initial coefficient values, recursively compute (for periods t, $t+1$, $t+2$, ...) the following two series:

$$y_{1,t}^* = [\omega_1(B) B^{b_1}/\delta_1(B)] x_{1,t}$$
$$= \delta_{1,1} y_{1,t-1}^* + \cdots + \delta_{1,r_1} y_{1,t-r_1}^* + \omega_{1,0} x_{1,t-b_1} + \cdots + \omega_{1,h_1} x_{1,t-b_1-h_1} \quad (9.1.3)$$

$$y_{2,t}^* = [\omega_2(B) B^{b_2}/\delta_2(B)] x_{2,t}$$
$$= \delta_{2,1} y_{2,t-1}^* + \cdots + \delta_{2,r_2} y_{2,t-r_2}^* + \omega_{2,0} x_{2,t-b_2} + \cdots + \omega_{2,h_2} x_{2,t-b_2-h_2} \quad (9.1.4)$$

These two series are the *outputs* from the $M = 2$ *transfer function components* of model (9.1.2), *given* the initial values of the coefficients. In other words $y_{1,t}^*$ is the set of values predicted by the transfer function $[\omega_1(B) B^{b_1}/\delta_1(B)] x_{1,t}$ in (9.1.2), conditional on the initial coefficient values; and $y_{2,t}^*$ is the set of values predicted by the transfer function $[\omega_2(B) B^{b_2}/\delta_2(B)] x_{2,t}$ in (9.1.2), also conditional on the initial coefficient values.

Because of the recursive nature of (9.1.3), computation of $y_{1,t}^*$ requires r_1 *starting values* for $y_{1,t}^*$ prior to $t = k_1$, where $k_1 = \max[r_1 + 1, b_1 + h_1 + 1]$. Similarly, computation of $y_{2,t}^*$ in (9.1.4) requires r_2 starting values for $y_{2,t}^*$ prior to $t = k_2$, where $k_2 = \max[r_2 + 1, b_2 + h_2 + 1]$. For example, suppose $r_1 = 0$, $b_1 = 0$, and $h_1 = 3$. Then from (9.1.3) we have

$$y_{1,t}^* = \omega_{1,0} x_{1,t} + \omega_{1,1} x_{1,t-1} + \omega_{1,2} x_{1,t-2} + \omega_{1,3} x_{1,t-3} \quad (9.1.5)$$

Thus $y_{1,t}^*$ can be computed starting from $t = k_1 = \max[r_1 + 1, b_1 + h_1 + 1] = \max[1, 4] = 4$. According to (9.1.5), to compute $y_{1,1}^*$ we need $x_{1,1}, x_{1,0}, x_{1,-1}$, and $x_{1,-2}$, but the latter three values are not observed. To compute $y_{1,2}^*$ we need $x_{1,2}, x_{1,1}, x_{1,0}$, and $x_{1,-1}$, but the latter two values are not observed, and so forth. The current and lagged values of $x_{1,t}$ needed to compute $y_{1,t}^*$ are all observed only for $t \geq k_1 = 4$. Because $r_1 = 0$ in this example, there are no lagged values of $y_{1,t}^*$ on the right side of (9.1.5). Thus (9.1.5) is not recursive, and $r_1 = 0$ starting values for $y_{1,t}^*$ are needed; that is, no values of $y_{1,t}^*$ are needed for $t < 4$.

Alternatively, suppose $r_1 = 1$, $b_1 = 1$, and $h_1 = 1$. Then from (9.1.3) we have

$$y_{1,t}^* = \delta_{1,1} y_{1,t-1}^* + \omega_{1,0} x_{1,t-1} + \omega_{1,1} x_{1,t-2} \quad (9.1.6)$$

Thus $y_{1,t}^*$ can be computed starting from $t = k_1 = \max[r_1 + 1, b_1 + h_1 + 1] = \max[2, 3] = 3$ onward. According to (9.1.6), to compute $y_{1,1}^*$ we need $y_{1,0}^*$, $x_{1,0}$, and $x_{1,-1}$, none of which is observed. To compute $y_{1,2}^*$ we need $y_{1,1}^*$, $x_{1,1}$, and $x_{1,0}$, the first and third of which are not observed, and so forth. The current and lagged $x_{1,t}$ values needed to compute $y_{1,t}^*$ are all observed only for $t \geq k_1 = 3$; thus $y_{1,3}^*$ is the first $y_{1,t}^*$ value that we can compute. Since there is only one lagged value of $y_{1,t}^*$ on the right side of (9.1.6), $r_1 =$

1 starting value for $y^*_{1,t}$ is needed: $y^*_{1,2}$ must be assigned a value in order to compute $y^*_{1,t}$ starting with $y^*_{1,3}$.

9.1.4 Disturbance and Residuals

The next step in estimation is to compute the stationary disturbance series as

$$\hat{n}_t = y_t - y^*_{1,t} - y^*_{2,t} \qquad (9.1.7)$$

where $\hat{n}_t = \nabla^D_s \nabla^d \hat{N}_t$ can be computed only from $t = k_0 = \max[k_1, k_2]$ onward. Next compute the residual series recursively as

$$\begin{aligned}\hat{a}_t &= [\phi(B)/\theta(B)]\hat{n}_t \\ &= \hat{n}_t - \phi_1 \hat{n}_{t-1} - \cdots - \phi_p \hat{n}_{t-p} + \theta_1 \hat{a}_{t-1} + \cdots + \theta_q \hat{a}_{t-q}\end{aligned} \qquad (9.1.8)$$

Due to the recursive nature of (9.1.8), computing $\hat{a}_t$ requires q *starting values* for $\hat{a}_t$ prior to $t = k_0 + p$. For example, suppose that $p = 2$ and $q = 1$, with $k_1 = 1$ and $k_2 = 0$. Then $\hat{a}_t$ can be computed from $t = k_0 + p = \max[k_1, k_2] + p = \max[1, 0] + 2 = 1 + 2 = 3$ onward. The values of $\hat{a}_t$ in this example cannot be computed for $t < 3$ because the required values of $\hat{n}_t$ are not available. Since there is one lagged value of $\hat{a}_t$ on the right side of (9.1.8) in this example, $q = 1$ starting value for $\hat{a}_t$ is needed: $\hat{a}_2$ must be assigned a value in order to compute $\hat{a}_t$ starting with $\hat{a}_3$.

9.1.5 Search to Minimize Sum of Squared Residuals

The next step in estimation is to compute the SSR (sum of squared residuals):

$$\text{SSR} = \sum_{t=k_0+p}^{n} \hat{a}_t^2,$$

where n is the last time period for which data are available. Then, typically a nonlinear least-squares optimizing routine similar to that given by Marquardt (1963) is used to choose better model coefficients (those giving a smaller SSR). Then we repeat all previous steps, starting with Section 9.1.2: We replace the previous coefficients with the new ones given by the optimizing routine; then we compute the new transfer function outputs, the new disturbance series, the new residual series, and the new SSR. Then we use the optimizing routine to select a new set of coefficients. These steps are repeated until the coefficient combination that minimizes the SSR is found.

9.1.6 Output Series Starting Values

As discussed in Section 9.1.3, we require *starting values* for the transfer function output series $y_{i,t}^*$, $i = 1, 2, \ldots, M$. Box and Jenkins (1976) and Wall (1976) propose setting the starting values for these series equal to zero. However, Liu (1984) presents evidence that this can lead to serious bias in the estimation of MA coefficients. There are a variety of other treatments of output series starting values. The interested reader may consult Liu (1984) and Newbold (1973), for example. The SCA System uses what Liu (1984) calls the "short-cut method" treatment of output series starting values. This method seems to give fairly stable estimates of the MA coefficients in a computationally efficient manner.

9.1.7 Residual Series Starting Values

As discussed in Section 9.1.4, we also require q *starting values* for the residual series $\hat{a}_t$. These starting values may be treated in several ways. Box and Jenkins (1976) discuss the *conditional* likelihood function for the estimation of ARIMA coefficients. Hillmer and Tiao (1979) discuss the *exact* likelihood function. In both of these approaches the observations of the input and the output series are treated as fixed. The conditional method in the SCA System assumes that the q starting values for the residual series are all zero ($\hat{a}_p = \cdots = \hat{a}_{p-q+1} = 0$), while the so-called exact method uses estimates of those values. Note that the exact method is exact for the MA terms only.

9.2 FORECASTING

Forecasts from DR models and the associated forecast error variances typically are produced by the same software used to identify and estimate the model. In this section we explain how DR model forecasts and their error variances may be obtained. Throughout this discussion we assume that the true DR coefficients are known. In practice we must replace unknown coefficients with sample estimates.

First we consider how point forecasts (single values for a given time period) are computed from a DR model. Then we discuss the forecast error variances and how we use this information to establish confidence limits (interval forecasts) around the point forecasts. Next we examine the matter of forecasting Y_t in the original metric when the data have been modeled in a transformed metric such as natural logarithms. Finally we establish that DR forecasts are minimum mean squared error forecasts, conditional on the data.

FORECASTING   329

9.2.1 Computation of Point Forecasts

In this section we explain how forecasts of future values of Y_t are produced from the following DR model with $M = 1$ input:

$$Y_t = \frac{\omega(B)}{\delta(B)} X_{t-b} + \frac{\theta(B)}{\nabla^d \phi(B)} a_t \qquad (9.2.1)$$

The following results are easily extended to include seasonal terms in the disturbance ARIMA model, a constant term, and additional inputs. To proceed, multiply through by $\delta(B)\nabla^d \phi(B)$ to write (9.2.1) as

$$\delta^*(B) Y_t = \omega^*(B) X_{t-b} + \theta^*(B) a_t \qquad (9.2.2)$$

where

$$\delta^*(B) = 1 - \delta_1^* B - \cdots - \delta_{p+d+r}^* B^{p+d+r} = \delta(B)\nabla^d \phi(B)$$
$$\omega^*(B) = \omega_0^* + \omega_1^* B + \cdots + \omega_{p+d+h}^* B^{p+d+h} = \nabla^d \phi(B)\omega(B)$$
$$\theta^*(B) = 1 - \theta_1^* B - \cdots - \theta_{q+r}^* B^{q+r} = \delta(B)\theta(B)$$

The starred (*) coefficients in (9.2.2) are found by equating coefficients of like powers of B on either side of each equation that defines a starred operator. This is similar to the procedure discussed in Chapter 4 for finding v weights from ω weights and δ weights.

Suppose the current time period from which forecasts are to be made is period $t = n$ (called the *forecast origin*). Suppose also that we want to forecast the future value Y_{n+l}, where $l \geq 1$ is called the *forecast lead time*. Using (9.2.2), write the value for Y_{n+l} as

$$Y_{n+l} = \delta_1^* Y_{n+l-1} + \cdots + \delta_{p+d+r}^* Y_{n+l-p-d-r} + \omega_0^* X_{n+l-b}$$
$$+ \cdots + \omega_{p+d+h}^* X_{n+l-b-p-d-h} - \theta_1^* a_{n+l-1}$$
$$- \cdots - \theta_{q+r}^* a_{n+l-q-r} + a_{n+l} \qquad (9.2.3)$$

Forecasts are made using only information available through the forecast origin $t = n$. Denote the information in the set of data available at time n $(Y_n, Y_{n-1}, \ldots; X_n, X_{n-1}, \ldots)$ as I_n. The forecast of Y_{n+l}, denoted $\hat{Y}_n(l)$, is the conditional mathematical expectation of Y_{n+l} given I_n. Denote a conditional expectation with square brackets. Then from (9.2.3) the DR model forecast of Y_{n+l} is

$$\hat{Y}_n(l) = E(Y_{n+l} | I_n) = [Y_{n+l}]$$
$$= \delta_1^* [Y_{n+l-1}] + \cdots + \delta_{p+d+r}^* [Y_{n+l-p-d-r}] + \omega_0^* [X_{n+l-b}]$$

Table 9.1 Ten Forecasts from Model (9.1.1) with Forecast Error Standard Deviations and Observed Values

Time	Forecast	Standard Error	Observed
141	257.0681	0.2177	257.3
142	257.5065	0.2328	257.5
143	259.5781	0.2471	259.6
144	260.8058	1.3782	261.1
145	263.0722	2.2116	262.9
146	263.1207	2.9824	263.3
147	262.5990	3.7044	262.8
148	261.4753	4.3815	261.1
149	261.8486	5.0166	262.2
150	262.5054	5.6131	262.7

$$+ \cdots + \omega^*_{p+d+h}[X_{n+l-b-p-d-h}] - \theta^*_1[a_{n+l-1}]$$
$$- \cdots - \theta^*_{q+r}[a_{n+l-q-r}] + [a_{n+l}] \qquad (9.2.4)$$

The values of $[Y_{n+j}]$ and $[X_{n+j}]$ for $j \leq 0$ are the *observed* values of each series. Similarly, the values of $[a_{n+j}]$ for $j \leq 0$ are *estimated by the residuals* ($\hat{a}_{n+j}$) of the DR model. On the other hand, the values of $[Y_{n+j}]$ and $[X_{n+j}]$ for $j > 0$ are their *forecasts* $\hat{Y}_n(j)$ and $\hat{X}_n(j)$, which are their respective conditional expected values. And for $j > 0$, $[a_{n+j}] = 0$: At time n we have no estimate of a_{n+j} in the form of the DR model residual $\hat{a}_{n+j}$, so its expected value is zero.

The forecasts $\hat{X}_n(j)$ may be obtained from a variety of sources, including an ARIMA model for X_t. In some cases the forecasts $\hat{X}_n(j)$ may be replaced by known values. This may occur if future X_t values are fixed by a policy decision, for example. Or, the future X_t values may be known deterministic values, such as seasonal dummy variables (discussed in Chapter 3) or trading day variables (discussed in Case 2).

If the DR model in rational form has a constant term (C), then (9.2.4) also has an additive constant term K on the right side, where

$$K = C\left(1 - \sum_{i=1}^{p} \phi_i\right)\left(1 - \sum_{i=1}^{P} \phi_{is}\right)\left(1 - \sum_{i=1}^{r_1} \delta_{i,1}\right) \cdots \left(1 - \sum_{i=1}^{r_M} \delta_{i,M}\right) \qquad (9.2.5)$$

where M is the number of input series in the DR model.

Example
Ten forecasts from model (9.1.1), with lead times $l = 1, 2, \ldots, 10$ from origin=140 are shown in Table 9.1. To illustrate how these forecasts are found, we apply (9.2.4) to the estimated sales and leading indicator model (9.1.1). For this one-input model, we have orders

FORECASTING 331

$$d = 1 \quad b = 3 \quad h = 0 \quad r = 1 \quad p = 0 \quad q = 1$$

and estimated coefficients

$$\omega_0 = 4.7156 \quad \delta_1 = 0.7249 \quad \theta_1 = 0.6197 \quad C = 0.0326$$

where the circumflexes are not shown to simplify the notation. The δ^* weights for this example are found from

$$1 - \delta_1^* B - \delta_2^* B^2 = (1 - \delta_1 B) \nabla$$
$$= 1 - (1 + \delta_1) B - (-\delta_1) B^2 \quad (9.2.6)$$

Equating coefficients of like powers of B on either side of (9.2.6), and inserting the estimated coefficients from the sales and leading indicator model (9.1.1), we have

$$\delta_1^* = 1 + \delta_1 = 1 + 0.7249 = 1.7249$$

and

$$\delta_2^* = -\delta_1 = -0.7249$$

The ω^* weights for this example are found from

$$\omega_0^* + \omega_1^* B = \nabla \omega_0$$
$$\omega_0 + (-\omega_0) B \quad (9.2.7)$$

Equating coefficients of like powers of B on either side of (9.2.7) and inserting the estimated coefficients, we find

$$\omega_0^* = \omega_0 = 4.7156 \quad \text{and} \quad \omega_1^* = -\omega_0 = -4.7156$$

The θ^* weights for this example are found from

$$1 - \theta_1^* B - \theta_2^* B^2 = (1 - \delta_1 B)(1 - \theta_1 B)$$
$$= 1 - (\delta_1 + \theta_1) B - (-\delta_1 \theta_1) B^2 \quad (9.2.8)$$

Equating coefficients of like powers of B on either side of (9.2.8), and inserting the estimated coefficients, we have

$$\theta_1^* = \delta_1 + \theta_1 = 0.7249 + 0.6197 = 1.3446$$

and

$$\theta_2^* = -\delta_1 \theta_1 = -(0.7249)(0.6197) = -0.4492$$

Inserting the estimated coefficients from the sales and leading indicator model into Equation (9.2.5), the constant K for this example is

$$K = C(1 - \delta_1) = 0.0326(1 - 0.7249) = 0.0090 \qquad (9.2.9)$$

Now we may use (9.2.4), with added constant K, to forecast from origin $t = 140$. We will show the computations for the first two lead times, $l = 1, 2$. For $l = 1$, the forecast is

$$\begin{aligned}\hat{Y}_{140}(1) &= K + \delta_1^*[Y_{140}] + \delta_2^*[Y_{139}] + \omega_0^*[X_{138}] + \omega_1^*[X_{137}] \\ &\quad - \theta_1^*[a_{140}] - \theta_2^*[a_{139}] + [a_{141}] \\ &= 0.0090 + 1.7249(257.6) - 0.7249(257.0) + 4.7156(13.10) \\ &\quad - 4.7156(13.32) - 1.3446(0.137) + 0.4492(1.041) + 0 \\ &= 257.0681\end{aligned}$$

This is the result printed by the SCA System. You may get a slightly different result due to rounding if you do the computations by hand. In these computations note that the conditional expectations $[Y_{140}]$, $[Y_{139}]$, $[X_{138}]$, and $[X_{137}]$ are all available at $t = n = 140$ as observed values. The conditional expectations $[a_{140}]$ and $[a_{139}]$ are replaced by estimated values, which are the corresponding residuals $\hat{a}_t$ from model (9.1.1). No model residual $\hat{a}_{141}$ is available when $t = n = 140$ is treated as the present time; that is, only data through $t = n = 140$ were used in model estimation. Thus the expected value $[a_{141}]$ is zero.

For $l = 2$ the forecast from origin $t = n = 140$ is

$$\begin{aligned}\hat{Y}_{140}(2) &= K + \delta_1^*[Y_{141}] + \delta_2^*[Y_{140}] + \omega_0^*[X_{139}] + \omega_1^*[X_{138}] \\ &\quad - \theta_1^*[a_{141}] - \theta_2^*[a_{140}] + [a_{142}] \\ &= 0.0090 + 1.7249(257.0681) - 0.7249(257.6) + 4.7156(13.27) \\ &\quad - 4.7156(13.10) - 1.3446(0) + 0.4492(0.137) + 0 \\ &= 257.5065\end{aligned}$$

Again, you may get a slightly different result due to rounding if you do the computations by hand. The expectations $[Y_{140}]$, $[X_{139}]$, and $[X_{138}]$ are available as observed values at $t = 140$. But $[Y_{141}]$ is not observed; we substitute the forecast $\hat{Y}_{140}(1) = 257.0681$, computed previously. [$\hat{Y}_{140}(2)$ is called a "bootstrap" forecast since it depends, in part, on earlier forecasts.] The expectation $[a_{140}]$ is replaced by its estimate, the residual from the estimated DR model at $t = n = 140$. The expectations $[a_{141}]$ and $[a_{142}]$ are neither

FORECASTING   333

observed nor estimated, given the data through $t = n = 140$; therefore each of these expected values is zero.

9.2.2 Forecast Error Variance and Interval Forecasts

To find the variance of the forecast errors, it is convenient first to state (9.2.1) in *random shock* form. To do this, we first write the ARIMA model for X_t as

$$X_t = [\theta_x(B)/\nabla^{d_x}\phi_x(B)]\alpha_t \qquad (9.2.10)$$

Use (9.2.10) to substitute for X_t in (9.2.1) to obtain the random shock form of (9.2.1):

$$Y_t = \eta(B)\alpha_t + \psi(B)a_t \qquad (9.2.11)$$

where

$$\eta(B) = \eta_0 + \eta_1 B + \eta_2 B^2 + \cdots = \omega(B)\theta_x(B)B^b/\delta(B)\nabla^{d_x}\phi_x(B)$$

and

$$\psi(B) = \psi_0 + \psi_1 B + \psi_2 B^2 + \cdots = \theta(B)/\nabla^d \phi(B)$$

where $\psi_0 = 1$. Both $\eta(B)$ and $\psi(B)$ may be of infinitely high order. Using the definition of $\eta(B)$, the η weights are found by equating coefficients of like powers of B on either side of this expression:

$$\delta(B)\nabla^{d_x}\phi_x(B)\eta(B) = \omega(B)\theta_x(B)B^b \qquad (9.2.12)$$

Similarly, using the definition of $\psi(B)$, the ψ weights are found by equating coefficients of like power of B on either side of this expression:

$$\nabla^d \phi(B)\psi(B) = \theta(B) \qquad (9.2.13)$$

Now use (9.2.11) to write the future value Y_{n+l}, where $t = n$ is the forecast origin, as

$$Y_{n+l} = \eta_0 \alpha_{n+l} + \eta_1 \alpha_{n+l-1} + \eta_2 \alpha_{n+l-2}$$
$$+ \cdots + a_{n+l} + \psi_1 a_{n+l-1} + \psi_2 a_{n+l-2} + \cdots \qquad (9.2.14)$$

Recall that the forecast of Y_{n+l} (the lead l forecast from origin $t = n$) is the conditional expectation of Y_{n+l}. Therefore, from (9.2.14), this forecast is

$$\hat{Y}_n(l) = [Y_{n+l}] = \eta_l \alpha_n + \eta_{l+1} \alpha_{n-1} + \cdots + \psi_l a_n + \psi_{l+1} a_{n-1} + \cdots \tag{9.2.15}$$

That is, any a_t or α_t value for $t > n$ is neither known nor estimated from a model residual, given the data available through $t = n$. The expected value of these terms is zero; (9.2.15) shows the remaining nonzero terms.

The lead l forecast error from origin n is $e_n(l) = Y_{n+l} - \hat{Y}_n(l)$. Thus, subtracting (9.2.15) from (9.2.14) gives the forecast error

$$e_n(l) = \eta_0 \alpha_{n+l} + \eta_1 \alpha_{n+l-1} + \cdots + \eta_{l-1} \alpha_{n+1}$$
$$+ a_{n+l} + \psi_1 a_{n+l-1} + \cdots + \psi_{l-1} a_{n+1} \tag{9.2.16}$$

The forecast error variance for lead l from origin n is the mathematical expectation $V(l) = E[e_n(l)]^2$, where the expected value of the forecast error is zero. Therefore squaring (9.2.16) and taking expected values gives

$$V(l) = \sigma_\alpha^2 \sum_{j=0}^{l-1} \eta_j^2 + \sigma_a^2 \sum_{j=0}^{l-1} \psi_j^2 \tag{9.2.17}$$

In finding this result we use the fact that a_t and α_t are mutually independent and not autocorrelated.

If a_t and α_t are normally distributed, then $e_n(l)$ is also normally distributed and we can form confidence intervals around the point forecasts. For example, the 95% confidence interval for Y_{n+l} is

$$\hat{Y}_n(l) \pm 1.96 [V(l)]^{1/2} \tag{9.2.18}$$

Example
Let's illustrate the computation of $V(l)$ for $l = 1, 2$ for the sales and leading indicator model (9.1.1). From (9.2.17), we need η_0 (for $l = 1$ and 2) and η_1 (for $l = 2$). The η weights are found by equating coefficients of like powers of B on either side of (9.2.12). For model (9.1.1) we have

$$\delta(B) = (1 - \delta_1 B) \qquad \nabla^{d_x} = \nabla = (1 - B) \qquad \phi_x(B) = 1$$
$$\omega(B) = \omega_0 \qquad \theta_x(B) = (1 - \theta_{x1} B) \qquad B^b = B^3$$

Substituting these expressions into (9.2.12) gives $(1 - \delta_1 B)(1 - B)(\eta_0 + \eta_1 B + \cdots) = \omega_0(1 - \theta_{x1} B) B^3$; expanding each side of this expression gives

$$\eta_0 + [\eta_1 - (1 + \delta_1) \eta_0] B + [\eta_2 - (1 + \delta_1) \eta_1 + \delta_1 \eta_0] B^2$$
$$+ \cdots = \omega_0 B^3 - \omega_0 \theta_{x1} B^4 \tag{9.2.19}$$

FORECASTING

Equating coefficients of B^0 and B^1 on either side of (9.2.19) gives $\eta_0 = 0$, and $\eta_1 - (1 + \delta_1)\eta_0 = 0$. Substituting $\eta_0 = 0$ into the latter expression and solving gives $\eta_1 = 0$.

With $\eta_0 = \eta_1 = 0$, we may ignore the first summation in (9.2.17) in computing $V(1)$ and $V(2)$ for this example. This makes sense because the first summation in (9.2.17) reflects uncertainty about the input series. Since $b = 3$, the first three forecasts of Y_t use only *observed* values of X_t; using known X_t values should not add to the uncertainty of the forecasts of Y_t. Only after $l = 3$, when we must use *forecasts* of X_t, should the first summation in (9.2.17) be nonzero.

The ψ weights are found by equating coefficients on either side of (9.2.13). For model (9.1.1), we have

$$(1 - B)^d = (1 - B) \qquad \phi(B) = 1 \qquad \theta(B) = (1 - \theta_1 B)$$

Substituting these expressions into (9.2.13) gives $(1 - B)(\psi_0 + \psi_1 B + \psi_2 B^2 + \cdots) = 1 - \theta_1 B$; expanding the left side of this expression gives

$$\psi_0 - (\psi_0 - \psi_1)B - (\psi_1 - \psi_2)B^2 - (\psi_2 - \psi_3)B^3 - \cdots = 1 - \theta_1 B \quad (9.2.20)$$

Equating coefficients of B^0 and B^1 on either side of (9.2.20) gives $\psi_0 = 1$, and $\psi_0 - \psi_1 = \theta_1$. Substituting $\psi_0 = 1$ into the latter expression and rearranging gives $\psi_1 = 1 - \theta_1$. Substituting the estimated value of θ_1, we have $\psi_1 = 1 - 0.6197 = 0.3803$. Thus from (9.2.17) we have

$$V(1) = \sigma_a^2 \psi_0^2 = (0.217622)^2 (1)^2 = 0.0473593$$
$$V(2) = \sigma_a^2 (\psi_0^2 + \psi_1^2) = (0.217622)^2 (1^2 + 0.3803^2) = 0.0542088$$

The square roots of $V(1)$ and $V(2)$ are 0.2177 and 0.2328, respectively; these are the standard errors of the one-step and two-step ahead forecasts, respectively, as shown in the third column of Table 9.1. Therefore an approximate 95% confidence interval around the one-step ahead point forecasts is

$$\hat{Y}_{140}(1) \pm 1.96[V(1)]^{1/2}$$

$$257.0681 \pm 1.96(0.2177)$$

$$(256.6414, \ 257.4948)$$

An approximate 95% interval for the two-step ahead forecast is constructed in a similar fashion. In this example both intervals contain the future observed values (shown in the fourth column of Table 9.1).

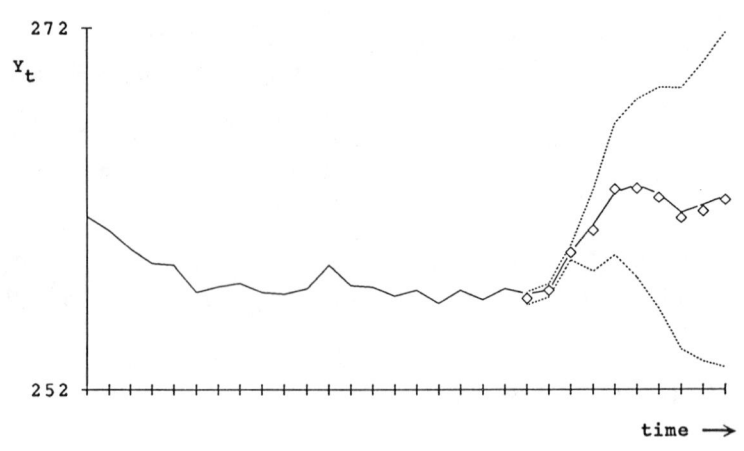

Figure 9.1 Forecasts in Table 9.1 with 90% confidence limits and observed values.

The solid line in Figure 9.1 shows the last 20 observed values of Y_t (for $t = 121, 122, \ldots, 140$) used for model estimation and the 10 future values ($t = 141, 142, \ldots, 150$). The 10 forecasts in Table 9.1 are represented in Figure 9.1 by the diamond symbols. The dashed lines are approximate 90% confidence limits. These limits widen sharply after $l = 3$. Given the DR model dead time $b = 3$, forecasts of Y_t after $l = 3$ are based on forecast rather than observed values of X_t; the uncertainty associated with the Y_t forecasts therefore increases substantially after $l = 3$.

9.2.3 Forecasting Transformed Series

Suppose series Y_t has been transformed to Y'_t using the Box–Cox transformation discussed in Chapter 2. Then we produce forecasts, denoted $\hat{Y}'_n(l)$, of the transformed series. Users of forecasts usually want forecasts stated in the original metric. For example, a manager will want a forecast of company sales, not of the natural log of sales.

Naive Retransformation
Retransforming the forecast $\hat{Y}'_n(l)$ into the original metric must be done with care. It may seem that we could find the original metric forecast by applying the simple (or "naive") inverse of the original transformation to the forecast $\hat{Y}'_n(l)$. Thus we might compute

$$\hat{Y}_n(l) = \exp[\hat{Y}'_n(l)] \qquad (9.2.21)$$

for the natural log transformation ($\lambda = 0$), or

$$\hat{Y}_n(l) = [\lambda \hat{Y}'_n(l) + 1]^{1/\lambda} \qquad (9.2.22)$$

for other Box–Cox transformations. If L is an upper or lower confidence limit in the transformed metric, then the original metric limit is $\exp(L)$ for $\lambda = 0$ or $(\lambda L + 1)^{1/\lambda}$ for $\lambda \neq 0$.

While the naive retransformation seems natural, if the transformed variable Y'_{n+l} is normally distributed (conditional on the data) as we typically assume, then the original metric variable Y_{n+l} is *not* normally distributed. For instance, under the log transformation Y_{n+l} is *log-normally* distributed; see Nelson (1973) for a discussion of this case. In general, the original metric distributions are skewed. It can be shown that the naive retransformation produces the *median* of the original metric distribution rather than its *mean* (expected value). In this sense the naive retransformation produces biased forecasts in the original metric.

Error Loss Function

Whether we want the mean, median, or some other point along the original metric distribution depends on the *error loss function* associated with the forecast. Presumably a zero forecast error entails no loss due to forecast error. However, both negative and positive forecast errors entail a loss of some kind. For example, underforecasting product sales could lead to insufficient product inventory, lost revenue, and less profit; overforecasting sales could lead to excess production, higher inventory carrying charges, and less profit.

Many forecast error loss functions are possible. We will consider two types:

1. *Absolute Error Loss.* In this case the loss function is linear and symmetrical around zero. Granger (1969) shows that the optimal original metric forecast for this loss function is the median of the distribution of Y_{n+l} obtained from the naive retransformation.

2. *Quadratic Error Loss.* In this case the cost of forecast error is proportional to the squared forecast error. Granger (1969) shows that the optimal original metric forecast for this loss function is the mean of the distribution of Y_{n+l}.

It seems reasonable that extreme forecast errors may often bring sharply increased costs. Then the quadratic error loss function may be a useful approximation in many situations, even when the true loss function is linear and asymmetric for small to moderate errors.

Retransformation to the Mean

Granger and Newbold (1976) show that the mean (M) of the conditional distribution of Y_{n+l} for the natural log transformation is

Table 9.2 Percent Difference Between m and M under Box–Cox Transformation Based on (9.2.24)

λ	$\|r\|$				
	0.02	0.05	0.10	0.15	0.20
3.00	0.0	0.0	0.1	0.3	0.5
2.00	0.0	0.0	0.1	0.3	0.5
0.75	−0.0	−0.1	−0.2	−0.5	−0.9
0.50	−0.0	−0.2	−1.0	−2.2	−3.8
0.25	−0.2	−1.5	−5.7	−12.0	−19.7
−0.25	−0.4	−2.5	−10.1	−23.1	−43.8
−0.50	−0.1	−0.8	−3.1	−7.1	−13.4
−0.75	−0.1	−0.4	−1.6	−3.7	−6.9
−1.00	−0.0	−0.3	−1.0	−2.4	−4.4
−2.00	−0.0	−0.1	−0.4	−0.9	−1.6
−3.00	−0.0	−0.1	−0.2	−0.5	−1.0

$$\hat{Y}_n(l) = \exp[\hat{Y}'_n(l) + \tfrac{1}{2} V(l)] \qquad (9.2.23)$$

where $V(l)$ is the lead l forecast error variance in (9.2.17).

For other Box–Cox transformations, Pankratz and Dudley (1987) give an algorithm for obtaining the mean (M) of the conditional distribution of Y_{n+l}:

$$M = m\left\{1 + \sum_{k=1}^{\infty} \lambda^{-1}(\lambda^{-1} - 1) \cdots [\lambda^{-1} - (2k-2)][\lambda^{-1} - (2k-1)] r^{2k}/2^k k!\right\} \qquad (9.2.24)$$

where m is the median (naive) retransformation obtained from (9.2.22), and $r = [V(l)]^{1/2}/[\hat{Y}'_n(l) + \lambda^{-1}]$. (If the simple transformation $Y'_t = Y_t^\lambda$ is used, then $r = [V(l)]^{1/2}/\hat{Y}'_n(l)$.) Pankratz and Dudley (1987) report that truncation of the sum in (9.2.24) after no more than eight terms gives good results for most combinations of λ and r that are likely to occur in practice. The appendix at the end of this chapter gives a short BASIC language routine that efficiently computes the nonbiasing factor (in curly brackets) in (9.2.24). Guerrero (1990) provides an alternative algorithm for the case $\lambda < 1$.

Often the percent difference between m and M [$=100\,(m - M)/M$] is quite small. Then m may be used for simplicity even when M is called for. Pankratz and Dudley (1987) report values of this percent difference under the Box–Cox transformation for various values of λ and $|r|$, shown in Table 9.2. The percent difference is quite small for many values of λ and $|r|$, though it is quite large as λ approaches zero and as $|r|$ increases.

It is difficult to generalize about the combinations of λ and $|r|$ that are likely to occur in practice. However, we may obtain some guidance in the case of economic data from a study by Nelson and Granger (1979). For three

FORECASTING

of their series the values of λ and $|r|$ for lead times $l = 1$ through $l = 10$ are as follows:

Series	λ	Range for $|r|$
Ratio, price to unit labor cost in manufacturing	-1.52	0.014–0.052
Index of industrial materials prices	0.14	0.003–0.020
Gross national product in current dollars	0.63	0.004–0.023

For these combinations of λ and $|r|$, Table 9.2 shows that the percent difference between m and M is generally a small fraction of -1%. Thus, in many cases the naive retransformed forecast m may be adequate in practice.

Upper and lower confidence limit values in the original metric are computed using the naive procedure even if the point forecast is retransformed to the mean. Thus forecast confidence intervals in the original metric are asymmetric.

9.2.4 Minimum Mean Squared Error DR Forecasts

We can show that the l-step ahead forecast of Y_{n+l} from origin n given by (9.2.4), denoted $\hat{Y}_n(l)$, is the minimum mean squared error forecast conditional on I_n. To show this it is convenient to work with the random shock form of the future and forecast values. Suppose from (9.2.15) that the forecast is

$$\hat{Y}_n(l) = \sum_{j=0}^{\infty} \eta^*_{l+j} \alpha_{n-j} + \sum_{j=0}^{\infty} \psi^*_{l+j} a_{n-j} \qquad (9.2.25)$$

Subtracting forecast (9.2.25) from future value (9.2.14) and rearranging we have

$$Y_{n+l} - \hat{Y}_n(l) = \sum_{j=0}^{l-1} (\eta_j \alpha_{n+l-j} + \psi_j a_{n+l-j})$$

$$+ \sum_{j=0}^{\infty} [(\eta_{l+j} - \eta^*_{l+j}) \alpha_{n-j} + (\psi_{l+j} - \psi^*_{l+j}) a_{n-j}] \qquad (9.2.26)$$

Squaring (9.2.26) and taking the expected value gives

$$E[Y_{n+l} - \hat{Y}_n(l)]^2 = (\eta_0^2 + \eta_1^2 + \cdots + \eta_{l-1}^2)\sigma_\alpha^2 + (1 + \psi_1^2 + \cdots + \psi_{l-1}^2)\sigma_a^2$$
$$+ \sum_{j=0}^{\infty} [(\eta_{l+j} - \eta_{l+j}^*)^2 \sigma_\alpha^2 + (\psi_{l+j} - \psi_{l+j}^*)^2 \sigma_a^2] \quad (9.2.27)$$

Inspection of (9.2.27) shows that it is minimized when $\eta_{l+j}^* = \eta_{l+j}$ and $\psi_{l+j}^* = \psi_{l+j}$. Therefore the minimum mean squared error forecast of Y_{n+l} from origin n is the conditional expectation of Y_{n+l} at $t = n$; that is, it is equal to the DR forecast given by (9.2.4). This expectation is conditional on the input and output series going back into the infinite past. However, in practice the δ^* weights, ω^* weights, and θ^* weights tend to become small rather quickly as we move further into the past.

QUESTIONS AND PROBLEMS

1. How is (9.1.8) modified when seasonal AR and MA terms are present? What is the first time period for which $\hat{a}_t$ may be found? How many starting values of $\hat{a}_t$ are needed?

2. How is (9.2.2) modified when two inputs are present? When seasonal AR and MA terms are present?

3. Find $V(l)$ for $l = 3$ for model (9.1.1). Find approximate 95% confidence limits for the forecast of Y_{143} shown in Table 9.1.

4. In Case 3 we found original metric forecasts (Table C3.3) using the naive retransformation (9.2.21). Find the forecasts when retransformation to the mean(s) is appropriate. How much is the percent difference between these forecasts and those shown in Case 3?

5. Find $V(l)$ for $l = 1, 2, 3$ for model (C4.3.4) in Case 4. Find the corresponding approximate 95% confidence intervals for the values Y_{121}, Y_{122}, and Y_{123}.

APPENDIX 9A. A BASIC ROUTINE FOR COMPUTING THE NONBIASING FACTOR IN (9.2.24)

The following BASIC language routine may be used to compute efficiently the nonbiasing factor [in curly brackets in Equation (9.2.24)] for a given forecast origin n and lead time l. Variable G is the nonbiasing factor; k is the number of terms used in the series in (9.2.24); $c = \lambda^{-1}$; $d =$

APPENDIX 9A. A BASIC ROUTINE FOR COMPUTING THE NONBIASING FACTOR

$r^2/2$, for the Box–Cox transformation where r is defined and for the simple transformation $Y'_t = Y^\lambda_t$ following Equation (9.2.24).

```
100 k = 8  \  G = 1
300 G = 1 + G*(c − 2*j + 2)*(c − 2*j + 1)*(d/j) for j = k to 1 step −1
```

CHAPTER 10

Dynamic Regression Models in a Vector ARMA Framework

A dynamic regression model is a special type of *vector ARMA* model. Vector ARMA models are the next important class of time series models after univariate ARIMA models and DR models. Vector ARMA models involve a set of K equations that show how each of a set of K time series is linearly related to *its own past* and *the past of the other $K - 1$ series in the set*. Thus, vector ARMA models are richer than univariate ARIMA or DR models since vector ARMA models can include both *multiple series* and *feedback effects*.

If any of the equations in a vector ARMA model involve *no feedback effects* elsewhere in the set of equations, then this equation separately is a DR model or a univariate ARIMA model. However, to put a DR model embedded in a vector ARMA model into the familiar form of DR models as we have studied them in this text may require substantial algebraic rearrangements; this is especially so when there are one or more *contemporaneous relationships* within the DR model.

A full discussion of vector ARMA models is beyond the scope of this text. For more detailed introductions see Tiao and Box (1981), Jenkins and Alavi (1981), Liu (1986b), Granger and Newbold (1986), and Wei (1990). In this chapter we will first set out some basic ideas about vector ARMA models. Then we will consider the vector AR (VAR) form of a vector ARMA model. Next we will show how a DR model can be written as a VAR model. Finally, we will show how we can use an estimated VAR model to check for (1) feedback, (2) a contemporaneous relationship between an input and the output in a possible DR model, and (3) the associated dead time b.

10.1 VECTOR ARMA PROCESSES

In this section we introduce the class of vector ARMA processes with known coefficients. The ideas also apply to vector ARMA models with estimated

VECTOR ARMA PROCESSES

coefficients. Vector ARMA processes and models are an extension of the univariate ARIMA framework to include the possibility of *time-lagged feedback among two or more series*. Vector ARMA model coefficients do not *explicitly* reflect *contemporaneous relationships* among the series. However, as we will see, a vector ARMA model provides *indirect* information about contemporaneous relationships.

Let's start with a simple example. Suppose we have two series, $Z_{1,t}$ and $Z_{2,t}$. Suppose each series is a linear function of its own past (lag 1) and the past of the other series (also lag 1), plus a random shock term:

$$Z_{1,t} = \phi_{11} Z_{1,t-1} + \phi_{12} Z_{2,t-1} + a_{1,t}$$
$$Z_{2,t} = \phi_{21} Z_{1,t-1} + \phi_{22} Z_{2,t-1} + a_{2,t} \quad (10.1.1)$$

where the ϕ coefficients link current values of $Z_{1,t}$ and $Z_{2,t}$ to past values of these two series, and $a_{1,t}$ and $a_{2,t}$ are independent, zero-mean, and normally distributed white-noise series with variances $\sigma_{a_1}^2$ and $\sigma_{a_2}^2$, respectively. These equations could also have constant terms; for simplicity we assume the constants are zero. The first subscript on each ϕ coefficient refers to the left-side ("dependent") variable in its equation; the second subscript refers to the right-side ("independent") variable to which the ϕ coefficient is attached. Thus ϕ_{12} tells us how $Z_{1,t}$ (series 1) is related to a past value of $Z_{2,t}$ (series 2), while ϕ_{22} tells us how $Z_{2,t}$ (series 2) is related to its own past (also series 2).

If ϕ_{12} and ϕ_{21} were zero, the equations in (10.1.1) would be a pair of univariate (single series) AR(1) processes for series $Z_{1,t}$ and $Z_{2,t}$. Since these two coefficients are not zero, the equations in (10.1.1) contain *feedback* from a past value of $Z_{2,t}$ to $Z_{1,t}$, and from a past value of $Z_{1,t}$ to $Z_{2,t}$. Taken together, these two equations are a vector autoregressive (VAR) process.

Now let's consider a more complicated example, still with $K = 2$ series. Suppose the equations in (10.1.1) are correct as far as they go, but incomplete: Suppose that $Z_{1,t}$ is also related to $Z_{2,t-2}$, and to its own past shock $a_{1,t-1}$. And suppose that $Z_{2,t}$ is also related to $Z_{2,t-2}$ and to $a_{1,t-1}$. This may be written as

$$Z_{1,t} = \phi_{111} Z_{1,t-1} + \phi_{112} Z_{2,t-1} + \phi_{212} Z_{2,t-2} - \theta_{111} a_{1,t-1} + a_{1,t}$$
$$Z_{2,t} = \phi_{121} Z_{1,t-1} + \phi_{122} Z_{2,t-1} + \phi_{222} Z_{2,t-2} - \theta_{121} a_{1,t-1} + a_{2,t} \quad (10.1.2)$$

The coefficient subscripts in (10.1.2) are more complicated than those in (10.1.1) to allow us to distinguish lag lengths as well as dependent (left-side) and independent (right-side) terms. The first subscript on each coefficient in (10.1.2) indicates the lag length. [We omitted this subscript from (10.1.1) for simplicity since all lags were one period.] The second subscript corresponds to the dependent variable on the left side of (10.1.2). The third subscript corresponds to the independent variable on the right side of (10.1.2). Thus, coefficient ϕ_{212} tells how $Z_{1,t}$ is related to past values of $Z_{2,t}$ with a lag of two time periods, while θ_{121} tells how $Z_{2,t}$ is related to past random shocks from the equation for $Z_{1,t}$ with a lag of one time period.

Generalizing from these two examples, the main idea of a vector ARMA process is as follows. We can have K series $(Z_{1,t}, Z_{2,t}, \ldots, Z_{K,t})$. Each series [on the left side of equations like (10.1.1) or (10.1.2)] may be related to its own past or to the past of any of the other $K - 1$ series in the group. These terms from the past [on the right side of equations like (10.1.1) or (10.1.2)] may be either AR terms (past Z values with ϕ coefficients) or MA terms (past random shocks with θ coefficients).

It is often convenient to write a vector ARMA process in matrix form and with the backshift operator. For example, (10.1.2) may be written as

$$\left(\begin{bmatrix} 1 & 0 \\ 0 & 1 \end{bmatrix} - \begin{bmatrix} \phi_{111} & \phi_{112} \\ \phi_{121} & \phi_{122} \end{bmatrix} B - \begin{bmatrix} 0 & \phi_{212} \\ 0 & \phi_{222} \end{bmatrix} B^2 \right) \begin{bmatrix} Z_{1,t} \\ Z_{2,t} \end{bmatrix}$$

$$= \left(\begin{bmatrix} 1 & 0 \\ 0 & 1 \end{bmatrix} - \begin{bmatrix} \theta_{111} & 0 \\ \theta_{121} & 0 \end{bmatrix} B \right) \begin{bmatrix} a_{1,t} \\ a_{2,t} \end{bmatrix} \qquad (10.1.3)$$

Performing the matrix multiplication indicated in (10.1.3) and using the definition of the backshift operator yields the set of equations in (10.1.2). Notice in (10.1.3) that some of the ϕ or θ coefficients may be zero. [What would be the subscripts of the lower left lag 2 ϕ coefficient in (10.1.3) if that coefficient were not zero?)

Now let's consider a general matrix form for vector ARMA processes. A vector ARMA(p, q) process of autoregressive order p (maximum AR lag) and moving average order q (maximum MA lag) is written as

$$\boldsymbol{\phi}(B)\mathbf{Z}_t = \mathbf{C} + \boldsymbol{\theta}(B)\mathbf{a}_t \qquad (10.1.4)$$

where

1. $\mathbf{Z}_t = (Z_{1,t}, Z_{2,t}, \ldots, Z_{K,t})'$ is a time-ordered sequence of $K \times 1$ vectors of K times series, where the prime denotes matrix transposition.
2. $\mathbf{C}$ is a $K \times 1$ vector of constants.
3. $\boldsymbol{\phi}(B) = (\mathbf{I}_K - \boldsymbol{\phi}_1 B - \boldsymbol{\phi}_2 B^2 - \cdots - \boldsymbol{\phi}_p B^p)$ is an autoregressive matrix operator of order p.
4. $\mathbf{I}_K$ is the $K \times K$ identity matrix.
5. $\boldsymbol{\phi}_i$ is a $K \times K$ matrix of autoregressive coefficients.
6. $\boldsymbol{\theta}(B) = (\mathbf{I}_K - \boldsymbol{\theta}_1 B - \boldsymbol{\theta}_2 B^2 - \cdots - \boldsymbol{\theta}_q B^q)$ is a moving average matrix operator of order q.
7. $\boldsymbol{\theta}_i$ is a $K \times K$ matrix of moving average coefficients.
8. $\mathbf{a}_t = (a_{1,t}, a_{2,t}, \ldots, a_{K,t})'$ is a time-ordered sequence of $K \times 1$ vectors

of random shocks, independent of Z_{t-i}, $i > 0$, that are normally, identically, and independently distributed with a mean vector of zeros and covariance matrix V.

The diagonal elements of V are the variances of the random shocks, and the off-diagonal elements are the covariances of the random shocks. For $K = 2$ series, for example, V looks like

$$V = \begin{bmatrix} \text{var}(a_{1,t}) & \text{cov}(a_{1,t}, a_{2,t}) \\ \text{cov}(a_{2,t}, a_{1,t}) & \text{var}(a_{2,t}) \end{bmatrix}$$

Like all covariance matrices, V is symmetric since $\text{cov}(a_{i,t}, a_{j,t}) = \text{cov}(a_{j,t}, a_{i,t})$. For (10.1.4) to be stationary and invertible, all roots of the determinantal polynomials $|\boldsymbol{\phi}(B)|$ and $|\boldsymbol{\theta}(B)|$, respectively, must lie outside the unit circle. This framework can be expanded to include multiplicative seasonal elements.

10.2 THE VECTOR AR (π WEIGHT) FORM

As discussed in Chapters 2 and 8, an invertible univariate ARIMA process for series z_t may be written in autoregressive (π weight) form. That is, we can state how z_t is related to its own past values, with all differencing, AR, and MA terms restated as AR-type terms. If MA terms are present, then the order of this π-weight form is not finite. Invertible vector ARMA models may also be written in VAR form. Unlike a univariate ARIMA model in π-weight form, however, the order of a vector ARMA model in π-weight form might be finite even when MA terms are present; see Wei (1990, Chapter 14). While the AR form of the vector ARMA model may not be parsimonious, it is useful in checking for feedback, as discussed in Section 10.4.

To find the VAR form of (10.1.4), with $C = 0$ for simplicity, multiply both sides of (10.1.4) by $\boldsymbol{\theta}^{-1}(B)$ to get

$$\boldsymbol{\pi}(B)Z_t = a_t \tag{10.2.1}$$

where $\boldsymbol{\pi}(B) = (I - \boldsymbol{\pi}_1 B - \boldsymbol{\pi}_2 B^2 - \cdots) = \boldsymbol{\theta}^{-1}(B)\boldsymbol{\phi}(B)$ is a matrix operator whose order may not be finite. With $K = 2$ series, the typical $\boldsymbol{\pi}_i$-coefficient matrix looks like

$$\boldsymbol{\pi}_i = \begin{bmatrix} \pi_{i11} & \pi_{i12} \\ \pi_{i21} & \pi_{i22} \end{bmatrix}$$

where the subscripts are interpreted as they were in (10.1.2). Thus, π_{i12}

indicates how series 1 is related to series 2 at lag i. The coefficient values in the arrays are found from the $\boldsymbol{\phi}_i$ and $\boldsymbol{\theta}_i$ coefficient matrices by equating coefficient matrices of like powers of B in the expression $\boldsymbol{\theta}(B)\boldsymbol{\pi}(B) = \boldsymbol{\phi}(B)$. For $K = 2$ series, writing out (10.2.1) in full gives

$$\left(\begin{bmatrix} 1 & 0 \\ 0 & 1 \end{bmatrix} - \begin{bmatrix} \pi_{111} & \pi_{112} \\ \pi_{121} & \pi_{122} \end{bmatrix} B - \begin{bmatrix} \pi_{211} & \pi_{212} \\ \pi_{221} & \pi_{222} \end{bmatrix} B^2 - \cdots - \right) \begin{bmatrix} Z_{1,t} \\ Z_{2,t} \end{bmatrix} = \begin{bmatrix} a_{1,t} \\ a_{2,t} \end{bmatrix}$$
(10.2.2)

Now define the following four operators in B:

$$\pi_{11}(B) = 1 - \pi_{111}B - \pi_{211}B^2 - \cdots$$
$$\pi_{12}(B) = \pi_{012} - \pi_{112}B - \pi_{212}B^2 - \cdots$$
$$\pi_{21}(B) = \pi_{021} - \pi_{121}B - \pi_{221}B^2 - \cdots$$
$$\pi_{22}(B) = 1 - \pi_{122}B - \pi_{222}B^2 - \cdots$$

where $\pi_{012} = \pi_{021} = 0$. Using these definitions, Equation (10.2.2) may be written compactly as

$$\begin{bmatrix} \pi_{11}(B) & \pi_{12}(B) \\ \pi_{21}(B) & \pi_{22}(B) \end{bmatrix} \begin{bmatrix} Z_{1,t} \\ Z_{2,t} \end{bmatrix} = \begin{bmatrix} a_{1,t} \\ a_{2,t} \end{bmatrix}$$
(10.2.3)

Equations (10.2.1), (10.2.2), and (10.2.3) are equivalent expressions for a VAR model with two series.

10.3 DR MODELS IN VAR FORM

In this section we show that a DR model can be written in VAR form. There is a complication in doing so: While a DR model can explicitly include a *contemporaneous* relationship (v_0 may be nonzero), a VAR model does not explicitly include contemporaneous relationships among the variables within its coefficient matrices. However, VAR analysis of a DR model yields useful information about contemporaneous relationships *in the covariance matrix* **V**.

For simplicity, suppose there is one input and that Y_t and X_t are stationary, nonseasonal, and zero-mean series. The following results can be generalized to other cases. Consider the following DR model for Y_t, along with an ARIMA model for X_t:

DR MODELS IN VAR FORM 347

$$Y_t = v(B)X_t + \frac{\theta(B)}{\phi(B)}u_t$$

$$X_t = \frac{\theta_x(B)}{\phi_x(B)}w_t$$

(10.3.1)

where u_t and w_t are each a sequence of zero-mean and normally distributed white-noise random shocks, uncorrelated with each other. If there is dead time in the first equation, then the first b coefficients in $v(B)$ are zero.

The two equations in the DR model (10.3.1) can each be rewritten, with Y_t a function of its own past and the past of X_t, and with X_t a function of its own past. To do this, multiply each equation in (10.3.1) by its AR operator and divide by its MA operator:

$$\frac{\phi(B)}{\theta(B)}Y_t = \frac{\phi(B)v(B)}{\theta(B)}X_t + u_t$$

$$\frac{\phi_x(B)}{\theta_x(B)}X_t = w_t$$

(10.3.2)

Now (10.3.2) may be written in matrix form as

$$\begin{bmatrix} \frac{\phi(B)}{\theta(B)} & -\frac{\phi(B)v(B)}{\theta(B)} \\ 0 & \frac{\phi_x(B)}{\theta_x(B)} \end{bmatrix} \begin{bmatrix} Y_t \\ X_t \end{bmatrix} = \begin{bmatrix} u_t \\ w_t \end{bmatrix}$$

(10.3.3)

(Note the negative sign in front of the upper right coefficient ratio.) If you perform the multiplication in (10.3.3) you will see that it is just a restatement of system (10.3.2). Inspection shows that the DR model in AR form (10.3.3) appears to be identical to the VAR model (10.2.3), with the following substitutions:

$$(Y_t, X_t)' = (Z_{1,t}, Z_{2,t})' \qquad (u_t, w_t)' = (a_{1,t}, a_{2,t})'$$

$$\phi(B)/\theta(B) = \pi_{11}(B) \qquad -\phi(B)v(B)/\theta(B) = \pi_{12}(B)$$

$$\pi_{i21} = 0, \, i \geq 0 \qquad \phi_x(B)/\theta_x(B) = \pi_{22}(B)$$

There is one possible difference, however, between a DR model in form (10.3.3) and the VAR model (10.2.3). This difference arises when there is *no dead time* in the DR model (i.e., when $b = 0$); that is, it arises when Y_t is *contemporaneously* related to X_t.

To understand the relationship between a DR model in form (10.3.3) and a VAR model, consider that we can find the relationship between the individual π weights and the other coefficients (v's, ϕ's, θ's, ϕ_x's, and θ_x's) in (10.3.3) by equating coefficients of like powers of B on either side of the relevant equation. Thus, the π weights in $\pi_{11}(B)$ are found by expanding $\phi(B) = \theta(B)\pi_{11}(B)$ and equating coefficients of B on either side of the resulting equation. The coefficients in $\pi_{22}(B)$ are found in a similar way from $\theta_x(B)\pi_{22}(B) = \phi_x(B)$, and those in $\pi_{12}(B)$ are found from $\theta(B)\pi_{12}(B) = -\phi(B)v(B)$. One result that emerges from this procedure is

$$\pi_{012} = -v_0$$

which is zero if $b < 0$ but *nonzero if $b = 0$*.

Substituting the π-weight operators into (10.3.3), and writing out the π-weight operators in full and gathering terms gives this alternative expression for the DR model in AR form:

$$\left(\begin{bmatrix} 1 & \pi_{012} \\ 0 & 1 \end{bmatrix} - \begin{bmatrix} \pi_{111} & \pi_{112} \\ 0 & \pi_{122} \end{bmatrix} B - \begin{bmatrix} \pi_{211} & \pi_{212} \\ 0 & \pi_{222} \end{bmatrix} B^2 - \cdots - \right)\begin{bmatrix} Y_t \\ X_t \end{bmatrix} = \begin{bmatrix} u_t \\ w_t \end{bmatrix}$$
(10.3.4)

Now compare the DR model in form (10.3.4) and the VAR model in form (10.2.2). If $b > 0$, then in (10.3.4) we have $\pi_{012} = -v_0 = 0$, and (10.3.4) and (10.2.2) are identical (with $\pi_{i21} = 0, i \geq 0$). But if $b = 0$, then in (10.3.4) $\pi_{012} = -v_0$ is nonzero, Y_t is *contemporaneously related* to X_t, and (10.3.4) and (10.2.2) are not identical.

Since vector ARMA (and VAR) models do not capture contemporaneous relationships explicitly in the coefficient matrices, the DR model in form (10.3.4) is not directly equivalent to the VAR model (10.2.2) when $b = 0$. However, we can alter (10.3.4) to achieve equivalence between (10.3.4) and (10.2.2), as follows. Multiply (10.3.4) from the left by the inverse of the first coefficient matrix, which is

$$\begin{bmatrix} 1 & \pi_{012} \\ 0 & 1 \end{bmatrix}^{-1} = \begin{bmatrix} 1 & -\pi_{012} \\ 0 & 1 \end{bmatrix}$$

The result of this multiplication is

$$\left(\begin{bmatrix} 1 & 0 \\ 0 & 1 \end{bmatrix} - \begin{bmatrix} \pi_{111} & \pi^*_{112} \\ 0 & \pi_{122} \end{bmatrix} B - \begin{bmatrix} \pi_{211} & \pi^*_{212} \\ 0 & \pi_{222} \end{bmatrix} B^2 - \cdots - \right)\begin{bmatrix} Y_t \\ X_t \end{bmatrix} = \begin{bmatrix} a_{1,t} \\ a_{2,t} \end{bmatrix}$$
(10.3.5)

where

$$\pi_{i12}^* = \pi_{i12} - \pi_{012}\pi_{i22} \quad i > 0$$

$$a_{1,t} = u_t - \pi_{012}w_t \quad (10.3.6)$$

$$a_{2,t} = w_t$$

Inspection shows that (10.3.5) is equivalent to (10.2.2), with $\mathbf{Z}_t = (Y_t, X_t)'$, and with each $\pi_{i21} = 0$. In other words the DR model system (10.3.2) has a corresponding VAR form (10.3.5), for any $b \geq 0$. If $b = 0$, then the contemporaneous relationship between Y_t and X_t is reflected in the contemporaneous relationship between $a_{1,t}$ and $a_{2,t}$, as shown in (10.3.6); that is, $a_{1,t}$ is related to $w_t = a_{2,t}$ via the (nonzero) coefficient $\pi_{012} = -v_0$.

10.4 FEEDBACK CHECK

Now let's consider a practical feature of these results: We can estimate a possible DR model in VAR form to check for feedback. Suppose we have two observed series Y_t and X_t whose relationship *might* be described by the DR system (10.3.2). By estimating a VAR model like (10.3.5) for these two series, we can check to see if a unidirectional (no feedback) system like (10.3.2) characterizes the relationship of these series. To check for feedback, we examine the lower left estimated coefficients, $\hat{\pi}_{i21}$ for $i > 0$, in the coefficient matrices in (10.3.5). These are the coefficients that link X_t to past values of Y_t. *If there is no feedback* from past values of Y_t to X_t, then we expect these coefficients to be insignificant (small relative to their standard errors), and each coefficient matrix in (10.3.5) is upper triangular. (These estimated coefficients and standard errors will be identical to those found using the single-equation regression feedback check discussed in Chapter 5.) This check should not be applied mechanically. In practice, when estimating a large number of π weights in a VAR model, we may be willing to attribute the significance of some $\hat{\pi}_{i21}$ coefficients to sampling variation. Tsay (1985) discusses an overall test that may be used in addition to the tests on the individual $\hat{\pi}_{i21}$ coefficients. If we are satisfied that there is no feedback, then we can apply the LTF identification method set out in Chapter 5. If there is feedback, we may instead attempt to identify a parsimonious vector ARMA model using methods discussed by, for example, Liu (1986b) or some other simultaneous equation model.

Deciding whether feedback is present is problematic when there are *no significant* $\hat{\pi}_{i21}$ coefficients in the estimated VAR model, but when there *is a contemporaneous relationship* between Y_t and X_t. The presence of a contemporaneous relationship may be checked using the VAR estimation results, as discussed in Section 10.5, or using the LTF identification method. A contemporaneous relationship between Y_t and X_t could indicate contemporaneous causality from X_t to Y_t, or from Y_t to X_t, or both. Contemporaneous causality from X_t to Y_t is not a problem since we can represent this in a DR model by setting $b = 0$. Causation from Y_t to X_t, or causal interaction between X_t and Y_t, however, will cause our estimated single equation DR model to give misleading information about the response of Y_t to changes in X_t. When Y_t and X_t are contemporaneously related, we must be willing to ascribe this to causality from X_t to Y_t in order to properly proceed with a single-equation DR model.

The VAR feedback check procedure is applicable with more than one input. Suppose the output Y_t is the first series ($Z_{1,t}$) in the set of K series being considered; then the possible inputs are series $Z_{2,t}, Z_{3,t}, \ldots, Z_{K,t}$. To check for feedback from Y_t to the inputs, we check the significance of estimated coefficients $\hat{\pi}_{ik1}$, for $k = 2, 3, \ldots, K$, $i = 1, 2, \ldots, p_{\max}$, where $p_{\max}$ is the order of the estimated VAR model. If these coefficients are insignificant, we may conclude that there is no feedback from the output to the inputs.

10.4.1 Order Selection for the VAR Model

To implement the VAR feedback check procedure, the first step is to choose the order $p_{\max}$ of the VAR approximation of a possible DR model such as (10.3.2). That is, we want to choose the highest power of B in the VAR model such as (10.3.5). This VAR model is only an approximation since we must choose a finite order $p_{\max}$ to represent a system that might, in theory, require an infinite number of π-weight matrices to represent model (10.3.2).

We may choose the value of $p_{\max}$ based on experience and judgment. For example, a chemical engineer may not know the exact dynamic response of the amount of a chemical output to changes in the amount of an input, but experience with similar processes may lead her to be fairly sure that the response is complete within two hours. If readings are taken every 15 minutes, she could reasonably choose $p_{\max} = 8$.

The choice of $p_{\max}$ may also depend on the sample size; a small data set will require a relatively small value for $p_{\max}$. Consider the case of one output and two input series, with $n = 100$ observations. With three series ($K = 3$), each coefficient matrix in the VAR model has $K^2 = 9$ elements. If a model of order $p_{\max} = 10$ is called for, we must estimate $K^2 \times p_{\max} = 9 \times 10 = 90$ parameters, leaving few degrees of freedom with $n = 100$. This difficulty is less serious when the sample size is large, but many practical problems involve only moderate sample sizes.

We may also choose the value of $p_{\max}$ more formally. There are several formal order selection criteria for VAR models. We will discuss the *Akaike information criterion* (*AIC*) due to Akaike (1973, 1974). The AIC statistic is

$$\text{AIC}(p) = \ln|\hat{\mathbf{V}}| + 2pK^2/n \tag{10.4.1}$$

where K is the number of series in the model; ln denotes natural logarithm;

FEEDBACK CHECK 351

Table 10.1 AIC Values for VAR Models (Orders 1 to 15) for the Sales and Leading Indicator Data ($n = 140$)

Order	1	2	3	4	5
AIC	−1.891	−2.081	−4.940	−4.912	−5.061
Order	6	7	8	9	10
AIC	−5.174	−5.195	−5.214	−5.290	−5.287
Order	11	12	13	14	15
AIC	−5.258	−5.210	−5.198	−5.194	−5.150

n is the number of observations; p is the order of the fitted AR model; the vertical lines ($|\ |$) denote the determinant of an array; and $\hat{V}$ is the least-squares residual covariance matrix of the fitted model. In practice we choose some maximum value for p, denoted p'. Then we compute the AIC statistic stepwise for VAR models of order $p = 1, 2, \ldots, p'$, and choose the model of order p_{max} ($0 \leq p_{max} \leq p'$) that *minimizes* the AIC statistic. The choice of p' is somewhat arbitrary; if we have some prior knowledge about the nature of the mechanism generating the data, we can choose a value for p' that seems large enough to cover the maximum anticipated time lag.

The AIC was originally developed for stationary series. But Tsay (1984) has shown that the asymptotic distribution of the order that minimizes the AIC (given in Shibata, 1976) also holds when the series are homogeneously nonstationary. Thus we can apply the AIC to a VAR model *without differencing* the data.

10.4.2 Application 1

Let's use the AIC to choose the order of a VAR model for the sales and leading indicator data (shown in Figures 1.4 and 1.5) using the first $n = 140$ observations, with $p' = 15$. Then we can check for feedback. The needed AIC values may be found using the SCA System. All AIC values for orders $p = 1, 2, \ldots, 15$ for these data are shown in Table 10.1. The minimum AIC value is -5.290 for a VAR model of order 9, so we choose $p_{max} = 9$.

Next we estimate a VAR model of order $p_{max} = 9$ using the SCA System. We use only the first $n = 140$ observations in the estimation. The results of this VAR estimation (Table 10.2) give us a direct test for feedback. Since $Z_t = (Y_t, X_t)'$, if there is any important time-lagged feedback from Y_t to X_t, then some of the estimated $\hat{\pi}_{i21}$ coefficients (row = 2, column = 1) should be significant. Recall that coefficient π_{i21} tells us how variable 2 ($Z_{2,t} = Y_t$ in this case) at time t is related to variable 1 ($Z_{1,t} = X_t$ in this case) at time $t - i$. The indicator symbols (+ or −) in Table 10.2 tell us which coefficients are significant. A + symbol denotes a coefficient that is more than two standard errors above zero; a − symbol denotes a coefficient that is more

Table 10.2 VAR(9) Estimation Results for the Sales and Leading Indicator Data

Lag	AR Coefficient Matrices		Standard Error Matrices		Indicator Symbols	
1	0.468	−0.041	0.083	0.067	+	·
	0.111	0.498	0.107	0.087	·	+
2	0.260	0.047	0.092	0.075	+	·
	−0.106	0.312	0.119	0.096	·	+
3	0.105	4.780	0.095	0.077	·	+
	0.063	0.128	0.123	0.099	·	·
4	0.061	1.253	0.095	0.404	·	+
	0.083	−0.349	0.123	0.521	·	·
5	0.237	−0.498	0.094	0.394	+	·
	−0.048	−0.030	0.122	0.508	·	·
6	0.050	−0.740	0.091	0.386	·	·
	−0.175	−0.170	0.117	0.498	·	·
7	−0.230	−0.888	0.061	0.379	−	−
	0.100	−0.669	0.078	0.489	·	·
8	0.012	−1.581	0.022	0.383	·	−
	0.041	−0.071	0.029	0.494	·	·
9	−0.009	−1.506	0.014	0.390	·	−
	−0.032	0.593	0.018	0.503	·	·

than two standard errors below zero. None of the $\hat{\pi}_{i21}$ estimated coefficients in Table 10.2 is significant: no + or − symbols occur in the (2, 1) positions. We conclude that there is no time-lagged feedback from Y_t to X_t for these data.

10.4.3 Application 2

The data plotted in Figure 10.1 are annual sales, from 1907 through 1960, of Lydia Pinkham's vegetable compound, a mixture of herbs and alcohol (Series 21 in the Data Appendix). Figure 10.2 shows annual advertising outlays for the same company (Series 22 in the Data Appendix). Changes in advertising outlays (X_t) might lead to changes in sales (Y_t), perhaps with a distributed lag, as advertising increases people's knowledge of the product. But advertising outlays might also be affected by sales, as higher sales are plowed back into higher advertising outlays. These data have been studied frequently over the years; they are often used to illustrate a distributed lag relationship. Among the studies are Clarke (1976), Caines et al. (1977), Helmer and Johansson (1977), Kyle (1978), Vandaele (1983), and Wei (1990).

Since these are annual data, it seems reasonable to choose a fairly small value for $p_{\max}$. That is, it seems the time-lagged interaction between sales and advertising would not extend past a few years. We may also choose $p_{\max}$

FEEDBACK CHECK 353

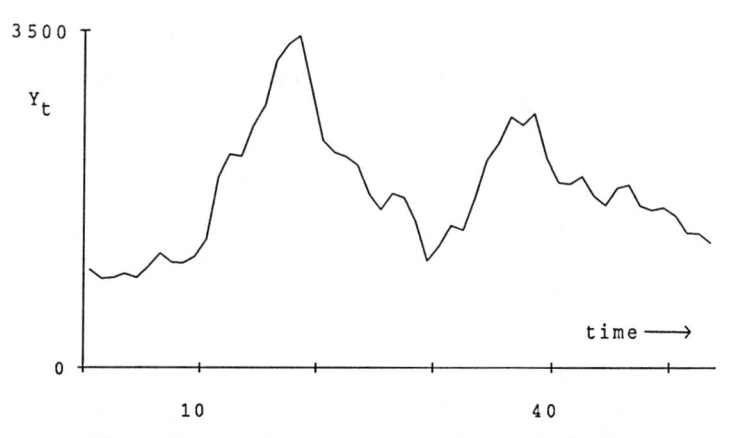

Figure 10.1 Lydia Pinkham annual sales, 1907–1960.

more formally using the AIC criterion. To do this, we use the SCA System to find the minimum AIC according to (10.4.1). Setting $p' = 5$, the minimum AIC is 20.764 for $p = p_{max} = 3$. Therefore we estimate a VAR(3) model to check for feedback.

The results of the VAR(3) estimation are shown in Table 10.3. We are especially interested in the $\hat{\pi}_{i21}$ coefficients; they should tend to be insignificant if there is no time-lagged feedback from sales to advertising outlays. In Table 10.3 we see that $\hat{\pi}_{121} = 0.586$ is positive and more than twice its standard error (0.127), so its indicator symbol is +. And $\hat{\pi}_{221} = -0.536$ is negative and more than twice its standard error in absolute value, so its indicator symbol is −. This is strong evidence of feedback from Y_t to X_t; therefore it is probably not appropriate to use a single-equation DR model. A similar result based on prewhitening and cross correlation analysis (see Appendix 5B) is reported by Wei (1990).

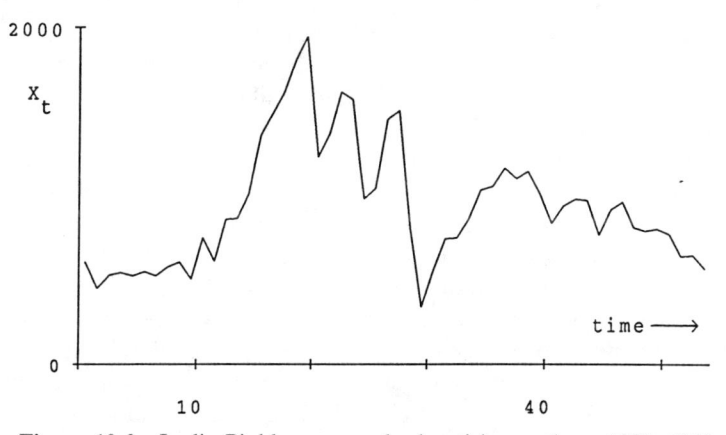

Figure 10.2 Lydia Pinkham annual advertising outlays, 1907–1960.

Table 10.3 VAR(3) Estimation Results for Lydia Pinkham Data

Lag	AR Coefficient Matrices		Standard Error Matrices		Indicator Symbols	
1	1.366	−0.162	0.158	0.167	+	·
	0.586	0.519	0.127	0.134	+	+
2	−0.307	−0.280	0.239	0.196	·	·
	−0.536	−0.339	0.191	0.157	−	−
3	0.004	0.097	0.168	0.175	·	·
	0.264	0.274	0.135	0.141	·	·

The feedback we have discovered could result from the fact that the data have been aggregated over time, as discussed by Tiao and Wei (1976) and Wei (1982, 1990). That is, annual sales are the sum of daily sales during the year, and annual advertising outlays are the sum of outlays at various times during the year. It might be that less highly aggregated data would show no feedback. But the feedback apparent in Table 10.3 says that a single-equation DR model with Y_t = sales and X_t = advertising outlays should not be used with the annual data since it is likely to give inconsistent estimates of the DR model parameters.

10.5 CHECK FOR CONTEMPORANEOUS RELATIONSHIP AND DEAD TIME

We may use the VAR model estimation results to check for a *contemporaneous relationship* between Y_t and X_t and to determine the dead time b. (To simplify notation, let $a_t = a_{1,t}$ and $\alpha_t = a_{2,t}$.) When $b > 0$, so that $\pi_{012} = -v_0 = 0$ (no contemporaneous relationship), then from (10.3.6) we have $\pi^*_{i12} = \pi_{i12}$; $a_t = u_t$; and $\alpha_t = w_t$. Since u_t and w_t are uncorrelated, we should find that the residual series $\hat{a}_t$ and $\hat{\alpha}_t$ are not significantly correlated if $b > 0$. In this case the estimated variance–covariance matrix of the shock terms (**V**) should be a diagonal matrix: The off-diagonal element (the estimated covariance between a_t and α_t) should be insignificant.

On the other hand, when $b = 0$, so that $\pi_{012} = -v_0 \neq 0$ (there is a contemporaneous relationship between Y_t and X_t), then we should find that the estimated correlation between a_t and α_t is significant. That is, from (10.3.6) we have $a_t = u_t - \pi_{012}w_t$, and $\alpha_t = w_t$. When $b = 0$, we have $\pi_{012} \neq 0$, and the residual series $\hat{a}_t$ and $\hat{\alpha}_t$ should thus be contemporaneously correlated. In this case the contemporaneous relationship between Y_t and X_t is reflected in the estimate of the variance–covariance matrix **V**: The off-diagonal element (the estimated covariance between a_t and α_t) should be significant. The idea of looking for contemporaneous relationships in **V** is discussed by Tsay (1985), Öller (1985), Hillmer et al. (1983), and Melicher et al. (1983).

For the sales and leading indicator data the estimated covariance between

CHECK FOR CONTEMPORANEOUS RELATIONSHIP AND DEAD TIME

a_t and α_t is -0.002893. The correlation between the residual series is found as

$$r = \text{cov}(\hat{a}_t, \hat{\alpha}_t)/(\hat{\sigma}_a^2 \hat{\sigma}_\alpha^2)^{1/2} \tag{10.5.1}$$

For the sales and leading indicator data,

$$r = -0.002893/[(0.043436)(0.072199)]^{1/2} = -0.052$$

Under the null hypothesis that the population correlation between a_t and α_t is zero, the following test statistic is t distributed:

$$t = r[(n - p_{\max} - 2)/(1 - r^2)]^{1/2} \tag{10.5.2}$$

with $n - p_{\max} - 2$ degrees of freedom. For the sales and leading indicator example,

$$t = -0.052\{(140 - 9 - 2)/[1 - (-0.052)^2]\}^{1/2} = -0.588$$

which has $n - p_{\max} - 2 = 140 - 9 - 2 = 129$ degrees of freedom. Here $n = 140$ is the original number of observations, and $n - p_{\max} = 140 - 9 = 131$ is the effective sample size. Since $t = -0.588$ is insignificant at the 5% level for 129 degrees of freedom, we conclude in this example that the input and the output series are not contemporaneously related. This is the same conclusion we reached when studying the $\hat{v}$ weights using the LTF identification method in Chapter 5.

If there is a contemporaneous relationship between an input and the output, then the dead time is $b = 0$. When there is no contemporaneous relationship, the estimated π weights may be used to choose the dead time. The estimated π_{i12} coefficients tell us about the relationship of Y_t to past values of X_t. For the sales and leading indicator data, in Table 10.2 the $\hat{\pi}_{i12}$ coefficients for lags 1 and 2 (coefficients $\hat{\pi}_{112}$ and $\hat{\pi}_{212}$) are not significant (no + or − symbols occur there). But coefficient $\hat{\pi}_{i12}$ for lag 3 ($\hat{\pi}_{312}$) has a + symbol. Since we have concluded that Y_t and X_t are not contemporaneously related in this example, we conclude that $v_0 = v_1 = v_2 = 0$. Apparently Y_t responds to a change in X_t starting at lag 3, so the dead time is $b = 3$. We found the same result using the LTF identification method for these data in Chapter 5.

Tsay (1985) also shows how the VAR estimation results may be used to obtain estimates of the v weights in a possible DR model, along with approximate standard errors. These estimates may then be used to identify a rational form transfer function for the DR model. However, estimates of the v weights and their standard errors are more easily found using the LTF identification method discussed in Chapter 5.

QUESTIONS AND PROBLEMS

1. In the Data Appendix are two series called U.K. Index of Consumption of Petrochemicals (Y_t, Series 15) and U.K. Index of Industrial Production (X_t, Series 16). The latter series has been seasonally adjusted. Using the VAR method, do you find evidence of feedback from Y_t to X_t? Do you find evidence of a contemporaneous relationship? Explain.

2. In the Data Appendix are two series called Percent Carbon Dioxide in Output from a Gas Furnace (Y_t, Series 17) and Methane Gas Input in Cubic Feet per Minute (X_t, Series 18). Using the VAR method, do you find evidence of feedback from Y_t to X_t? Do you find evidence of a contemporaneous relationship? Explain.

3. In the Data Appendix are two series called capital expenditures (Y_t, Series 19) and capital appropriations (X_t, Series 20). Both series are seasonally adjusted. Since funds must be appropriated before expenditures occur, and since capital construction projects take time to complete, it is expected that appropriations might lead expenditures.
   a. Using the VAR method, do you find evidence of feedback from Y_t to X_t when the data are in their original metrics? Do you find evidence of a contemporaneous relationship? Explain.
   b. Using the VAR method, do you find evidence of feedback from Y_t to X_t when the data are each in the natural log metric? Do you find evidence of a contemporaneous relationship? Explain.

APPENDIX

Tables

Table A Student's *t* Distribution[a]

The tabled values are two-tailed values $t(\alpha; \nu)$, such that

$$\text{prob}\{|t_\nu \text{ variate}| > t(\alpha; \nu)\} = \alpha$$

The entries in the table were computed on a CDC Cyber 172 computer at the University of Minnesota using IMSL subroutine MDSTI

ν	\multicolumn{5}{c}{α}
	0.200	0.100	0.050	0.010	0.001
1	3.08	6.31	12.71	63.66	636.62
2	1.89	2.92	4.30	9.92	31.60
3	1.64	2.35	3.18	5.84	12.92
4	1.53	2.13	2.78	4.60	8.61
5	1.48	2.02	2.57	4.03	6.87
6	1.44	1.94	2.45	3.71	5.96
7	1.41	1.89	2.36	3.50	5.41
8	1.40	1.86	2.31	3.36	5.04
9	1.38	1.83	2.26	3.25	4.78
10	1.37	1.81	2.23	3.17	4.59
11	1.36	1.80	2.20	3.11	4.44
12	1.36	1.78	2.18	3.05	4.32
13	1.35	1.77	2.16	3.01	4.22
14	1.35	1.76	2.14	2.98	4.14
15	1.34	1.75	2.13	2.95	4.07
16	1.34	1.75	2.12	2.92	4.01
17	1.33	1.74	2.11	2.90	3.97
18	1.33	1.73	2.10	2.88	3.92

Table A (*Continued*)

ν	\multicolumn{5}{c}{α}
	0.200	0.100	0.050	0.010	0.001
19	1.33	1.73	2.09	2.86	3.88
20	1.33	1.72	2.09	2.85	3.85
21	1.32	1.72	2.08	2.83	3.82
22	1.32	1.72	2.07	2.82	3.79
23	1.32	1.71	2.07	2.81	3.77
24	1.32	1.71	2.06	2.80	3.75
25	1.32	1.71	2.06	2.79	3.73
26	1.31	1.71	2.06	2.78	3.71
27	1.31	1.70	2.05	2.77	3.69
28	1.31	1.70	2.05	2.76	3.67
29	1.31	1.70	2.05	2.76	3.66
30	1.31	1.70	2.04	2.75	3.65
31	1.31	1.70	2.04	2.74	3.63
32	1.31	1.69	2.04	2.74	3.62
33	1.31	1.69	2.03	2.73	3.61
34	1.31	1.69	2.03	2.73	3.60
35	1.31	1.69	2.03	2.72	3.59
36	1.31	1.69	2.03	2.72	3.58
37	1.30	1.69	2.03	2.72	3.57
38	1.30	1.69	2.02	2.71	3.57
39	1.30	1.68	2.02	2.71	3.56
40	1.30	1.68	2.02	2.70	3.55
41	1.30	1.68	2.02	2.70	3.54
42	1.30	1.68	2.02	2.70	3.54
43	1.30	1.68	2.02	2.70	3.53
44	1.30	1.68	2.02	2.69	3.53
45	1.30	1.68	2.01	2.69	3.52
46	1.30	1.68	2.01	2.69	3.51
47	1.30	1.68	2.01	2.68	3.51
48	1.30	1.68	2.01	2.68	3.51
49	1.30	1.68	2.01	2.68	3.50
50	1.30	1.68	2.01	2.68	3.50
60	1.30	1.67	2.00	2.66	3.46
70	1.29	1.67	1.99	2.65	3.44
80	1.29	1.66	1.99	2.64	3.42
90	1.29	1.66	1.99	2.63	3.40
100	1.29	1.66	1.98	2.63	3.39
120	1.29	1.66	1.98	2.62	3.37
∞	1.28	1.64	1.96	2.58	3.29

[a]Source: Weisberg (1980). Reprinted by permission.

Table B χ^2 Critical Points[a]

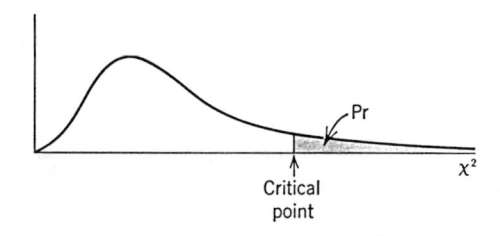

d.f. \ Pr	.250	.100	.050	.025	.010	.005	.001
1	1.32	2.71	3.84	5.02	6.63	7.88	10.8
2	2.77	4.61	5.99	7.38	9.21	10.6	13.8
3	4.11	6.25	7.81	9.35	11.3	12.8	16.3
4	5.39	7.78	9.49	11.1	13.3	14.9	18.5
5	6.63	9.24	11.1	12.8	15.1	16.7	20.5
6	7.84	10.6	12.6	14.4	16.8	18.5	22.5
7	9.04	12.0	14.1	16.0	18.5	20.3	24.3
8	10.2	13.4	15.5	17.5	20.1	22.0	26.1
9	11.4	14.7	16.9	19.0	21.7	23.6	27.9
10	12.5	16.0	18.3	20.5	23.2	25.2	29.6
11	13.7	17.3	19.7	21.9	24.7	26.8	31.3
12	14.8	18.5	21.0	23.3	26.2	28.3	32.9
13	16.0	19.8	22.4	24.7	27.7	29.8	34.5
14	17.1	21.1	23.7	26.1	29.1	31.3	36.1
15	18.2	22.3	25.0	27.5	30.6	32.8	37.7
16	19.4	23.5	26.3	28.8	32.0	34.3	39.3
17	20.5	24.8	27.6	30.2	33.4	35.7	40.8
18	21.6	26.0	28.9	31.5	34.8	37.2	42.3
19	22.7	27.2	30.1	32.9	36.2	38.6	32.8
20	23.8	28.4	31.4	34.2	37.6	40.0	45.3
21	24.9	29.6	32.7	35.5	38.9	41.4	46.8
22	26.0	30.8	33.9	36.8	40.3	42.8	48.3
23	27.1	32.0	35.2	38.1	41.6	44.2	49.7
24	28.2	33.2	36.4	39.4	32.0	45.6	51.2
25	29.3	34.4	37.7	40.6	44.3	46.9	52.6
26	30.4	35.6	38.9	41.9	45.6	48.3	54.1
27	31.5	36.7	40.1	43.2	47.0	49.6	55.5
28	32.6	37.9	41.3	44.5	48.3	51.0	56.9
29	33.7	39.1	42.6	45.7	49.6	52.3	58.3
30	34.8	40.3	43.8	47.0	50.9	53.7	59.7
40	45.6	51.8	55.8	59.3	63.7	66.8	73.4
50	56.3	63.2	67.5	71.4	76.2	79.5	86.7
60	67.0	74.4	79.1	83.3	88.4	92.0	99.6
70	77.6	85.5	90.5	95.0	100	104	112
80	88.1	96.6	102	107	112	116	125
90	98.6	108	113	118	124	128	137
100	109	118	124	130	136	140	149

[a] Source: Ronald J. Wonnacott and Thomas H. Wonnacott, *Econometrics*, 2nd ed., John Wiley and Sons, 1979. Reprinted by permission.

Table C F Critical Points[a]

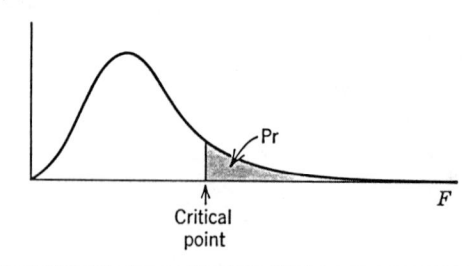

Critical point

		\multicolumn{10}{c}{Degrees of freedom for numerator}
	Pr	1	2	3	4	5	6	8	10	20	40	∞
1	.25	5.83	7.50	8.20	8.58	8.82	8.98	9.19	9.32	9.58	9.71	9.85
	.10	39.9	49.5	53.6	55.8	57.2	58.2	59.4	60.2	61.7	62.5	63.3
	.05	161	200	216	225	230	234	239	242	248	251	254
2	.25	2.57	3.00	3.15	3.23	3.28	3.31	3.35	3.38	3.43	3.45	3.48
	.10	8.53	9.00	9.16	9.24	9.29	9.33	9.37	9.39	9.44	9.47	9.49
	.05	18.5	19.0	19.2	19.2	19.3	19.3	19.4	19.4	19.4	19.5	19.5
	.01	98.5	99.0	99.2	99.2	99.3	99.3	99.4	99.4	99.4	99.5	99.5
	.001	998	999	999	999	999	999	999	999	999	999	999
3	.25	2.02	2.28	2.36	2.39	2.41	2.42	2.44	2.44	2.46	2.47	2.47
	.10	5.54	5.46	5.39	5.34	5.31	5.28	5.25	5.23	5.18	5.16	5.13
	.05	10.1	9.55	9.28	9.12	9.10	8.94	8.85	8.79	8.66	8.59	8.53
	.01	34.1	30.8	29.5	28.7	28.2	27.9	27.5	27.2	26.7	26.4	26.1
	.001	167	149	141	137	135	133	131	129	126	125	124
4	.25	1.81	2.00	2.05	2.06	2.07	2.08	2.08	2.08	2.08	2.08	2.08
	.10	4.54	4.32	4.19	4.11	4.05	4.01	3.95	3.92	3.84	3.80	3.76
	.05	7.71	6.94	6.59	6.39	6.26	6.16	6.04	5.96	5.80	5.72	5.63
	.01	21.2	18.0	16.7	16.0	15.5	15.2	14.8	14.5	14.0	13.7	13.5
	.001	74.1	61.3	56.2	53.4	51.7	50.5	49.0	48.1	46.1	45.1	44.1
5	.25	1.69	1.85	1.88	1.89	1.89	1.89	1.89	1.89	1.88	1.88	1.87
	.10	4.06	3.78	3.62	3.52	3.45	3.40	3.34	3.30	3.21	3.16	3.10
	.05	6.61	5.79	5.41	5.19	5.05	4.95	4.82	4.74	4.56	4.46	4.36
	.01	16.3	13.3	12.1	11.4	11.0	10.7	10.3	10.1	9.55	9.29	9.02
	.001	47.2	37.1	33.2	31.1	29.8	28.8	27.6	26.9	25.4	24.6	23.8
6	.25	1.62	1.76	1.78	1.79	1.79	1.78	1.77	1.77	1.76	1.75	1.74
	.10	3.78	3.46	3.29	3.18	3.11	3.05	2.98	2.94	2.84	2.78	2.72
	.05	5.99	5.14	4.76	4.53	4.39	4.28	4.15	4.06	3.87	3.77	3.67
	.01	13.7	10.9	9.78	9.15	8.75	8.47	8.10	7.87	7.40	7.14	6.88
	.001	35.5	27.0	23.7	21.9	20.8	20.0	19.0	18.4	17.1	16.4	15.8
7	.25	1.57	1.70	1.72	1.72	1.71	1.71	1.70	1.69	1.67	1.66	1.65
	.10	3.59	3.26	3.07	2.96	2.88	2.83	2.75	2.70	2.59	2.54	2.47
	.05	5.59	4.74	4.35	4.12	3.97	3.87	3.73	3.64	3.44	3.34	3.23
	.01	12.2	9.55	8.45	7.85	7.46	7.19	6.84	6.62	6.16	5.91	5.65
	.001	29.3	21.7	18.8	17.2	16.2	15.5	14.6	14.1	12.9	12.3	11.7
8	.25	1.54	1.66	1.67	1.66	1.66	1.65	1.64	1.63	1.61	1.59	1.58
	.10	3.46	3.11	2.92	2.81	2.73	2.67	2.59	2.54	2.42	2.36	2.29
	.05	5.32	4.46	4.07	3.84	3.69	3.58	3.44	3.35	3.15	3.04	2.93
	.01	11.3	8.65	7.59	7.01	6.63	6.37	6.03	5.81	5.36	5.12	4.86
	.001	25.4	18.5	15.8	14.4	13.5	12.9	12.0	11.5	10.5	9.92	9.33
9	.25	1.51	1.62	1.63	1.63	1.62	1.61	1.60	1.59	1.56	1.55	1.53
	.10	3.36	3.01	2.81	2.69	2.61	2.55	2.47	2.42	2.30	2.23	2.16
	.05	5.12	4.26	3.86	3.63	3.48	3.37	3.23	3.14	2.94	2.83	2.71
	.01	10.6	8.02	6.99	6.42	6.06	5.80	5.47	5.26	4.81	4.57	4.31
	.001	22.9	16.4	13.9	12.6	11.7	11.1	10.4	9.89	8.90	8.37	7.81

Degrees of freedom for denominator

[a]Source: Ronald J. Wonnacott and Thomas H. Wonnacott, *Econometrics*, 2nd ed., John Wiley and Sons, 1979. Reprinted by permission

Table C (*Continued*)

			\multicolumn{10}{c}{Degrees of freedom for numerator}
		Pr	1	2	3	4	5	6	8	10	20	40	∞
Degrees of freedom for denominator	10	.25	1.49	1.60	1.60	1.59	1.59	1.58	1.56	1.55	1.52	1.51	1.48
		.10	3.28	2.92	2.73	2.61	2.52	2.46	2.38	2.32	2.20	2.13	2.06
		.05	4.96	4.10	3.71	3.48	3.33	3.22	3.07	2.98	2.77	2.66	2.54
		.01	10.0	7.56	6.55	5.99	5.64	5.39	5.06	4.85	4.41	4.17	3.91
		.001	21.0	14.9	12.6	11.3	10.5	9.92	9.20	8.75	7.80	7.30	6.76
	12	.25	1.56	1.56	1.56	1.55	1.54	1.53	1.51	1.50	1.47	1.45	1.42
		.10	3.18	2.81	2.61	2.48	2.39	2.33	2.24	2.19	2.06	1.99	1.90
		.05	4.75	3.89	3.49	3.26	3.11	3.00	2.85	2.75	2.54	2.43	2.30
		.01	9.33	6.93	5.95	5.41	5.06	4.82	4.50	4.30	3.86	3.62	3.36
		.001	18.6	13.0	10.8	9.63	8.89	8.38	7.71	7.29	6.40	5.93	5.42
	14	.25	1.44	1.53	1.53	1.52	1.51	1.50	1.48	1.46	1.43	1.41	1.38
		.10	3.10	2.73	2.52	2.39	2.31	2.24	2.15	2.10	1.96	1.89	1.80
		.05	4.60	3.74	3.34	3.11	2.96	2.85	2.70	2.60	2.39	2.27	2.13
		.01	8.86	5.51	5.56	5.04	4.69	4.46	4.14	3.94	3.51	3.27	3.00
		.001	17.1	11.8	9.73	8.62	7.92	7.43	6.80	6.40	5.56	5.10	4.60
	16	.25	1.42	1.51	1.51	1.50	1.48	1.48	1.46	1.45	1.40	1.37	1.34
		.10	3.05	2.67	2.46	2.33	2.24	2.18	2.09	2.03	1.89	1.81	1.72
		.05	4.49	3.63	3.24	3.01	2.85	2.74	2.59	2.49	2.28	2.15	2.01
		.01	8.53	6.23	5.29	4.77	4.44	4.20	3.89	3.69	3.26	3.02	2.75
		.001	16.1	11.0	9.00	7.94	7.27	6.81	6.19	5.81	4.99	4.54	4.06
	18	.25	1.41	1.50	1.49	1.48	1.46	1.45	1.43	1.42	1.38	1.35	1.32
		.10	3.01	2.62	2.42	2.29	2.20	2.13	2.04	1.98	1.84	1.75	1.66
		.05	4.41	3.55	3.16	2.93	2.77	2.66	2.51	2.41	2.19	2.06	1.92
		.01	8.29	6.01	5.09	4.58	4.25	4.01	3.71	3.51	3.08	2.84	2.57
		.001	15.4	10.4	8.49	7.46	6.81	6.35	5.76	5.39	4.59	4.15	3.67
	20	.25	1.40	1.49	1.48	1.46	1.45	1.44	1.42	1.40	1.36	1.33	1.29
		.10	2.97	2.59	2.38	2.25	2.16	2.09	2.00	1.94	1.79	1.71	1.61
		.05	4.35	3.49	3.10	2.87	2.71	2.60	2.45	2.35	2.12	1.99	1.84
		.01	8.10	5.85	4.94	4.43	4.10	3.87	3.56	3.37	2.94	2.69	2.42
		.001	14.8	9.95	8.10	7.10	6.46	6.02	5.44	5.08	4.29	3.86	3.38
	30	.25	1.38	1.45	1.44	1.42	1.41	1.39	1.37	1.35	1.30	1.27	1.23
		.10	2.88	2.49	2.28	2.14	2.05	1.98	1.88	1.82	1.67	1.57	1.46
		.05	4.17	3.32	2.92	2.69	2.53	2.42	2.27	2.16	1.93	1.79	1.62
		.01	7.56	5.39	4.51	4.02	3.70	3.47	3.17	2.98	2.55	2.30	2.01
		.001	13.3	8.77	7.05	6.12	5.53	5.12	4.58	4.24	3.49	3.07	2.59
	40	.25	1.36	1.44	1.42	1.40	1.39	1.37	1.35	1.33	1.28	1.24	1.19
		.10	2.84	2.44	2.23	2.09	2.00	1.93	1.83	1.76	1.61	1.51	1.38
		.05	4.08	3.23	2.84	2.61	2.45	2.34	2.18	2.08	1.84	1.69	1.51
		.01	7.31	5.18	4.31	3.83	3.51	3.29	2.99	2.80	2.37	2.11	1.80
		.001	12.6	8.25	6.60	5.70	5.13	4.73	4.21	3.87	3.15	2.73	2.23
	60	.25	1.35	1.42	1.41	1.38	1.37	1.35	1.32	1.30	1.25	1.21	1.15
		.10	2.79	2.39	2.18	2.04	1.95	1.87	1.77	1.71	1.54	1.44	1.29
		.05	4.00	3.15	2.76	2.53	2.37	2.25	2.10	1.99	1.75	1.59	1.39
		.01	7.08	4.98	4.13	3.65	3.34	3.12	2.82	2.63	2.20	1.94	1.60
		.001	12.0	7.76	6.17	5.31	4.76	4.37	3.87	3.54	2.83	2.41	1.89
	120	.25	1.34	1.40	1.39	1.37	1.35	1.33	1.30	1.28	1.22	1.18	1.10
		.10	2.75	2.35	2.13	1.99	1.90	1.82	1.72	1.65	1.48	1.37	1.19
		.05	3.92	3.07	2.68	2.45	2.29	2.17	2.02	1.91	1.66	1.50	1.25
		.01	6.85	4.79	3.95	3.48	3.17	2.96	2.66	2.47	2.03	1.76	1.38
		.001	11.4	7.32	5.79	4.95	4.42	4.04	3.55	3.24	2.53	2.11	1.54
	∞	.25	1.32	1.39	1.37	1.35	1.33	1.31	1.28	1.25	1.19	1.14	1.00
		.10	2.71	2.30	2.08	1.94	1.85	1.77	1.67	1.60	1.42	1.30	1.00
		.05	3.84	3.00	2.60	2.37	2.21	2.10	1.94	1.83	1.57	1.39	1.00
		.01	6.63	4.61	3.78	3.32	3.02	2.80	2.51	2.32	1.88	1.59	1.00
		.001	10.8	6.91	5.42	4.62	4.10	3.74	3.27	2.96	2.27	1.84	1.00

Data Appendix

Series 1. Valve Shipments (monthly, 1/84–5/88; n = 53; read across):

39377	39417	39475	39843	40223	40105	40502	40726
41444	41256	41803	42028	41653	41706	41629	41648
41153	41744	41522	41282	43115	42253	42113	42231
42583	43106	42800	43103	42850	43653	43079	42361
42070	40892	40361	40895	41201	40712	40883	40464
40507	40260	40325	40575	40455	41040	41643	41726
42245	42586	43301	43299	43687			

Series 2. Valve Orders (monthly, 1/84–6/88; n = 54; read across):

34662	34165	34127	33917	33900	34618	35463	33067
34095	34443	34868	35835	36207	37769	36883	36861
36365	35854	36278	35862	36603	36960	36954	36957
37547	37631	37751	36428	38490	37795	37547	37997
36639	36171	36044	35695	34710	34813	34584	35540
35992	36014	36187	36599	37355	36421	37133	36880
36368	39150	39101	39331	40203	38916		

Series 3. U.S. Saving Rate (quarterly, 1955 I–1979 IV; n = 100; read across). Source: Pankratz (1983):

4.9	5.2	5.7	5.7	6.2	6.7	6.9	7.1	6.6	7.0	6.9	6.4	6.6
6.4	7.0	7.3	6.0	6.3	4.8	5.3	5.4	4.7	4.9	4.4	5.1	5.3
6.0	5.9	5.9	5.6	5.3	4.5	4.7	4.6	4.3	5.0	5.2	6.2	5.8
6.7	5.7	6.1	7.2	6.5	6.1	6.3	6.4	7.0	7.6	7.2	7.5	7.8
7.2	7.5	5.6	5.7	4.9	5.1	6.2	6.0	6.1	7.5	7.8	8.0	8.0
8.1	7.6	7.1	6.6	5.6	5.9	6.6	6.8	7.8	7.9	8.7	7.7	7.3
6.7	7.5	6.4	9.7	7.5	7.1	6.4	6.0	5.7	5.0	4.2	5.1	5.4
5.1	5.3	5.0	4.8	4.7	5.0	5.4	4.3	3.5				

Series 4. Sales (n = 150; read across). Source: Box and Jenkins (1976):

200.1	199.5	199.4	198.9	199.0	200.2	198.6	200.0	200.3
201.2	201.6	201.5	201.5	203.5	204.9	207.1	210.5	210.5
209.8	208.8	209.5	213.2	213.7	215.1	218.7	219.8	220.5
223.8	222.8	223.8	221.7	222.3	220.8	219.4	220.1	220.6
218.9	217.8	217.7	215.0	215.3	215.9	216.7	216.7	217.7
218.7	222.9	224.9	222.2	220.7	220.0	218.7	217.0	215.9
215.8	214.1	212.3	213.9	214.6	213.6	212.1	211.4	213.1
212.9	213.3	211.5	212.3	213.0	211.0	210.7	210.1	211.4
210.0	209.7	208.8	208.8	208.8	210.6	211.9	212.8	212.5
214.8	215.3	217.5	218.8	220.7	222.2	226.7	228.4	233.2
235.7	237.1	240.6	243.8	245.3	246.0	246.3	247.7	247.6
247.8	249.4	249.0	249.9	250.5	251.5	249.0	247.6	248.8
250.4	250.7	253.0	253.7	255.0	256.2	256.0	257.4	260.4
260.0	261.3	260.4	261.6	260.8	259.8	259.0	258.9	257.4
257.7	257.9	257.4	257.3	257.6	258.9	257.8	257.7	257.2
257.5	256.8	257.5	257.0	257.6	257.3	257.5	259.6	261.1
262.9	263.3	262.8	261.8	262.2	262.7			

Series 5. Leading Indicator (n = 150; read across). Source: Box and Jenkins (1976):

10.01	10.07	10.32	9.75	10.33	10.13	10.36	10.32	10.13
10.16	10.58	10.62	10.86	11.20	10.74	10.56	10.48	10.77
11.33	10.96	11.16	11.70	11.39	11.42	11.94	11.24	11.59
10.96	11.40	11.02	11.01	11.23	11.33	10.83	10.84	11.14
10.38	10.90	11.05	11.11	11.01	11.22	11.21	11.91	11.69
10.93	10.99	11.01	10.84	10.76	10.77	10.88	10.49	10.50
11.00	10.98	10.61	10.48	10.53	11.07	10.61	10.86	10.34
10.78	10.80	10.33	10.44	10.50	10.75	10.40	10.40	10.34
10.55	10.46	10.82	10.91	10.87	10.67	11.11	10.88	11.28
11.27	11.44	11.52	12.10	11.83	12.62	12.41	12.43	12.73
13.01	12.74	12.73	12.76	12.92	12.64	12.79	13.05	12.69
13.01	12.90	13.12	12.47	12.47	12.94	13.10	12.91	13.39
13.13	13.34	13.34	13.14	13.49	13.87	13.39	13.59	13.27
13.70	13.20	13.32	13.15	13.30	12.94	13.29	13.26	13.08
13.24	13.31	13.52	13.02	13.25	13.12	13.26	13.11	13.30
13.06	13.32	13.10	13.27	13.64	13.58	13.87	13.53	13.41
13.25	13.50	13.58	13.51	13.77	13.40			

Series 6. Change in Business Inventories (quarterly, 1955 I–1969 IV; n = 60; read across). Source: Pankratz (1983):

4.4	5.8	6.7	7.1	5.7	4.1	4.6	4.3	2.0	2.2	3.6
−2.2	−5.1	−4.9	0.1	4.1	3.8	9.9	0.0	6.5	10.8	4.1
2.7	−2.9	−2.9	1.5	5.7	5.0	7.9	6.8	7.1	4.1	5.5
5.1	8.0	5.6	4.5	6.1	6.7	6.1	10.6	8.6	11.6	7.6
10.9	14.6	14.5	17.4	11.7	5.8	11.5	11.7	5.0	10.0	8.9
7.1	8.3	10.2	13.3	6.2						

Series 7. U.S. Federal Government Net Receipts ($ millions; monthly, 1/76–12/87; n = 144; read across). Source: U.S. Commerce Department, Business Statistics *and* Survey of Current Business:

25634	20845	20431	33348	22679	36807	22589	27349	31748
21018	25694	29470	29954	24182	24817	39832	27549	43948
24967	29683	36647	24130	27597	32796	33201	26922	25233
42545	35090	47657	29194	35040	42591	28745	33227	37477
38364	32639	31144	52238	38287	53910	33268	39353	47302
33099	38320	42617	43429	37866	33351	61097	36071	59055
37348	44259	53545	38923	39175	48904	52214	38129	44623
74464	38243	70429	47829	47669	60279	45150	44016	56822
55269	43042	45291	75777	36753	66353	44675	44924	59694
40539	42007	54498	57505	38818	43504	66234	33755	66517
43948	49683	63556	45157	46202	58044	62537	46886	44464
80180	37459	69282	52017	55209	68019	52251	51494	62404
70454	54021	49606	94593	39794	72151	57970	55776	73808
57886	51163	68193	76710	53370	49557	91438	46246	77024
62974	56523	78013	59012	52967	78035	81771	55463	56515
122897	47691	82945	64223	60213	92410	62354	56987	85525

Series 8. Demand for Repair Parts for a Large Heavy Equipment Manufacturer in Iowa ($ thousands; monthly, 1/72–10/79; n = 94; read across). Source: Abraham and Ledolter (1983):

954	765	867	940	913	1014	801	990	712	959	828
965	915	891	991	971	1129	1091	1195	1295	1046	1121
1033	1222	1199	1012	1404	1137	1421	1162	1639	1545	1420
1916	1491	1295	1764	1727	1654	1811	1520	1635	1984	1898
1853	2015	1709	1667	1625	1562	2121	1783	1474	1657	1746
1763	1517	1457	1388	1501	1227	1342	1666	2091	1629	1451
1727	1736	1952	1420	1345	842	1576	1485	1928	2072	1887
2175	2199	1961	2091	1993	1595	1372	1607	1871	2594	1743
2267	2602	2565	2567	2344	2805					

DATA APPENDIX

Series 9. Housing Permits in the United States (quarterly seasonally adjusted; 1947 I–1967 II; n = 82; read across). Source: Pankratz (1983):

83.3	83.2	105.3	117.7	104.6	108.8	93.9	86.1	83.0	102.4	119.6
141.4	158.6	161.3	158.2	136.1	121.9	97.7	103.3	92.7	106.8	102.1
110.3	114.1	109.1	105.4	97.6	100.7	102.7	110.9	120.2	131.3	138.9
130.9	123.1	110.8	108.8	103.8	97.0	93.2	89.7	89.9	90.2	89.6
85.8	96.9	112.7	122.7	119.8	117.4	111.9	104.7	98.3	94.9	93.3
90.9	91.9	97.2	104.7	107.7	108.2	110.7	113.2	114.6	112.2	120.2
122.1	126.6	122.3	115.9	116.9	110.1	110.4	108.9	112.1	117.6	112.2
96.0	78.0	66.9	83.5	95.8	107.7	113.7				

Series 10. Retail Sales of Nondurable Goods Stores (millions of dollars; monthly, 1/73–12/83; n = 132; read across). Source: U.S. Commerce Department, Survey of Current Business:

23921	22971	26627	26295	27583	28267	27264	28796	27497	
28650	30414	36408	26521	25337	28933	29472	31172	30436	
30513	32916	30085	31873	33123	38728	29401	27590	31556	
30906	34642	32869	33560	35230	32880	34809	35140	43636	
32824	30549	33749	35240	35445	35659	36916	36225	35632	
37539	38052	48168	33750	32804	37098	38498	38431	38629	
39495	39684	38988	40246	42203	53247	36171	35619	41981	
40690	42927	43496	42805	44473	43984	44351	47192	59031	
41306	40078	46783	45532	48218	48969	47497	51113	48723	
50933	54614	65457	47963	47665	51727	51507	54949	52617	
53852	56518	53057	57156	58364	71757	53704	50127	56006	
57892	59556	58451	59538	59956	57950	61571	61294	76386	
54759	52008	57709	59127	60338	59354	61948	60731	59471	
62602	63692	79508	55871	53181	61235	61867	63486	63911	
64633	65455	63907	65803	68406	85505				

Series 11. Housing Starts (quarterly, 1965 I–1975 IV; n = 44; read across). Source: these data are the moving sums (without overlap) of every 3 months' observations in Series 23.

181.504	296.690	266.195	219.299	180.158	258.919	197.834	
141.714	147.023	254.774	244.250	197.818	179.990	266.213	
248.998	204.226	171.150	259.024	214.492	165.935	136.609	
231.690	228.713	215.877	204.727	348.586	321.559	276.148	
263.849	386.877	370.893	287.613	255.801	366.918	306.056	
203.259	177.831	297.895	243.940	168.399	142.248	260.843	
267.954	221.122						

Series 12. Index of Kilowatt Hours Used (monthly, 1/70–12/84; n = 180; read across):

74.5	64.1	64.2	60.6	61.8	73.1	85.0	81.3	70.3
70.1	70.9	78.6	82.6	70.6	72.7	65.0	62.0	84.9
84.9	89.7	82.5	78.4	79.0	85.9	93.1	84.7	80.8
76.1	83.9	90.9	98.5	114.6	88.0	91.4	94.3	103.6
104.9	89.5	91.1	83.7	88.1	100.0	116.5	122.7	99.8
98.4	94.7	97.2	102.0	87.6	90.7	83.3	89.4	93.6
130.3	111.2	93.3	96.9	99.1	105.9	110.7	100.4	100.1
95.3	99.0	110.7	143.6	131.5	102.6	104.5	103.1	117.4
122.7	107.4	110.2	96.7	98.9	119.8	144.1	143.7	118.0
114.5	120.4	140.4	143.6	114.7	111.4	105.3	115.4	124.4
152.5	130.6	119.6	116.4	121.7	137.6	146.4	130.5	123.5
115.8	115.1	133.6	147.0	157.7	148.5	124.9	130.4	144.1
157.7	143.6	133.8	128.8	124.5	130.5	153.6	157.5	135.1
131.3	142.6	149.4	154.0	145.6	142.3	126.6	124.3	139.6
178.3	162.4	144.8	138.6	140.1	154.5	158.9	144.8	138.4
130.8	129.1	145.1	179.5	172.8	150.8	147.1	146.1	161.9
174.7	157.7	149.7	145.2	143.3	144.2	182.8	190.8	158.3
157.5	157.3	170.3	173.2	153.5	162.1	146.9	144.6	166.3
223.1	236.3	181.8	165.7	168.7	199.0	193.8	174.1	182.9
158.8	165.0	186.8	207.3	238.8	175.6	183.0	181.9	194.7

Series 13. Heating Degree Days (monthly, 1/70–12/84; n = 180; read across):

2038	1509	1289	638	269	61	18	12	222	520	1047	1572
1986	1471	1296	614	351	21	31	34	184	391	994	1504
1963	1649	1238	763	237	95	42	77	237	723	1090	1812
1654	1356	859	731	383	22	3	16	214	378	979	1676
1747	1488	1191	593	379	85	0	54	335	518	1020	1428
1687	1502	1474	819	213	65	21	8	296	412	890	1421
1751	1148	1107	507	307	24	1	28	244	770	1296	1860
2246	1310	876	402	98	40	1	65	155	589	1069	1702
2063	1714	1226	652	275	72	12	22	154	626	1159	1761
2238	1798	1283	737	346	34	0	51	133	586	1056	1302
1662	1564	1285	589	223	57	0	12	196	715	956	1535
1544	1285	981	538	331	46	23	29	255	673	941	1630
2084	1478	1215	779	178	140	0	52	193	503	1045	1298
1579	1241	1095	829	437	93	9	0	227	602	1017	2087
1762	1238	1389	651	413	43	25	20	305	510	1030	1533

DATA APPENDIX

Series 14. Cooling Degree Days (monthly, 1/70–12/84; n = 180; read across):

0	0	0	13	37	134	179	116	51	0	0	0
0	0	0	0	0	193	88	109	78	10	0	0
0	0	0	0	65	99	115	157	12	0	0	0
0	0	0	0	0	109	173	170	33	10	0	0
0	0	0	0	11	57	267	86	6	0	0	0
0	0	0	0	60	125	247	172	16	11	0	0
0	0	0	8	0	111	246	162	32	0	0	0
0	0	0	16	73	87	226	60	23	0	0	0
0	0	0	0	34	99	130	129	98	0	0	0
0	0	0	0	19	105	218	148	50	0	0	0
0	0	0	12	61	107	232	145	36	0	0	0
0	0	0	0	0	66	158	100	11	0	0	0
0	0	0	0	23	27	225	148	47	0	0	0
0	0	0	0	0	107	256	241	70	6	0	0
0	0	0	0	0	65	102	178	20	0	0	0

Series 15. U.K. Consumption of Petrochemicals (quarterly, 1958 I–1976 IV; n = 76; read across). Source: McLeod (1982):

31.1	31.3	29.6	32.0	31.9	35.7	34.5	38.2	39.7	40.6	37.1
39.2	38.7	40.6	37.0	39.1	40.6	42.4	39.9	42.9	43.0	47.1
46.5	50.9	52.1	55.9	51.4	56.1	56.6	57.5	55.6	58.7	61.7
63.8	58.7	61.4	63.1	66.4	61.3	68.1	74.3	74.7	72.3	77.3
79.4	79.9	73.5	80.1	78.9	84.6	76.9	85.3	84.8	83.7	80.6
85.7	83.5	91.1	87.5	95.0	101.8	104.9	100.1	105.6	99.9	104.0
98.8	97.3	90.5	90.8	87.4	95.7	101.9	99.9	96.8	101.2	

Series 16. U.K. Index of Industrial Production (quarterly, seasonally adjusted, 1958 I–1976 IV; n = 76; read across). Source: McLeod (1982):

61.25	60.23	59.86	60.88	61.34	63.19	64.02	67.62	68.91	69.01
69.10	69.75	69.19	70.02	69.47	68.64	68.64	69.75	70.30	69.38
68.36	70.95	73.35	75.10	77.69	77.69	78.43	80.65	80.46	80.74
80.55	81.85	83.14	82.77	83.23	80.55	81.85	82.68	82.86	84.43
87.39	88.41	89.15	89.88	90.90	92.66	92.19	92.19	91.92	91.92
92.47	93.12	91.73	92.56	92.29	91.82	90.53	94.13	95.15	98.57
102.08	102.17	103.00	102.82	98.85	101.71	101.52	97.92	98.01	92.84
92.10	92.75	94.23	95.61	95.52	96.26				

Series 17. Percent Carbon Dioxide in Gas Furnace Output (recorded every 9 seconds; n = 296; read across). Source: Box and Jenkins (1976):

53.8	53.6	53.5	53.5	53.4	53.1	52.7	52.4	52.2	52.0	52.0
52.4	53.0	54.0	54.9	56.0	56.8	56.8	56.4	55.7	55.0	54.3
53.2	52.3	51.6	51.2	50.8	50.5	50.0	49.2	48.4	47.9	47.6
47.5	47.5	47.6	48.1	49.0	50.0	51.1	51.8	51.9	51.7	51.2
50.0	48.3	47.0	45.8	45.6	46.0	46.9	47.8	48.2	48.3	47.9
47.2	47.2	48.1	49.4	50.6	51.5	51.6	51.2	50.5	50.1	49.8
49.6	49.4	49.3	49.2	49.3	49.7	50.3	51.3	52.8	54.4	56.0
56.9	57.5	57.3	56.6	56.0	55.4	55.4	56.4	57.2	58.0	58.4
58.4	58.1	57.7	57.0	56.0	54.7	53.2	52.1	51.6	51.0	50.5
50.4	51.0	51.8	52.4	53.0	53.4	53.6	53.7	53.8	53.8	53.8
53.3	53.0	52.9	53.4	54.6	56.4	58.0	59.4	60.2	60.0	59.4
58.4	57.6	56.9	56.4	56.0	55.7	55.3	55.0	54.4	53.7	52.8
51.6	50.6	49.4	48.8	48.5	48.7	49.2	49.8	50.4	50.7	50.9
50.7	50.5	50.4	50.2	50.4	51.2	52.3	53.2	53.9	54.1	54.0
53.6	53.2	53.0	52.8	52.3	51.9	51.6	51.6	51.4	51.2	50.7
50.0	49.4	49.3	49.7	50.6	51.8	53.0	54.0	55.3	55.9	55.9
54.6	53.5	52.4	52.1	52.3	53.0	53.8	54.6	55.4	55.9	55.9
55.2	54.4	53.7	53.6	53.6	53.2	52.5	52.0	51.4	51.0	50.9
52.4	53.5	55.6	58.0	59.5	60.0	60.4	60.5	60.2	59.7	59.0
57.6	56.4	55.2	54.5	54.1	54.1	54.4	55.5	56.2	57.0	57.3
57.4	57.0	56.4	55.9	55.5	55.3	55.2	55.4	56.0	56.5	57.1
57.3	56.8	55.6	55.0	54.1	54.3	55.3	56.4	57.2	57.8	58.3
58.6	58.8	58.8	58.6	58.0	57.4	57.0	56.4	56.3	56.4	56.4
56.0	55.2	54.0	53.0	52.0	51.6	51.6	51.1	50.4	50.0	50.0
52.0	54.0	55.1	54.5	52.8	51.4	50.8	51.2	52.0	52.8	53.8
54.5	54.9	54.9	54.8	54.4	53.7	53.3	52.8	52.6	52.6	53.0
54.3	56.0	57.0	58.0	58.6	58.5	58.3	57.8	57.3	57.0	

Series 18. Methane Gas Input to Furnace (recorded every 9 seconds; n = 296; read across). Source: Box and Jenkins (1976):

−0.109	0.000	0.178	0.339	0.373	0.441	0.461	0.348
0.127	−0.180	−0.588	−1.055	−1.421	−1.520	−1.302	−0.814
−0.475	−0.193	0.088	0.435	0.771	0.866	0.875	0.891
0.987	1.263	1.775	1.976	1.934	1.866	1.832	1.767
1.608	1.265	0.790	0.360	0.115	0.088	0.331	0.645
0.960	1.409	2.670	2.834	2.812	2.483	1.929	1.485
1.214	1.239	1.608	1.905	2.023	1.815	0.535	0.122
0.009	0.164	0.671	1.019	1.146	1.155	1.112	1.121
1.223	1.257	1.157	0.913	0.620	0.255	−0.280	−1.080
−1.551	−1.799	−1.825	−1.456	−0.944	−0.570	−0.431	−0.577
−0.960	−1.616	−1.875	−1.891	−1.746	−1.474	−1.201	−0.927
−0.524	0.040	0.788	0.943	0.930	1.006	1.137	1.198

DATA APPENDIX 369

1.054	0.595	−0.080	−0.314	−0.288	−0.153	−0.109	−0.187
−0.255	−0.229	−0.007	0.254	0.330	0.102	−0.423	−1.139
−2.275	−2.594	−2.716	−2.510	−1.790	−1.346	−1.081	−0.910
−0.876	−0.885	−0.800	−0.544	−0.416	−0.271	0.000	0.403
0.841	1.285	1.607	1.746	1.683	1.485	0.993	0.648
0.577	0.577	0.632	0.747	0.900	0.993	0.968	0.790
0.399	−0.161	−0.553	−0.603	−0.424	−0.194	−0.049	0.060
0.161	0.301	0.517	0.566	0.560	0.573	0.592	0.671
0.933	1.337	1.460	1.353	0.772	0.218	−0.237	−0.714
−1.099	−1.269	−1.175	−0.676	0.033	0.556	0.643	0.484
0.109	−0.310	−0.697	−1.047	−1.218	−1.183	−0.873	−0.336
0.063	0.084	0.000	0.001	0.209	0.556	0.782	0.858
0.918	0.862	0.416	−0.336	−0.959	−1.813	−2.378	−2.499
−2.473	−2.330	−2.053	−1.739	−1.261	−0.569	−0.137	−0.024
−0.050	−0.135	−0.276	−0.534	−0.871	−1.234	−1.439	−1.422
−1.175	−0.813	−0.634	−0.582	−0.625	−0.713	−0.848	−1.039
−1.346	−1.628	−1.619	−1.149	−0.488	−0.160	−0.007	−0.092
−0.620	−1.086	−1.525	−1.858	−2.029	−2.024	−1.961	−1.952
−1.794	−1.302	−1.030	−0.918	−0.798	−0.867	−1.047	−1.123
−0.876	−0.395	0.185	0.662	0.709	0.605	0.501	0.603
0.943	1.223	1.249	0.824	0.102	0.025	0.382	0.922
1.032	0.866	0.527	0.093	−0.458	−0.748	−0.947	−1.029
−0.928	−0.645	−0.424	−0.276	−0.158	−0.033	0.102	0.251
0.280	0.000	−0.493	−0.759	−0.824	−0.740	−0.528	−0.204
0.034	0.204	0.253	0.195	0.131	0.017	−0.182	−0.262

Series 19. Capital Expenditures (quarterly, 1953 I–1974 IV; n = 88; read across). Source: Judge, et al. (1982, Chapter 27):

2072	2077	2078	2043	2062	2067	1964	1981	1914	1991	2129
2309	2614	2896	3058	3309	3446	3466	3435	3183	2697	2338
2140	2012	2071	2192	2240	2421	2639	2733	2721	2640	2513
2448	2429	2516	2534	2494	2596	2572	2601	2648	2840	2937
3136	3299	3514	3815	4040	4274	4565	4838	5222	5406	5705
5871	5953	5868	5573	5672	5543	5526	5750	5761	5943	6212
6631	6828	6645	6703	6659	6337	6165	5875	5798	5921	5772
5874	5872	6159	6583	6961	7449	8093	9013	9752	10704	11597

Series 20. Capital Appropriations (quarterly, 1953 I–1974 IV; n = 88; read across). Source: Judge, et al. (1982, Chapter 27):

1767	2061	2289	2047	1856	1842	1866	2279	2688	3264
3896	4014	4041	3710	3383	3431	3613	3205	2426	2330
1954	1936	2201	2233	2690	2940	3127	3131	2872	2515
2271	2711	2394	2457	2720	2703	2992	2516	2817	3153
2756	3269	3657	3941	4123	4656	4906	4344	5080	5539
5583	6147	6545	6770	5955	6015	6029	5975	5894	5951

5952	5723	6351	6636	6799	7753	7595	7436	6679	6475
6319	5860	5705	5521	5920	5937	6570	7087	7206	8431
9718	10921	11672	12199	12865	14985	16378	12680		

Series 21. Lydia Pinkham Sales (thousands of dollars; annual, 1907–60; n = 54; read across). Source: Vandaele (1983):

1016	921	934	976	930	1052	1184	1089	1087	1154
1330	1980	2223	2203	2514	2726	3185	3351	3438	2917
2359	2240	2196	2111	1806	1644	1814	1770	1518	1103
1266	1473	1423	1767	2161	2336	2602	2518	2637	2177
1920	1910	1984	1787	1689	1866	1896	1684	1633	1657
1569	1390	1387	1289						

Series 22. Lydia Pinkham Advertising (thousands of dollars; annual, 1907–1960; n = 54; read across). Source: Vandaele (1983):

608	451	529	543	525	549	525	578	609	504
752	613	862	866	1016	1360	1482	1608	1800	1941
1229	1373	1611	1568	983	1046	1453	1504	807	339
562	745	749	862	1034	1054	1164	1102	1145	1012
836	941	981	974	766	920	964	811	789	802
770	639	644	564						

Series 23. Housing Starts (monthly, 1/65–12/75; n = 132; read across). Source: Abraham and Ledolter (1983):

52.149	47.205	82.150	100.931	98.408	97.351	96.489	88.830
80.876	85.750	72.351	61.198	46.561	50.361	83.236	94.343
84.748	79.828	69.068	69.362	59.404	53.530	50.212	37.972
40.157	40.274	66.592	79.839	87.341	87.594	82.344	83.712
78.194	81.704	69.088	47.026	45.234	55.431	79.325	97.983
86.806	81.424	86.398	82.522	80.078	85.560	64.819	53.847
51.300	47.909	71.941	84.982	91.301	82.741	73.523	69.465
71.504	68.039	55.069	42.827	33.363	41.367	61.879	73.835
74.848	83.007	75.461	77.291	75.961	79.393	67.443	69.041
54.856	58.287	91.584	116.013	115.627	116.946	107.747	111.663
102.149	102.882	92.904	80.362	76.185	76.306	111.358	119.840
135.167	131.870	119.078	131.324	120.491	116.990	97.428	73.195
77.105	73.560	105.136	120.453	131.643	114.822	114.746	106.806
84.504	86.004	70.488	46.767	43.292	57.593	76.946	102.237
96.340	99.318	90.715	79.782	73.443	69.460	57.898	41.041
39.791	39.959	62.498	77.777	92.782	90.284	92.782	90.655
84.517	93.826	71.646	55.650				

DATA APPENDIX

Series 24. Housing Sales (monthly, 1/65–12/75; n = 132; read across).
Source: Abraham and Ledolter (1983):

```
38  44  53  49  54  57  51  58  48  44  42  37  42  43  53  49  49  40  40  36  29
31  26  23  29  32  41  44  49  47  46  47  43  45  34  31  35  43  46  46  43  41
44  47  41  40  32  32  34  40  43  42  43  44  39  40  33  32  31  28  34  29  36
42  43  44  44  48  45  44  40  37  45  49  62  62  58  59  64  62  50  52  50  44
51  56  60  65  64  63  63  72  61  65  51  47  54  58  66  63  64  60  53  52  44
40  36  28  36  42  53  53  55  48  47  43  39  33  30  23  29  33  44  54  56  51
51  53  45  45  44  38
```

Series 25. Industrial Production (monthly, 1/47–9/80; n = 405; read across).
Source: U.S. Commerce Department, Survey of Current Business:

```
 38.6   39.3   39.5   38.7   38.5   38.8   36.8   39.3   40.3   41.5   41.2
 40.4   40.6   41.1   40.5   40.1   40.4   41.2   39.3   41.6   42.3   43.2
 41.8   40.5   40.0   40.1   39.3   38.5   37.7   37.9   36.0   39.0   40.0
 39.2   39.0   38.8   39.8   40.3   41.6   42.6   43.0   44.7   43.4   48.3
 48.7   50.0   48.3   48.2   48.5   49.6   50.0   49.4   48.6   49.1   45.5
 48.0   49.1   49.5   48.9   48.2   49.0   50.1   50.4   49.1   48.3   48.4
 44.9   50.6   53.4   54.5   54.5   53.6   54.2   55.6   56.3   55.8   55.8
 55.9   53.2   55.8   55.5   55.8   53.2   51.0   51.3   52.0   51.8   51.2
 51.3   51.9   48.8   51.3   52.3   53.5   53.6   53.4   54.0   56.2   57.5
 57.9   58.4   59.1   55.8   58.6   60.1   61.9   61.3   60.4   60.5   61.1
 61.2   61.6   60.7   61.0   55.2   60.4   62.6   63.8   62.6   62.0   62.0
 63.5   63.6   62.3   61.8   63.1   59.4   62.8   63.0   62.6   60.4   57.8
 57.0   56.4   55.8   54.7   55.0   57.6   54.9   58.7   60.7   61.3   62.1
 60.8   61.9   63.8   65.2   66.4   67.2   68.5   62.9   63.3   64.5   64.6
 63.6   65.9   67.8   68.2   68.0   67.5   67.3   67.5   63.4   65.6   66.6
 66.9   64.5   61.7   62.0   62.7   63.5   65.1   66.1   68.0   64.6   67.9
 69.5   71.1   70.5   69.4   68.9   71.0   72.1   72.4   72.4   73.4   69.7
 71.9   74.7   75.0   73.6   71.8   72.4   74.7   75.7   76.4   77.1   78.6
 73.4   75.5   78.9   80.2   78.7   76.5   77.5   79.8   80.2   81.5   81.9
 83.2   78.3   81.3   84.6   83.9   84.5   83.4   84.8   87.0   88.8   88.8
 89.5   91.7   86.7   89.4   92.4   94.5   93.0   91.4   92.7   95.3   97.3
 97.1   97.8  100.0   93.8   97.2  101.7  102.8  100.0   97.5   98.1   99.1
 99.0   99.6   98.7  100.9   94.4   99.6  102.7  103.4  103.1  101.4  101.8
104.5  105.6  104.9  106.5  109.3  102.5  105.5  109.6  110.1  109.6  106.2
107.3  110.4  111.6  110.6  110.5  114.0  107.3  111.6  115.1  115.1  112.0
108.3  106.5  109.1  109.4  108.8  108.6  110.8  104.5  108.0  110.4  108.0
105.1  104.1  105.5  108.3  108.6  108.8  109.5  112.5  105.4  108.8  113.5
113.9  111.6  108.5  111.5  115.6  116.8  118.7  118.4  121.8  114.2  120.5
125.5  126.8  125.2  121.8  122.7  128.1  128.8  128.6  129.6  133.0  126.4
130.3  134.8  135.3  132.9  126.7  126.3  129.8  130.8  129.9  131.7  135.3
127.3  131.4  135.5  133.1  125.5  114.9  111.8  113.0  111.8  113.0  113.8
119.2  114.5  121.4  125.9  125.4  123.8  119.8  122.4  128.7  129.1  129.1
130.5  134.0  127.2  132.5  135.1  134.8  133.0  129.1  130.0  134.7  136.6
137.1  138.3  142.6  135.3  139.6  143.5  143.9  140.4  135.8  135.7  140.5
142.4  145.3  144.9  149.8  142.9  148.2  153.0  153.4  150.5  147.1  146.6
152.3  154.0  151.1  152.5  156.5  149.0  152.7  157.1  156.2  152.4  147.7
148.0  152.7  153.2  148.1  143.5  145.0  137.2  142.9  148.3
```

Series 26. Standard and Poor's 500 Stock Price Index (monthly, 1/47–9/80; n = 405; read across). Source: U.S. Commerce Department, Business Conditions Digest:

15.21	15.80	15.16	14.60	14.34	14.84	15.77	15.46
15.06	15.45	15.27	15.03	14.83	14.10	14.30	15.40
16.15	16.82	16.42	15.94	15.76	16.19	15.29	15.19
15.36	14.77	14.91	14.89	14.78	13.97	14.76	15.29
15.49	15.89	16.11	16.54	16.88	17.21	17.35	17.84
18.44	18.74	17.38	18.43	19.08	19.87	19.83	19.75
21.21	22.00	21.63	21.92	21.93	21.55	21.93	22.89
23.48	23.36	22.71	23.41	24.19	23.75	23.81	23.74
23.73	24.38	25.08	25.18	24.78	24.26	25.03	26.04
26.18	25.86	25.99	24.71	24.84	23.95	24.29	24.39
23.27	23.97	24.50	24.83	25.46	26.02	26.57	27.63
28.73	28.96	30.13	30.73	31.45	32.18	33.44	34.97
35.60	36.79	36.50	37.76	37.60	39.78	42.69	42.43
44.34	42.11	44.95	45.37	44.15	44.43	47.49	48.05
46.54	46.27	48.78	48.49	46.84	46.24	45.76	46.44
45.43	43.47	44.03	45.05	46.78	47.55	48.51	45.84
43.98	41.24	40.35	40.33	41.12	41.26	42.11	42.34
43.70	44.75	45.98	47.70	48.96	50.95	52.50	53.49
55.62	54.77	56.15	57.10	57.96	57.46	59.74	59.40
57.05	57.00	57.23	59.06	58.03	55.78	55.02	55.73
55.22	57.26	55.84	56.51	54.81	53.73	55.47	56.80
59.72	62.17	64.12	65.83	66.50	65.62	65.44	67.79
67.26	68.00	71.08	71.74	69.07	70.22	70.29	68.05
62.99	55.63	56.97	58.52	58.00	56.17	60.04	62.64
65.06	65.92	65.67	68.76	70.14	70.11	69.07	70.98
72.85	73.03	72.62	74.17	76.45	77.39	78.80	79.94
80.72	80.24	83.22	82.00	83.41	84.85	85.44	83.96
86.12	86.75	86.83	87.97	89.28	85.04	84.91	86.49
89.38	91.39	92.15	91.73	93.32	92.69	88.88	91.60
86.78	86.06	85.84	80.65	77.81	77.13	80.99	81.33
84.45	87.36	89.42	90.96	92.59	91.43	93.01	94.49
95.81	95.66	92.66	95.30	95.04	90.75	89.09	95.67
97.87	100.53	100.30	98.11	101.34	103.76	105.40	106.48
102.04	101.46	99.30	101.26	104.62	99.14	94.71	94.18
94.51	95.52	96.21	91.11	90.31	87.16	88.65	85.95
76.06	75.59	75.72	77.92	82.58	84.37	84.28	90.05
93.49	97.11	99.60	103.04	101.64	99.72	99.00	97.24
99.40	97.29	92.78	99.17	103.30	105.24	107.69	108.81
107.65	108.01	107.21	111.01	109.39	109.56	115.05	117.50
118.42	114.16	112.42	110.27	107.22	104.75	105.83	103.80
105.61	109.84	102.03	94.78	96.11	93.45	97.44	92.46

DATA APPENDIX

89.67	89.79	82.82	76.03	68.12	69.44	71.74	67.07
72.56	80.10	83.78	84.72	90.10	92.40	92.49	85.71
84.67	88.57	90.07	88.70	96.86	100.64	101.08	101.93
101.16	101.77	104.20	103.29	105.45	101.89	101.19	104.66
103.81	100.96	100.57	99.05	98.76	99.29	100.18	97.75
96.23	93.74	94.28	93.82	90.25	88.98	88.82	92.71
97.41	97.66	97.19	103.92	103.86	100.58	94.71	96.11
99.71	98.23	100.11	102.07	99.73	101.73	102.71	107.36
108.60	104.47	103.66	107.78	110.87	115.34	104.69	102.97
107.69	114.55	119.83	123.50	126.51			

Series 27. Vendor Performance (monthly, 1/47–9/80; n = 405; read across).
Source: U.S. Commerce Department, Business Conditions Digest:

37	42	34	25	20	24	22	25	30	31	38	41	35	34	26	36
31	30	36	36	38	38	32	17	16	12	10	14	12	12	22	38
53	60	58	50	54	62	60	60	66	64	88	94	96	88	87	84
84	85	74	58	46	38	34	38	50	50	34	31	28	22	18	19
23	34	50	47	46	46	44	40	37	37	40	38	36	34	30	30
25	22	20	20	21	24	27	30	35	36	40	41	47	53	52	50
54	60	66	71	70	65	70	72	72	72	66	56	48	46	49	50
39	40	56	52	47	44	46	36	33	28	25	28	30	29	38	34
32	36	29	25	28	28	32	34	38	38	44	49	57	58	58	52
58	62	62	62	62	62	60	62	64	64	56	50	44	30	27	28
32	34	36	40	41	39	38	38	38	40	40	47	48	48	49	52
55	55	51	53	56	56	55	48	46	42	44	44	48	48	48	48
50	52	54	60	58	54	42	48	52	48	48	46	55	54	60	60
63	55	59	65	74	72	70	66	68	72	66	72	70	66	62	64
62	60	66	72	74	85	86	82	75	69	70	73	72	70	64	57
48	51	38	39	36	38	41	43	44	50	51	48	50	55	54	52
52	52	56	46	46	52	60	56	62	61	61	68	69	70	66	68
66	65	62	64	56	58	50	52	72	69	50	45	45	38	36	36
38	44	46	52	53	50	48	49	48	50	48	51	52	52	58	58
60	60	63	63	65	73	70	77	78	84	88	90	92	89	88	88
90	90	91	88	85	88	88	84	79	76	72	68	52	46	32	22
18	16	17	22	24	26	30	36	44	45	44	39	42	50	52	58
58	62	60	64	60	50	48	45	44	55	56	58	56	58	59	58
56	56	50	56	55	64	67	64	64	66	56	65	66	68	66	68
69	77	78	76	76	70	60	55	51	50	47	49	48	42	45	40
32	28	32	34	39											

Series 28. Cincinnati Average Daily Calls to Directory Assistance (hundreds; monthly, 1/62–12/76; n = 180; read across). Source: McCleary and Hay (1980):

350	339	351	364	369	331	331	340	346	341	357	398
381	367	383	375	353	361	375	371	373	366	382	429
406	403	429	425	427	409	402	409	419	404	429	463
428	449	444	467	474	463	432	453	462	456	474	514
489	475	492	525	527	533	527	522	526	513	564	599
572	587	599	601	611	620	579	582	592	581	630	663
638	631	645	682	601	595	521	521	516	496	538	575
537	534	542	538	547	540	526	548	555	545	594	643
625	616	640	625	637	634	621	641	654	649	662	699
672	704	700	711	715	718	652	664	695	704	733	772
716	712	732	755	761	748	748	750	744	731	782	810
777	816	840	868	872	811	810	762	634	626	649	697
657	549	162	177	175	162	161	165	170	172	178	186
178	178	189	205	202	185	193	200	196	204	206	227
225	217	219	236	253	213	205	210	216	218	235	241

Series 29. Perceptual Speed Scores for a Schizophrenic Patient (daily; n = 120; drug regimen started on day 61; read across). Source: McCleary and Hay (1980):

55	56	48	46	56	46	59	60	53	58	73	69	72	51	72	69	68	69	79	77	53	63
80	65	78	64	72	77	82	77	35	79	71	73	77	76	83	73	78	91	70	88	88	85
77	63	91	94	72	83	88	78	84	78	75	75	86	79	76	87	66	73	62	27	52	47
65	59	77	47	51	47	49	54	58	56	50	54	45	66	39	51	39	27	39	37	43	41
27	29	27	26	29	31	28	38	37	26	31	45	38	33	33	25	24	29	37	35	32	31
28	40	31	37	34	43	38	33	28	35												

Series 30. Air Carrier Freight (ton-miles; monthly, 1/69–12/80; n = 144; read across). Source: Pankratz (1983, Figure C9.1 and Table C9.1):

1299	1148	1345	1363	1374	1533	1592	1687	1384	1388	1295
1489	1403	1243	1466	1434	1520	1689	1763	1834	1494	1439
1327	1554	1405	1252	1424	1517	1483	1605	1775	1840	1573
1617	1485	1710	1563	1439	1669	1651	1654	1847	1931	2034
1705	1725	1687	1842	1696	1534	1814	1796	1822	2008	2088
2230	1843	1848	1736	1826	1766	1636	1921	1882	1910	2034
2047	2195	1765	1818	1634	1818	1698	1520	1820	1689	1775
1968	2110	2241	1803	1899	1762	1901	1839	1727	1954	1991
1988	2146	2301	2338	1947	1990	1832	2066	1952	1747	2098
2057	2060	2240	2425	2515	2128	2255	2116	2315	2143	1948
1460	2344	2363	2630	2811	2972	2515	2536	2414	2545	2445
2275	2857	2601	2593	2939	3149	3333	2650	2764	2608	2668

DATA APPENDIX

2536 2415 2883 2635 2665 2914 3050 3236 2540 2629 2379
2590

Series 31. Shipments of Consumer Product (monthly index; 1/76–5/86; n = 125; read across):

100	98	93	109	121	114	136	134	156	120	145
153	122	148	169	159	144	152	191	158	194	175
209	190	197	178	229	205	193	223	246	223	232
257	313	273	260	241	229	279	240	285	310	308
319	317	299	324	349	310	371	295	331	331	344
352	362	368	438	334	352	350	362	381	378	421
394	406	426	390	436	449	435	411	458	447	461
470	449	503	533	497	538	531	501	570	482	523
564	528	540	552	598	587	602	702	481	588	571
588	615	626	641	659	642	633	644	723	545	637
609	633	636	660	693	665	644	660	676	822	572
694	692	720	744							

References

Abraham, B. and J. Ledolter (1983). *Statistical Methods for Forecasting*, New York: Wiley.

Akaike, H. (1973). "Information Theory and an Extension of the Maximum Likelihood Principle," in B. N. Petrov and F. Csaki (eds.), *Proceedings, 2nd International Symposium on Information Theory*, Budapest: Akademiai Kiado, pp. 267–281.

Akaike, H. (1974). "A New Look at Statistical Model Identification," *IEEE Transactions on Automatic Control*, AC-19, 716–723.

Baker, G. A., Jr. (1975). *Essentials of Padé Approximants*, New York: Academic Press.

Bartlett, M. S. (1946). "On the Theoretical Specification and Sampling Properties of Autocorrelated Time Series," *Journal of the Royal Statistical Society, Series B*, 8, 27–41.

Bartlett, M. S. (1966). *An Introduction to Stochastic Processes with Special Reference to Methods and Applications*, 2nd ed., Cambridge, England: Cambridge University Press.

Bell, W. R. and S. C. Hillmer (1983). "Modeling Time Series with Calendar Variation," *Journal of the American Statistical Association*, 78, 526–534.

Box, G. E. P. and D. R. Cox (1964). "An Analysis of Transformations," *Journal of the Royal Statistical Society, Series B*, 26, 211–243; discussion 244–252.

Box, G. E. P. and G. M. Jenkins (1976). *Time Series Analysis: Forecasting and Control*, rev. ed., San Francisco: Holden-Day.

Box, G. E. P. and J. F. MacGregor (1974). "The Analysis of Closed-Loop Dynamic-Stochastic Systems," *Technometrics*, 16, 391–398.

Box, G. E. P. and G. C. Tiao (1975). "Intervention Analysis with Applications to Economic and Environmental Problems," *Journal of the American Statistical Association*, 70, 70–79.

Caines, P. E., S. P. Sethi, and T. W. Brotherton (1977). "Impulse Response Identification and Causality Detection for the Lydia Pinkham Data," *Annals of Economic and Social Measurement*, 6, 147–163.

Chang, I. (1982). "Outliers in Time Series," unpublished Ph.D. Dissertation, Department of Statistics, University of Wisconsin, Madison.

Chang, I. and G. C. Tiao (1983a). "Effect of Exogenous Interventions on the Estimation of Time Series Parameters," *Proceedings of the Business and Economic Statistics Section*, Washington, D.C.: American Statistical Association, pp. 532–537.

REFERENCES

Chang, I. and G. C. Tiao (1983b). "Estimation of Time Series Parameters in the Presence of Outliers," Technical Report 8, Statistical Research Center, University of Chicago.

Chang, I., G. C. Tiao, and C. Chen (1988). "Estimation of Time Series Parameters in the Presence of Outliers," *Technometrics*, 30, 193–204.

Chen, C., L.-M. Liu, and G. C. Tiao (1990). "Robust Estimation in Time Series: A Parametric Approach," manuscript.

Clarke, D. G. (1976). "Econometric Measurement of the Duration of the Advertising Effect on Sales," *Journal of Marketing Research*, 13, 345–357.

Denby, L. and R. D. Martin (1979). "Robust Estimation of the First-Order Autoregressive Parameter," *Journal of the American Statistical Association*, 74, 140–146.

Dhrymes, P. J. (1981). *Distributed Lags: Problems of Estimation and Formulation*, 2nd ed., New York: North-Holland.

Granger, C. W. J. (1969). "Prediction with a Generalized Cost of Error Function," *Operational Research Quarterly*, 20, 199–207.

Granger, C. W. J. and P. Newbold (1976). "Forecasting Transformed Series," *Journal of the Royal Statistical Society, Series B*, 38, 189–203.

Granger, C. W. J. and P. Newbold (1986). *Forecasting Economic Time Series*, 2nd ed., San Diego: Academic Press.

Guerrero, V. M. (1990). "Using Power Transformations in Applied Time Series Analysis," manuscript, Department of Statistics, Instituto Tecnológico Autónomo de México (ITAM), México 01000, D.F.

Gujarati, D. N. (1988). *Basic Econometrics*, New York: McGraw-Hill.

Guttman, I. and G. C. Tiao (1978). "Effect of Correlation on the Estimation of a Mean in the Presence of Spurious Observations," *The Canadian Journal of Statistics*, 6, 229–247.

Helmer, R. M. and J. K. Johansson (1977) . "An Exposition of the Box–Jenkins Transfer Function Analysis with an Application to the Advertising–Sales Relationship," *Journal of Marketing Research*, 14, 227–239.

Hillmer, S. C. (1982). "Forecasting Time Series with Trading Day Variation," *Journal of Forecasting*, 1, 385–395.

Hillmer, S. C. (1984). "Monitoring and Adjusting Forecasts in the Presence of Additive Outliers," *Journal of Forecasting*, 3, 205–215.

Hillmer, S. C. and G. C. Tiao (1979). "Likelihood Function of Stationary Multiple Autoregressive-Moving Average Models," *Journal of the American Statistical Association*, 74, 652–660.

Hillmer, S. C., D. F. Larcker, and D. A. Schroeder. (1983). "Forecasting Accounting Data: A Multiple Time-Series Analysis," *Journal of Forecasting*, 2, 389–404.

Jenkins, G. M. and A. S. Alavi (1981). "Some Aspects of Modelling and Forecasting Multivariate Time Series," *Journal of Time Series Analysis*, 2, 1–47.

Jenkins, G. M. and D. W. Watts (1968). *Spectral Analysis and its Applications*, San Francisco: Holden-Day.

Johnston, J. (1984). *Econometric Methods*, New York: McGraw-Hill.

Jorgenson, D. W. (1966). "Rational Distributed Lag Functions, " *Econometrica*, 32, 135–149.

Judge, G. G., R. C. Hill, W. E. Griffiths, H. Lütkepohl, and T.-C. Lee (1982). *Introduction to the Theory and Practice of Econometrics*, New York: Wiley.

Kennedy, P. (1985). *A Guide to Econometrics*, Cambridge, MA: MIT Press.

Kmenta, J. (1986). *Elements of Econometrics*, 2nd ed., New York: Macmillan.

Koch, P. D. and R. H. Rasche (1988). "An Examination of the Commerce Department Leading Indicator Approach," *Journal of Business & Economic Statistics*, 6, 167–187.

Koreisha, S. G. and S. A. Taylor (1985). "Identification of Transfer Function Models: An Asymptotic Test of Significance for the Corner Method," *Communications in Statistics, Part A—Theory and Methods*, 14, 159–173.

Koyck, L. M. (1954). *Distributed Lags and Investment Analysis*, New York: North-Holland.

Kyle, P. W. (1978). "Lydia Pinkham Revisited: A Box–Jenkins Approach," *Journal of Advertising Research*, 18, 32–39.

Ledolter, J. (1988). "The Effect of Additive Outliers on the Forecasts from ARIMA Models," *International Journal of Forecasting*, 5, 231–240.

Ledolter, J. and B. Abraham (1981). "Parsimony and Its Importance in Time Series Forecasting," *Technometrics*, 23, 411–414.

Liu, L.-M. (1984). "Estimation of Rational Transfer Function Models," *Communications in Statistics—Simulation and Computation*, 13, 775–784.

Liu, L.-M. (1986a). "Identification of Time Series Models in the Presence of Calendar Variation," *International Journal of Forecasting*, 3, 357–372.

Liu, L.-M. (1986b). "Multivariate Time Series Analysis Using Vector ARMA Models," Lincoln Center, Suite 106, 4513 Lincoln Avenue, Lisle, IL: Scientific Computing Associates.

Liu, L.-M. and D. M. Hanssens (1982). "Identification of Multiple Input Transfer Function Models," *Communications in Statistics, Part A—Theory and Methods*, 11, 297–314.

Liu, L.-M. and G. B. Hudak (1986). *The SCA Statistical System, Reference Manual for Forecasting and Time Series Analysis, Version III*, Lincoln Center, Suite 106, 4513 Lincoln Avenue, Lisle, IL: Scientific Computing Associates.

Ljung, G. M. and G. E. P. Box (1978). "On a Measure of Lack of Fit in Time Series Models," *Biometrika*, 65, 297–303.

McCleary, R. and R. A. Hay, Jr. (1980). *Applied Time Series Analysis for the Social Sciences*, Beverly Hills, CA: Sage Publications.

McLeod, G. (1982). *Box Jenkins in Practice*, Lancaster, UK: A GJP Publication.

Marquardt, D. W. (1963). "An Algorithm for Least Squares Estimation of Nonlinear Parameters," *Journal of the Society for Industrial and Applied Mathematics*, 11, 431–441.

Melicher, R. W., J. Ledolter, and L. J. D'Antonio (1983). "A Time Series Analysis of Aggregate Merger Activity," *Review of Economics and Statistics*, 65, 423–429.

Neftçi, S. N. (1979). "Lead-lag Relations, Exogeneity and Prediction of Economic Time Series," *Econometrica*, 47, 101–113.

Nelson, C. R. (1973). *Applied Time Series Analysis for Managerial Forecasting*, San Francisco: Holden-Day.

Nelson, H. R. and C. W. J. Granger (1979). "Experience with Using the Box–Cox Transformation when Forecasting Economic Time Series," *Journal of Econometrics*, 10, 57–69.

Nerlove, M. (1958). *Distributed Lags and Demand Analysis for Agricultural and Other Commodities*, Agricultural Handbook No. 141, Washington, D.C.: U.S. Department of Agriculture.

Newbold, P. (1973). "Bayesian Estimation of Box–Jenkins Transfer Function-Noise Models," *Journal of the Royal Statistical Society, Series B*, 35, 323–336.

Öller, L.-E. (1985). "Macroeconomic Forecasting with a Vector ARIMA Model: A Case Study of the Finnish Economy," *International Journal of Forecasting*, 1, 143–150.

Pankratz, A. (1983). *Forecasting with Univariate Box–Jenkins Models: Concepts and Cases*, New York: Wiley.

Pankratz, A. and U. Dudley (1987). "Forecasts of Power-Transformed Series," *Journal of Forecasting*, 6, 239–248.

Parzen, E., ed. (1984). *Time Series Analysis of Irregularly Observed Data*, Lecture Notes in Statistics, Vol. 25, New York: Springer, Verlag.

Pierce, D. A. (1979). "R^2 Measures for Time Series," *Journal of the American Statistical Association*, 74, 901–910.

Pindyck, R. S. and D. R. Rubinfeld (1991). *Econometric Models and Economic Forecasts*, 3rd ed., New York: McGraw-Hill.

Rao, C. R. (1965). *Linear Statistical Inference and Its Applications*, New York: Wiley.

Shibata, R. (1976). "Selection of the Order of an Autoregressive Model by Akaike's Information Criterion," *Biometrika*, 63, 117–126.

Tiao, G. C. and G. E. P. Box (1981). "Modeling Multiple Time Series with Applications," *Journal of the American Statistical Association*, 76, 802–816.

Tiao, G. C. and W. W. S. Wei (1976). "Effect of Temporal Aggregation on the Dynamic Relationship of Two Time Series Variables," *Biometrika*, 63, 513–523.

Tsay, R. S. (1984). "Order Selection in Nonstationary Autoregressive Models," *Annals of Statistics*, 12, 513–523.

Tsay, R. S. (1985). "Model Identification in Dynamic Regression (Distributed Lag) Models," *Journal of Business & Economic Statistics*, 3, 228–237.

Tsay, R. S. (1986). "Time Series Model Specification in the Presence of Outliers," *Journal of the American Statistical Association*, 81, 132–141.

Tsay, R. S. (1988). "Outliers, Level Shifts, and Variance Changes in Time Series," *Journal of Forecasting*, 7, 1–20.

Tsay, R. S. and G. C. Tiao (1984). "Consistent Estimates of Autoregressive Parameters and Extended Sample Autocorrelation Function for Stationary and Nonstationary ARMA Models," *Journal of the American Statistical Association*, 79, 84–96.

Vandaele, W. (1983). *Applied Time Series and Box–Jenkins Models*, New York: Academic Press.

Wall, K. D. (1976). "FIML Estimation of Rational Distributed Lag Structural Form Models," *Annals of Economic and Social Measurement*, 5, 53–64.

Wei, W. W. S. (1982). "The Effects of Systematic Sampling and Temporal Aggregation on Causality—a Cautionary Note," *Journal of the American Statistical Association*, 77, 316–319.

Wei, W. W. S. (1990). *Time Series Analysis*, Redwood City, CA; Addison-Wesley.

Weisberg, S. (1980). *Applied Linear Regression*, New York: Wiley.

Wichern, D. W. and R. H. Jones (1977) . "Assessing the Impact of Market Disturbances Using Intervention Analysis," *Management Science*, 24, 329–337.

Zimring, F. E. (1975). "Firearms and Federal Law: The Gun Control Act of 1968," *Journal of Legal Studies*, 4, 133–198.

Index

Abraham, B., 17, 135, 217, 364, 370, 371, 376, 378
ACF, *see* Autocorrelation function, theoretical
Additive outlier (AO), 290. *See also* Intervention effects, pulse
   defined and explained, 293
   example, 301, 302, 317
   initial effect of, 299
   test for occurrence of, 299, 301, 312, 313
Akaike, H., 350, 376
Akaike information criterion:
   example, 351, 353
   formula, 350
Alavi, A. S., 11, 342, 377
ARIMA models, 11, 25, 39
   backshift notation for, 53
   as baseline for comparison, 27, 134, 169, 219, 234
   and dynamic regression models, 26, 235
   and prewhitening, 199
   as proxy for disturbance series in dynamic regression, 175, 176, 180, 198, 240, 264, 265, 318
ARIMA process, *see* Autoregressive integrated moving average (ARIMA) process
Autocorrelation coefficients, 34
   estimated:
      defined and explained, 36
      formula, 36
      standard error of, 36
   theoretical:
      defined and explained, 35
      formula, 36
Autocorrelation function:
   estimated:
      and dynamic regression disturbance series, 174, 178
      graphical form, 37
      and stationarity of mean, 37, 47
   theoretical, 39
      AR(1), 40
      AR(2), 42
      $AR(1)_{s=4}$, 56
      ARMA(1,1), 45
      MA(1), 44
      MA(2), 45
      $MA(1)_{s=4}$, 57
Autoregressive integrated moving average (ARIMA) process, 11, 25, 39
   AR(1), 39
   AR(2), 39
   ARIMA(0,1,1), 46
   $ARIMA(p,d,q)(P,D,Q)_s$, 58
   ARMA(1,1), 43
   MA(1), 43
   MA(2), 43

Backshift notation, *see* Notation
   for ARIMA models, 52
   for c-weight function, 299
   for linear distributed lag transfer functions, 150
   for π-weight function, 299
   for VARMA models, 344
Baker, G. A., Jr., 156, 194, 376
Bartlett, M. S., 36, 199, 376
Bell, W. R., 115, 117, 376
Binary indicator variables, *see* Deterministic inputs; Dummy variables; Intervention analysis; Intervention effects
   defined and illustrated, 4
   example, 254, 259
Box, G. E. P., 11, 15, 24, 40, 60, 158, 160, 165, 169, 173, 175, 187, 190, 197, 199, 204, 207, 253, 266, 324, 328, 338, 342, 363, 368, 376, 378, 379
Box–Cox transformation:
   and forecasting in original metric, 336, 338, 340
   formula, 29
   to induce stationary variance, 29, 169
   and natural log transformation, 29, 69
Brotherton, T. W., 376

Caines, P. E., 352, 376
Causation, 96, 171
Chang, I., 266, 299, 313, 314, 376, 377
Change in business inventories data, 25, 34

381

382　INDEX

Chen, C., 314, 377
Clarke, D. G., 352, 377
Contamination in time series, 293, 296. *See also* Intervention analysis; Intervention effects; Outliers
Corner table:
　derivation of elements, 195
　explained and illustrated, 194
Correlation, 17. *See also* Autocorrelation function; Correlation matrix of parameter estimates; Partial autocorrelation function
　population coefficient, 18
　sample coefficient, 20
　test for significance, 20
Correlation matrix of parameter estimates, 124, 140, 212, 224, 227, 246, 270, 281, 284–285
Covariance:
　population, 19
　and residual cross correlations, 204
　sample, 20
Cox, D. R., 29, 69, 169, 336, 338, 340, 376
Cross correlation function, 197
　estimated, 199
　theoretical, 198
Cross-section data, 82
c weights (in outlier detection), 299, 301, 308, 309, 315
Cycles, stochastic, 66, 78, 115, 142, 169, 236, 252. *See also* Trading day pattern

D'Antonio, L. J., 378
Dead time, 149
　examples, 153, 159, 243, 244
　identification of, 185, 187, 354
　and VAR model, 354
Decay rate (of distributed lag regression coefficients), *see* Impulse response function
　examples, 154, 159, 243, 244, 258
　explained, 152
　identification of, 185, 187
　initial values for estimation, 325
　and intervention models, 271, 272, 294, 298
　start-up values, 152, 154, 185, 188, 243, 244
　and steady state gain, 160
　and transfer function stability, 161, 258
δ weights, *see* Decay rate
Denby, L., 291, 373
Deterministic inputs, 7. *See also* Binary indicator variables; Dummy variables; Interventional analysis; Intervention effects
Deterministic (time) trend, *see also* Year-end loading
　case study and forecasts, 68
　interpretation of, 69
Dhrymes, P., 156, 377
Diagnostic checking stage, 16, 49, 202
　for ARIMA models:
　　chi-squared test, 51, 77, 125
　　Durbin-Watson statistic, 100
　　goodness of fit ($\bar{R}^2$), 89
　　and model reformulation, 76

　　normality of random shocks, 51, 77
　　residual autocorrelation function, 76, 101, 119, 121, 125, 136
　　residual extended autocorrelation function, 101, 122
　　residual partial autocorrelation function, 101
　for dynamic regression models, 202
　　chi-squared test, 206, 207
　　goodness of fit, 214
　　intervention models, 266
　　and model reformulation, 205
　　residual autocorrelation function, 206, 267
　　residual cross correlations, 203, 205, 226
　　residual extended autocorrelation function, 206
　　residual partial autocorrelation function, 206
Differencing:
　nonseasonal, 30
　　and dynamic regression disturbance series, 119
　　example, 136, 154
　　first-degree, 30, 75, 236
　　and integrated processes, 47
　　and mean of a series, 30, 72, 136, 177, 211
　　second-degree, 30
　seasonal, 32
　　example, 222
　　first-degree, 121
　　and mean of a series, 32, 72, 137
Disturbance series, 8, 99, 100, 103, 107, 115. *See also* ARIMA models, as proxy for disturbance series in dynamic regression
　assumed properties in regression, 84
　autocorrelated, 11, 14, 98, 165, 175
　computation at estimation stage, 327
　and heteroskedasticity, 98
　and innovational outliers, 290
　and seasonal differencing, 138
DR, *see* Dynamic regression model
Dudley, U., 338, 379
Dummy variables, *see* Deterministic inputs; Binary indicator variables; Intervention analysis; Intervention effects
　and intervention analysis, 253
　seasonal, 103
　joint significance test, 112
Durbin, J., 100
Durbin–Watson statistic, 100
Dynamic regression (DR) models, 7, 137, 147, 163. *See also* Rational distributed lag models; Transfer function; Vector ARMA models
　and ARIMA analysis, 26, 128, 147, 163, 169, 235
　complete model, 164
　contemporaneous relationship within, 347
　and intervention analysis, 290
　minimum mean squared error forecasts, 339
　model building, 15
　model identification, *see* Identification stage, dynamic regression models
　multiple inputs, 232, 325
　as ordinary regression model, 8, 147, 164

INDEX

EACF, *see* Extended autocorrelation function
Equating coefficients of like powers of backshift operator, 161, 331, 333
ESACF, *see* Extended autocorrelation function
Estimation stage, 16, 76, 118, 120, 121, 123, 124, 135, 137, 324
   and ARIMA models (example), 76
   and dynamic regression models, 202, 220, 245, 324–328
   and intervention models, 266, 281
   and least squares estimation, 87
η weights (in deriving forecast error variance), 333
Exponentially weighted moving average (EWMA) models, 46, 55, 127, 135, 142, 220
   forecasts from, 65, 127
Extended autocorrelation function (EACF), 43, 62, 63, 75, 179, 183, 223, 282

Federal government receipts case study:
   ARIMA model, 72
   dynamic regression model, 115
Federal government receipts data, 28
Feedback:
   in dynamic regression models, 170, 172, 173
   examples, 217, 236
   explanation, 10
   problems caused by, 10
   and vector ARMA models, 343, 349, 353–354
First differences, *see* Differencing
Forecast error variance, 334
Forecasting, *see also* Exponentially weighted moving average (EWMA) models, forecasts from
   with ARIMA models, 60
      with deterministic (time) trend component, 68
      nonstationary, 61
      stationary, 61
      with stochastic cycle component, 66
      as weighted average, 24, 64
   with dynamic regression models, 328, 339
      computations, 329
   error loss function, 337
   interval forecasts, 59
      computations, 60, 80, 336, 144
      and forecast error variance, 333
      with logarithmic data, 144
   point forecasts, 59
      computations, 60, 79, 112, 128, 143, 250, 330
      dynamic regression *vs.* ARIMA (example), 230
      with logarithmic data, 128, 143
   transformed series, 336
Forecast leads time, 329
Forecast origin, 329
Forward shift operator, 308
Free form distributed lag transfer function models, 164, 174, 175, 220. *See also* Rational distributed lag models
F statistic, 93, 109
   and test for feedback, 172, 236

Granger, C. W. J., 12, 171, 338, 342, 377, 379
Granger causality, 171
Grittiths, W. E., 378
Guerrero, V. M., 338, 377
Gujarati, D., 82, 88, 96, 103, 164, 165, 377
Guttman, I., 266, 377

Hanssens, D., 176, 194, 378
Hay, R. A., Jr., 276, 374, 378
Heights data, 17
Helmer, R. M., 352, 377
Hill, R. C., 378
Hillmer, S. C., 115, 117, 126, 217, 320, 325, 328, 376, 377
Housing starts data (quarterly), 104
Housing starts and sales case study, 217
Hudak, G., 11, 151, 176, 209, 313, 324, 378

Identification stage, 16, 49
   for ARIMA models, 27
      using extended sample autocorrelation function, 62, 75
      using sample autocorrelation function, 37–45, 49–50, 55–57, 72–73
      using sample partial autocorrelation function, 38–45, 49–50, 55–57, 72, 74
   for dynamic regression models, 167, 173
      using corner table, 194
      using cross correlations, 198
      using extended sample autocorrelation function of disturbance series, 179, 241
      and feedback, 170, 172, 236
      and inspection for outliers, 168
      using linear transfer function (LTF) method, 174, 237, 240, 244, 271, 280
      using prewhitening, 199
      and relevant theory, 170
      using rules for choosing rational lag transfer function, 184
      using sample autocorrelation function of disturbance series, 174, 178–179, 237, 240, 242
      using sample partial autocorrelation function of disturbance series, 174, 178, 179, 242
      and stationarity of mean, 177, 180
      and transformation for constant variance, 169
Impulse response function, 149, 150, 157. *See also* Decay rate
   and corner table, 195, 197
   dead time, 158, 185, 187
   decay rate (δ weights), 152, 160, 161, 185, 187
   decay start-up values, 152, 185, 188
   estimated, 156, 159, 182, 188, 223, 224, 242, 244, 246, 247, 248, 269, 280
   and forecast computation, 329
   and intervention models, 256, 257, 261, 262
   and Koyck model, 151
   and linear transfer function (LTF) identification method, 174, 175, 178, 181, 240, 244, 270, 284
   numerator coefficients (ω weights), 158, 161, 185, 189

Impulse response function (*Continued*)
  obtained from rational form transfer function, 161, 329
  and steady state gain, 160
  theoretical, 151, 158, 185–187, 192
  unpatterned values, 185, 189
Impulse response weights, *see* Impulse response function
Industrial production, stock prices, and vendor performance case study, 232
Innovational outlier (IO), 290
  defined and explained, 296
  example, 306, 317
  initial effects, 299
  test for occurrence of, 299, 301, 312, 313
Integrated processes, 46
Intervention analysis:
  ARIMA component, 264–266
  and detected events near end of series, 319
  and estimation stage, 266
  iterative detection procedure, 313, 321
  and linear transfer function (LTF) identification method, 268
  model building, 264, 291
  and outlier detection, 290, 291, 299, 301
  transfer function component, 264
Intervention effects, 253
  compound, 273, 274
  multiple, 272, 273
  pulse, 254, 272, 290
    defined and explained, 254
    dynamic response, 256, 258
    temporary response, 255, 294
  ramp, 263, 272
  step, 259, 271, 272, 290, 295
    defined and explained, 259
    multiperiod permanent response, 260
    one-period permanent response, 260
Invertibility:
  conditions for higher-ordered MA models, 59
  condition(s) for MA(1) and MA(2), 46
  and DR models, 211, 245, 285
  example MA(1), 123
  reason for requirement, 45

Jenkins, G. M., 11, 15, 24, 36, 40, 60, 158, 160, 165, 173, 175, 187, 190, 197, 199, 204, 266, 324, 328, 342, 363, 368, 376, 377
Johansson, J. K., 352, 377
Johnston, J., 29, 377
Jones, R. H., 253, 380
Jorgenson, D. W., 156, 377
Judge, G. G., 369

Kennedy, P., 23, 378
Kilowatt-hours used, case study, 131
Kmenta, J., 82, 195, 378
Koch, P. D., 192, 378
Koreisha, S. G., 195, 378
Koyck, L. M., 150, 378
Koyck model:
  examples, 154, 155, 162, 182, 184, 214
  explained, 150
  and pulse interventions, 258
  in rational form, 156
  and sequential identification of intervention models, 271
  stability conditions, 161
  and step interventions, 263
Kyle, P. W., 352, 378

Larcker, D. F., 377
Leading indicator, 6, 13. *See also* Industrial production, stock prices, and vendor performance case study, 232
Least-squares estimation:
  and autocorrelated disturbance, 98
  and colinearity, 97
  and dynamic regression analysis, 98, 100
  and excluded variables, 102
  formulas for regression estimators, 87
  and heteroskedastic disturbance, 98
  matrix notation, 95
  and outlier detection, 299, 308
  properties of regression estimators, 88
  variance of repeated sampling distribution of estimates, 89, 96
Ledolter, J., 17, 135, 217, 364, 370, 371, 376, 378
Lee, T. C., 378
Level shifts, 290. *See also* Intervention, step
  defined and explained, 295
  example, 304, 317
  gradual, 320
  initial effect, 299
  test for occurrence of, 299, 301, 312, 313
Likelihood ratio criteria:
  and outlier detection, 299, 312–314, 316, 319, 321
Linear transfer function identification method, *see* Identification stage, for dynamic regression models; Transfer function
Liu, L.-M., 11, 51, 115, 176, 194, 209, 313, 324, 328, 342, 377, 378
Ljung, G., 51, 207, 378
Ljung–Box approximate chi-squared statistic, 51, 207
Logarithmic transformation, and stationary variance, 28, 30, 133, 169, 234
Lütkepohl, H., 378
Lydia Pinkham's vegetable compound data, 352

McCleary, R., 276, 374, 378
McLeod, G., 367
MacGregor, J. F., 11, 376
Marquardt, D. W., 327, 378
Martin, R. D., 291, 377
Maximum likelihood estimation:
  conditional likelihood function, 324
  when moving average coefficients are near invertibility boundary, 324
  in the SCA System, 324
Mean, *see* Stationarity, mean
  population (process), 19

INDEX 385

sample (realization), 19
weighted, 24
Melicher, R. W., 354, 374
Minimum mean squared error forecasts, 339
Model building strategies:
  for ARIMA models, 49
  Box–Jenkins, 15, 25, 152, 167, 173, 266
  for dynamic regression models, 152, 167, 173
  linear transfer function (LTF) method, 173

Neftçi, S. N., 232, 378
Nelson, C. R., 337, 379
Nelson, H. R., 338, 379
Nerlove, M., 147, 379
Newbold, P., 12, 171, 328, 342, 377, 379
Notation, *see also* Backshift notation
  for ARIMA processes, 40, 43, 53, 54, 55, 58
  for impulse response function, 150
  for least squares regression, 95
  for rational distributed lag models, 157
  for time-ordered variables, 5
  for vector ARMA processes, 342

Öller, L.-E., 354, 379
ω weights, 157
  and impulse response weights, 158, 161, 185, 189, 243, 244
  initial values (at estimation stage), 325
  in intervention models, 275, 276, 294, 298, 308
  and point forecasts, 329
  and steady state gain, 160
  variance of estimates in outlier detection, 311, 312
Ordinary least squares regression (OLS), *see* Regression
Outlier detection, 290. *See also* Intervention analysis; Intervention effects; Outliers
Outliers:
  checks for, 207, 208
  defined and explained, 168
  and intervention analysis, 253, 291

PACF, *see* Partial autocorrelation function, theoretical
Padé approximations, 194
Pankratz, A., 24, 40, 60, 338, 362, 364, 365, 374, 379
Parsimony, 17, 49
  and dynamic regression models, 150, 210
Partial autocorrelation function:
  estimated, 38, 174, 178
    standard errors, 38
  theoretical, 38, 39, 40, 42, 44, 45, 56, 57
Parzen, E., 129, 379
Pierce, D. A., 90, 379
Pindyck, R. S., 171, 379
π weights, 299, 308, 309, 315, 351, 345, 353
  and vector AR model, 345
Population regression function (PRF):
  with multiple inputs, 93
  with one input, 82, 118
  prediction with, 85, 94
  standard assumptions, 84

Prewhitening, 197
  and cross correlations, 199
ψ weights, 333, 334
Pulse interventions, *see* Intervention analysis; Intervention effects, pulse

Rao, C. R., 88, 379
Rasche, R. H., 192, 378
Rational distributed lag transfer function models, 147, 156, 163, 174, 184. *See also* Free form distributed lag transfer function models; Impulse response function; Koyck model; Transfer function
  and corner table, 196
  defined and explained, 147
  examples, 223, 246
  identification rules, 184
  with multiple inputs, 163
Regression, 8–12, 82. *See also* Dynamic regression (DR) models; Least squares estimation; Population regression function (PRF); Sample regression function (SRF)
Residual autocorrelation function, 50, 119, 121, 124, 125, 136, 138, 206, 221, 222, 226, 227, 228, 248, 267, 282, 284. *See also* Diagnostic checking stage
Residual cross correlation function, *see* Diagnostic checking stage, dynamic regression models
  examples, 226, 228, 246, 247
  explained, 203
  computations, 203
  standard error, 204
Residual partial autocorrelation function, 101, 122, 139, 206. *See also* Diagnostic checking stage
Residuals:
  histograms and normal probability plots of standardized values, 77, 78, 79, 125, 126, 141, 207, 208, 229, 249, 268, 286
  and outliers, 207, 228, 285, 301, 304, 306, 313
  starting values in estimation, 327, 328
Residual standard deviation, 88, 137
Retransformation to original metric:
  and forecast error loss function, 337
  and forecasting, 336
  to mean, 337
  to median, 336
Root mean squared forecast error, 80
$\bar{R}^2$ statistic:
  computations, 91
  and goodness of fit, 89
Rubinfeld, D. R., 171, 379

SACF, *see* Autocorrelation function, estimated
Sample regression function (SRF):
  with multiple inputs, 94
  with one input, 86
Saving rate data (quarterly), 254
Schroeder, D. A., 377
Seasonal ARIMA models, 54
  differencing, 32
  theoretical autocorrelation function, 55

Season ARIMA models (*Continued*)
 theoretical partial autocorrelation function, 55
Second differences, *see* Differencing
Sethi, S. P., 376
Shibata, R., 351, 379
Single equation models, *see* Dynamic regression models
SPACF, *see* Partial autocorrelation function, estimated
Square root transformation, 28, 133, 169
Stationarity:
 of ARIMA process, 27
 and integrated processes, 46
 of mean, 27. *See also* Differencing
  condition(s) for AR(1) and AR(2), 41, 59
  conditions for higher-ordered AR models, 59
  in dynamic regression models, 177, 211
  and linear transfer function identification method, 177
  practical checks for, 30, 47, 137
  reason for requirement, 27
 and sample autocorrelation function behavior, 37
 strong form, 27
 of variance, 27, 169. *See also* Box–Cox transformation; Logarithmic transformation; Square root transformation
  in dynamic regression models, 211
  reason for requirement, 27
  transformation to achieve, 28, 133
 weak form, 27
Steady-state gain, 160, 258, 263
Step interventions, *see* Intervention analysis; Intervention effects, step
Sum of squared residuals, 327

Taylor, S. A., 195, 378
Temporal aggregates, 11, 171, 218
Tiao, G. C., 11, 16, 62, 171, 217, 253, 266, 325, 328, 342, 354, 376, 377, 379
Time-lagged relationships, 9
Time-ordered variables, 1
 examples, 2, 3
 as input, 7
 as output, 7
Time series analysis, 1, 7
Trading day pattern in monthly data:
 data transformation, 117
 defined and explained, 115
 test for significance in regression model, 127
Transfer function:
 defined and explained, 148
 and intervention analysis, 264, 268, 275, 280, 293, 318

Koyck model, 150, 271
and linear distributed lags, 149, 163
linear transfer function (LTF) identification method, 173, 174, 175, 220, 237, 240, 268, 271, 280, 283
 with multiple inputs and disturbance ARIMA model, 164
 rules for identification of rational form, 184
 stability conditions for, 161
 starting values for output series in estimation of, 325, 328
 and steady-state gain, 160
Tsay, R., 62, 194, 299, 320, 351, 355, 379
t value:
 of regression model coefficient, 14, 89, 123, 210
 of sample autocorrelation coefficient (warning values), 36
 of sample correlation coefficient, 21

Valve orders and shipments data, 2, 3
Vandaele, W., 352, 370, 379
VAR, *see* Vector ARMA models
Variance:
 population, 19
 sample, 20
 and stationarity, *see* Stationarity, of variance
Vector ARMA models, 342, 351, 352
 and contemporaneous relationship, 346–349, 354
 and dead time, 354
 and feedback check, 349
 order selection for vector AR models, 350, 351, 352
Vector ARMA processes, 342, 343
 and contemporaneous relationship, 347, 354
 and dynamic regression models, 346
 matrix form, 344
 π weight form, 345
 and time-lagged feedback, 343
 variance–covariance matrix of random shocks, 345

Wall, K. D., 328, 380
Watson, G. S., 100
Watts, D. W., 36, 377
Wei, W. W. S., 11, 62, 171, 342, 352, 354, 379, 380
Weisberg, S., 51, 362, 380
White noise, 9
Wichern, D. W., 253, 380

Year-end loading case study, 279

Zimring, F. E., 253, 380

Applied Probability and Statistics (Continued)

HOEL and JESSEN · Basic Statistics for Business and Economics, *Third Edition*
HOGG and KLUGMAN · Loss Distributions
HOLLANDER and WOLFE · Nonparametric Statistical Methods
HOSMER and LEMESHOW · Applied Logistic Regression
IMAN and CONOVER · Modern Business Statistics
JACKSON · A User's Guide to Principle Components
JESSEN · Statistical Survey Techniques
JOHN · Statistical Methods in Engineering and Quality Assurance
JOHNSON · Multivariate Statistical Simulation
JOHNSON and KOTZ · Distributions in Statistics
                                Discrete Distributions
                                Continuous Univariate Distributions—1
                                Continuous Univariate Distributions—2
                                Continuous Multivariate Distributions
JUDGE, GRIFFITHS, HILL, LÜTKEPOHL, and LEE · The Theory and Practice of Econometrics, *Second Edition*
JUDGE, HILL, GRIFFITHS, LÜTKEPOHL, and LEE · Introduction to the Theory and Practice of Econometrics, *Second Edition*
KALBFLEISCH and PRENTICE · The Statistical Analysis of Failure Time Data
KASPRZYK, DUNCAN, KALTON, and SINGH · Panel Surveys
KAUFMAN and ROUSSEEUW · Finding Groups in Data: An Introduction to Cluster Analysis
KEENEY and RAIFFA · Decisions with Multiple Objectives
KISH · Statistical Design for Research
KISH · Survey Sampling
KUH, NEESE, and HOLLINGER · Structural Sensitivity in Econometric Models
LAWLESS · Statistical Models and Methods for Lifetime Data
LEAMER · Specification Searches: Ad Hoc Inference with Nonexperimental Data
LEBART, MORINEAU, and WARWICK · Multivariate Descriptive Statistical Analysis: Correspondence Analysis and Related Techniques for Large Matrices
LEVY and LEMESHOW · Sampling of Populations: Methods and Applications
LINHART and ZUCCHINI · Model Selection
LITTLE and RUBIN · Statistical Analysis with Missing Data
McNEIL · Interactive Data Analysis
MAGNUS and NEUDECKER · Matrix Differential Calculus with Applications in Statistics and Econometrics
MAINDONALD · Statistical Computation
MALLOWS · Design, Data, and Analysis by Some Friends of Cuthbert Daniel
MANN, SCHAFER, and SINGPURWALLA · Methods for Statistical Analysis of Reliability and Life Data
MASON, GUNST, and HESS · Statistical Design and Analysis of Experiments with Applications to Engineering and Science
MILLER · Survival Analysis
MILLER, EFRON, BROWN, and MOSES · Biostatistics Casebook
MONTGOMERY and PECK · Introduction to Linear Regression Analysis, *Second Edition*
NELSON · Accelerated Testing, Statistical Models, Test Plans, and Data Analyses
NELSON · Applied Life Data Analysis
OCHI · Applied Probability and Stochastic Processes in Engineering and Physical Sciences
OSBORNE · Finite Algorithms in Optimization and Data Analysis
OTNES and ENOCHSON · Digital Time Series Analysis
PANKRATZ · Forecasting with Dynamic Regression Models
PANKRATZ · Forecasting with Univariate Box-Jenkins Models: Concepts and Cases
POLLOCK · The Algebra of Econometrics
RAO and MITRA · Generalized Inverse of Matrices and Its Applications
RÉNYI · A Diary on Information Theory
RIPLEY · Spatial Statistics
RIPLEY · Stochastic Simulation
ROSS · Introduction to Probability and Statistics for Engineers and Scientists
ROUSSEEUW and LEROY · Robust Regression and Outlier Detection
RUBIN · Multiple Imputation for Nonresponse in Surveys

*Now available in a lower priced paperback edition in the Wiley Classics Library.